AN INTRODUCTION TO MANY-VALUED AND FUZZY LOGIC

This volume is an accessible introduction to the subject of many-valued and fuzzy logic suitable for use in relevant advanced undergraduate and graduate courses. The text opens with a discussion of the philosophical issues that give rise to fuzzy logic—problems arising from vague language—and returns to those issues as logical systems are presented. For historical and pedagogical reasons, three-valued logical systems are presented as useful intermediate systems for studying the principles and theory behind fuzzy logic. The major fuzzy logical systems—Łukasiewicz, Gödel, and product logics—are then presented as generalizations of three-valued systems that successfully address the problems of vagueness. Semantic and axiomatic systems for three-valued and fuzzy logics are examined along with an introduction to the algebras characteristic of those systems. A clear presentation of technical concepts, this book includes exercises throughout the text that pose straightforward problems, ask students to continue proofs begun in the text, and engage them in the comparison of logical systems.

Merrie Bergmann is an emerita professor of computer science at Smith College. She is the coauthor, with James Moor and Jack Nelson, of *The Logic Book*.

AN INTRODUCTION TO

Many-Valued and Fuzzy Logic

SEMANTICS, ALGEBRAS, AND DERIVATION SYSTEMS

Merrie Bergmann

Emerita, Smith College

CAMBRIDGE
UNIVERSITY PRESS

CAMBRIDGE UNIVERSITY PRESS
Cambridge, New York, Melbourne, Madrid, Cape Town, Singapore, São Paulo, Delhi

Cambridge University Press
32 Avenue of the Americas, New York, NY 10013-2473, USA

www.cambridge.org
Information on this title: www.cambridge.org/9780521881289

First published 2008

Printed in the United States of America

A catalog record for this publication is available from the British Library.

Library of Congress Cataloging in Publication Data

Bergmann, Merrie.
An introduction to many-valued and fuzzy logic : semantics, algebras, and derivation systems / Merrie Bergmann.
p. cm.
Includes bibliographical references and index.
ISBN 978-0-521-88128-9 (hardback) – ISBN 978-0-521-70757-2
1. Fuzzy logic. 2. Many-valued logic. I. Title.
QA9.64.B47 2008
511.3′13 – dc22 2007007725

ISBN 978-0-521-88128-9 hardback
ISBN 978-0-521-70757-2 paperback

To my husband, Michael Thorpe, with love,

for his understanding and support during this project

Contents

Preface

Formal fuzzy logic has developed into an extensive, rigorous, and exciting discipline since it was first proposed by Joseph Goguen and Lotfi Zadeh in the midtwentieth century, and it is a wonderful topic for introducing students to the richness and fascination of formal logic and the philosophy thereof. This textbook grew out of an interdisciplinary course on fuzzy logic that I've taught at Smith College, a course that attracts philosophy, computer science, and mathematics majors. I taught the course for several years with only a course reader because the few existing texts devoted to fuzzy logic were too advanced for my undergraduate audience (and probably for some graduate audiences as well). Finally, after writing voluminous supplements for the course, I decided to write an accessible introductory textbook on many-valued and fuzzy logic. It is my hope that after working through this textbook, students will have the necessary background to tackle more advanced texts, such as Gottwald (2001), Hájek (1998b), and Novák, Perfilieva, and Močkoř (1999), along with the rest of the vast fuzzy literature.

This book opens with a discussion of the philosophical issues that give rise to fuzzy logic—problems and paradoxes arising from vague language—and returns to those issues as new logical systems are presented. There is a two-chapter review of classical logic to familiarize students and instructors with my terminology and notation, and to introduce formal logic to those who have no prior background. Three-valued logical systems are introduced as candidate logics for vagueness, ultimately to be rejected but interesting in their own right and serving as useful intermediate systems for studying the principles and theory that guide fuzzy logics. The major fuzzy logical systems—Łukasiewicz, Gödel, and product logics—are then presented as generalizations of three-valued systems, generalizations that fully address the problems of vagueness. The text ends with two chapters introducing further directions for study: extensions of basic fuzzy systems and definitions of fuzzy membership functions.

Throughout, I have included both semantic and axiomatic systems, along with introductions to the algebras characteristic of those systems. Many texts that have a chapter or so on fuzzy logic restrict their attention to semantics, but much of the interest of fuzzy logic lies in the rich axiomatic systems developed by Jan Pavelka and in the insights garnered from studying the algebras for these systems.

I've used semantic concepts that aren't featured in standard presentations of fuzzy logic, specifically, the concepts of degree-validity and n-degree-validity (these concepts were proposed in Machina (1976)). Degree-validity occurs when an argument's conclusion is guaranteed to be at least as true as the least true premise and is an obvious generalization of classical validity. N-degree-validity measures the slippage of truth going from premises to conclusion: how much less true than the premises can the conclusion of an argument be? The latter concept is particularly useful in analyzing Sorites arguments, and in comparing the performance of the three major fuzzy logical systems with respect to these arguments.

There are exercises throughout the text. Some pose straightforward problems for the student to solve, but many exercises also ask students to continue proofs begun in the text, to prove results analogous to those in the text, and to compare the various logical systems that are presented.

This textbook can be used as a complete basis for an introductory course on formal many-valued and fuzzy logics, at either the upper-level undergraduate or the graduate level, and it can also be used as a supplementary text in a variety of courses. There is considerable flexibility in either case. The truth-valued semantic chapters are independent of the algebraic and axiomatic ones, so that either of the latter may be skipped. Except for Section 13.3 of Chapter 13, the axiomatic chapters are also independent of the algebraic ones, and an instructor who chooses to skip the algebraic material can simply ignore the latter part of 13.3. Finally, Łukasiewicz fuzzy logic is presented independently of Gödel and product fuzzy logics, thus allowing an instructor to focus solely on the former.

I am indebted to my students at Smith College for making this course such a pleasure to teach, and for the many questions and comments that have informed my presentations throughout the text. Joseph Goguen and Petr Hájek, the two men whose work most largely generated my own appreciation of fuzzy logic, generously answered questions that I e-mailed as I was writing the text. It was with great sadness that I learned of Professor Goguen's passing at the age of sixty-five last summer; fuzzy logic as we know it owes much to his pioneering work.

I also thank my colleague Michael Albertson for a helpful analytic suggestion that I used in Chapter 14, and two anonymous reviewers of several chapters for their careful reading and thoughtful suggestions. Any inelegance or errors remain my responsibility alone. Finally, I thank Smith College for generous sabbatical release time.

Merrie Bergmann
August 2007

1 Introduction

1.1 Issues of Vagueness

Some people, like 6′ 7″ Gina Biggerly, are just plain tall. Other people, like 4′ 7″ Tina Littleton, are just as plainly not tall. But now consider Mary Middleford, who is 5′ 7″. Is she tall? Well, kind of, but not really—certainly not as clearly as Gina is tall. If Mary Middleford is kind of but not really tall, is the sentence *Mary Middleford is tall* true? No. Nor is the sentence false. The sentence *Mary Middleford is tall* is neither true nor false. This is a counterexample to the **Principle of Bivalence**, which states that every declarative sentence is either true, like the sentence *Gina Biggerly is tall*, or false, like the sentence *Tina Littleton is tall* (*bivalence* means having two values).[1] The counterexample arises because the predicate *tall* is **vague**: in addition to the people to whom the predicate (clearly) applies or (clearly) fails to apply, there are people like Mary Middleford to whom the predicate neither clearly applies nor clearly fails to apply. Thus the predicate is true of some people, false of some other people, and neither true nor false of yet others. We call the latter people (or, perhaps more strictly, their heights) ***borderline*** or ***fringe*** cases of tallness.

Vague predicates contrast with **precise** ones, which admit of no borderline cases in their domain of application. The predicates that mathematicians typically use to classify numbers are precise. For example, the predicate *even* has no borderline cases in the domain of positive integers. It is true of the positive integers that are multiples of 2 and false of all other positive integers. Consequently, for any positive integer *n* the statement *n is even* is either true or false: *1 is even* is false; *2 is even* is true; *3 is even* is false; *4 is even* is true; and so on, for every positive integer. Thus, *even* is a precise predicate. (We hasten to acknowledge that there are also vague predicates that are applicable to positive integers, e.g., *large*.)

Classical logic, the standard logic that is taught in philosophy and mathematics departments, assumes the Principle of Bivalence: every sentence is assumed to be either true or false. Vagueness thus presents a challenge to classical logic, for sentences containing vague predicates can fail to be true or false and therefore such

1 We will italicize sentences and terms in our text when we are mentioning, that is (in the standard logical vocabulary), talking about them. An alternative convention that we do not use in this text is to place quotation marks around mentioned sentences and terms. We also italicize for emphasis; the distinction should be clear from the context.

sentences cannot be adequately represented in classical logic. "All traditional logic," wrote the philosopher Bertrand Russell, "habitually assumes that precise symbols are being employed. It is therefore not applicable to this terrestrial life, but only to an imagined celestial existence."[2] **Fuzzy logic**, the ultimate subject of this text, was developed to accommodate sentences containing vague predicates (as well as other vague parts of speech). One of the defining characteristics of fuzzy logic is that it admits truth-values other than *true* and *false*; in fact it admits infinitely many truth-values. Fuzzy logic does not assume the Principle of Bivalence.

Some will say, *Why bother? Logic is the study of reasoning, and good reasoning—whether it be in the sciences or in the humanities—exclusively involves precise terms. So we are justified in pursuing classical logic alone, tossing aside as don't-cares any sentences that contain vague expressions.* But as Bertrand Russell pointed out in 1923, vagueness is the norm rather than the exception in much of our discourse. Max Black concluded in 1937 that vagueness must therefore be addressed in an adequate logic for studying natural language discourse, whether that discourse occurs in scientific endeavors or in everyday casual conversations:

> Deviations from the logical or mathematical standards of precision are all pervasive in symbolism; [and] to label them as subjective aberrations sets an impassable gulf between formal laws and experience and leaves the *usefulness* of the formal sciences an insoluble mystery.... [W]ith the provision of an adequate symbolism [that is, a formal system] the need is removed for regarding vagueness as a defect of language. The ideal standard of precision which those have in mind who use vagueness as a term of reproach... is the standard of scientific precision. But the indeterminacy which is characteristic of vagueness is present also in all scientific measurement.... Vagueness is a feature of scientific as of other discourse.[3]

And vague predicates do abound both within and outside academic discourse: *hot, round, red, audible, rich*, and so on. After sitting in my mug for a while, my previously hot coffee becomes a borderline case of *hot*; a couple of days before or after full moon the moon may be a borderline case of *round*; as we move away from red in the color spectrum toward either orange or purple we get borderline cases of *red*; slowly turning the dial on our stereo we can move from loud (also a vague term) music to borderline cases of *audible*; and wealthy people may once have dwelled in the borderline of *rich*. Even in the realm of numbers, which we take as the epitome of precision, we have noted that we may speak of *large* (and *small*) ones, employing predicates as vague as *hot* and *round*.

But even if vagueness weren't pervasive, there are other reasons for developing logics that can handle vague statements. It is an interesting and informative exercise to see what adjustments can and need to be made to classical logic when

2 Russell (1923), p. 88.

3 Black (1937), p. 429. Black's article is a gem, with its appreciation of the pervasiveness and usefulness of vague terms and its attempt to formalize foundations for a logic that includes vague terms.

the Principle of Bivalence is dropped, and to explore ways of addressing the logical challenges posed by vagueness. Consider the classical **Law of Excluded Middle**, the claim that every sentence of the form *A or not A* is true. In classical logic, where precision is the norm, the Law of Excluded Middle is taken as a given. But while we may agree that the two sentences *Either Gina Biggerly is tall or she isn't* and *Either Tina Littleton is tall or she isn't* are both true, indeed on purely logical grounds, we may balk when it comes to Mary Middleford. *Either Mary Middleford is tall or she isn't* doesn't seem to be true, precisely because it's not true that she's tall, and it's also not true that she's not tall.

Not surprisingly, there is a close connection between the Principle of Bivalence and the Law of Excluded Middle. Negation, expressed by *not*, forms a true sentence from a false one and a false sentence from a true one. So if every sentence is either true or false (Principle of Bivalence)—then for any sentence *A*, either *A* is true or *not A* is true (the latter arising when *A* is false). And if either *A* is true or *not A* is true then the sentence *either A or not A* is also true—and this is the Law of Excluded Middle.[4]

In addition to challenging fundamental principles of classical logic, vagueness leads to a family of paradoxes known as the *Sorites paradoxes*. We'll illustrate with a Sorites paradox using the predicate *tall*. As we noted, Gina Biggerly is tall. That is the first premise of the Sorites paradox. Moreover, it is clear that $\frac{1}{8}''$ can't make or break tallness; specifically, someone who is $\frac{1}{8}''$ less tall than a tall person is also tall. That is the second premise. But then it follows that 4′ 7″ Tina Littleton is also tall! For using the two premises we may reason as follows. Since Gina Biggerly is tall, it follows from the second premise that anyone whose height is $\frac{1}{8}''$ less than Gina Biggerly's is also tall; that is, that anyone who is 6′ $6\frac{7}{8}''$ is tall. But then, using the second premise again, we may conclude that anyone who is 6′ $6^6/_8''$ is tall, and again that anyone who is 6′ $6^5/_8''$ is tall, and so on, eventually leading us to the conclusion that Tina Litteleton, along with everyone else who is 4′ 7″, is tall.[5]

Sorites is the Greek word for "heap," and in a heap version of the paradox we have the premises that a large pile of sand—say, one that is 4′ deep—is a heap and that if you remove one grain of sand from a heap what is left is also a heap. Iterated reasoning eventually results in the conclusion that even a single grain of sand is a heap! (In fact, it looks like *no* grains of sand will also count as a heap.) The general pattern of a **Sorites paradox**, given a vague term T, is:

Premise 1 x is T (where x is something of which T is clearly true).

Premise 2 Some type of small change to a thing that is T results in something that is also T.

4 It *is* possible to retain the Law of Excluded Middle while rejecting bivalence; this is the case for *supervaluational* logics. For references see footnote 1 to Chapter 5.

5 Indeed, we may replace $\frac{1}{8}''$ with $^1/_{1000}''$ and conclude that *everyone* whose height is 6′ 7″ or less is tall. In fact, Joseph Goguen (1968–1969) pointed out that we can arrive at an even stronger conclusion: Certainly anyone who is $^1/_{1000}''$ *taller* than a tall person is also tall. So we can conclude that *everyone* is tall, given the existence of one tall person.

Conclusion y is T (where y is something of which T is clearly false, but which you can get to from a long chain of small changes of the sort in Premise 2 beginning with x).

For any vague predicate, a Sorites paradox can be formed.[6] Why are these called *paradoxes*? It is because they appear to be valid (truth-preserving) arguments with true premises, and that means that the conclusions should also be true; but the conclusions are clearly false. Sorites paradoxes are an additional motivation for developing logics to handle vague terms and statements—logics that do not lead us to the paradoxical conclusions of the Sorites paradoxes.[7]

There is a further troubling feature of the Sorites paradoxes. An obvious way out of these paradoxes in classical logic is to deny the truth of the second premise, which is sometimes called the Principle of Charity premise. For the *tall* version of the Sorites paradox just given, the classical logician can simply deny the claim that $^1/_8''$ can't make or break tallness. The paradox dissolves, because a valid argument with false premises need not have a true conclusion. But here's the trouble: when we deny a claim, we accept its negation. This means accepting the negation of the claim that $^1/_8''$ can't make or break tallness, namely, accepting that $^1/_8''$ *does* (at some point) make a difference. But that can't be right since it entails that there is some pair of heights that differ by $^1/_8''$, such that one is tall and the other is not. But where would that pair be? Is it, perhaps, the pair 6′ 2″ and 6′ $1^7/_8''$, so that 6′ 2″ is tall but 6′ $1^7/_8''$ isn't? To see how very unacceptable this is, change the second premise to one that states that $^1/_{1000}''$ doesn't make a difference. The conclusion, that Tina Littleton is tall, still follows. But if we deny the second premise we are saying that $^1/_{1000}''$ *does* make a difference, that there is some pair of heights differing by $^1/_{1000}''$ such that one is tall and the other isn't. That's ludicrous!

Some react to Sorites paradoxes as if they are jokes. They are not. The same type of reasoning with vague concepts, because it is so seductive, can be very dangerous in the world we live in. Consider the population of a country that has a reasonable living standard, including diet and housing, for all. Should we worry about population growth? Of course we should, because at some point population may

6 This may seem contentious for the following reason. Some terms exhibit what I shall call *multidimensional vagueness*. *Tall* exhibits one-dimensional vagueness insofar as tallness is a function of a single measure, height. The Sorites argument depends on small adjustments in that single measure. Other terms' vagueness turns on several factors. Max Black (1937) asks us to consider the word *chair*. There is a multiplicity of characteristics involved in being a chair, including being made of suitably solid material, being of a suitable size, having a suitable horizontal plane for a seat, and having a suitable number of legs (a stool is not a chair). In this respect *chair* exemplifies multidimensional vagueness. The reader is asked to consider whether Sorites arguments can always be constructed for terms that exhibit multidimensional vagueness, as claimed in the text, or whether they arise mainly in the case of one-dimensional vagueness.

7 Some theoreticians, most recently in the school of paraconsistent logics, choose to embrace Sorites paradoxes by concluding that their conclusions are indeed both true and false. See, for example, Hyde (1997) and Beall and Colyvan (2001).

ow, it seems reasonable to say that if
ng standard, then if the population
still be acceptable. It may also seem
se of .01 percent, but clearly this will

ıpplication] concerning which it is
erm] in question does, or does not,

hose things that are the *sort* of thing
ı of the term *tall* includes people and
People and buildings are the sort of
:hat *could* be tall. Integers and colors
e other hand, the field of application
's people, colors, and buildings.
Iary Middleford is tall or that Mary
means that it is not simply a matter
's height is and still find it impossible
ıot apply to her. This contrasts with
ın of ignorance. For example, is the
author's brother Barrie Bergmann tall? You probably can't say, because you have no idea what his height is. But Barrie is not in the fringe of this predicate—he is 6′ 3$^1/_2$″ and clearly tall. Your inability was not an *intrinsic* impossibility, as it is in the case of Mary Middleford. We call those objects within a term's field of application concerning which it is intrinsically impossible to say that the term does or does not apply ***borderline*** cases, and we call the collection of borderline cases the ***fringe*** of the term. Mary is in the fringe of the term *tall*; Gina, Tina, and Barrie are not.

The opposite of *vague* is ***precise***. The term *exactly 6′ 2″ tall* is precise. Given any object in its field of application, the term either does or does not apply, and so there is no intrinsic impossibility in saying whether it does or doesn't. It applies if the object is exactly 6′ 2″ tall and fails to apply otherwise. The term *even* (as applied to positive integers) is also precise. Given any positive integer, the term either applies or fails to apply—it applies if the integer is a multiple of 2 and fails to apply otherwise.

[8] Black (1937, p. 430). For the most part we will restrict our attention to terms that are *adjectives* (like *tall*), *common nouns* (like *chair*), and verbs (like to *smile*)—terms that can appear in predicate position in a sentence. Other parts of speech can also be vague; we will return to some of these in Chapter 16.

English speakers also use the word *vague* to describe terms that are not specific about the properties they connote. We will call such terms ***general*** rather than vague. The term *interesting* is general in this sense: what does it mean to say, for example, that a book is interesting? It could mean that the book contains little-known facts, that the book contains compelling arguments, that the style of writing is unusual, and so on. The term *interesting* is not very specific, unlike the term *tall*, which specifically connotes a magnitude of height (albeit underdetermined). Generality is not the source of borderline cases, which are the exclusive domain of vagueness. This is not to say that a general term cannot also be vague—indeed, this is frequently the case. For example, *interesting* is certainly vague as well as general. But it is important for our purposes to distinguish the two categorizations of terms.

Vagueness is also distinct from ambiguity. A term is **ambiguous** if it has two or more distinct meanings or connotations. For example, *light* is ambiguous: it can mean *light in color* or *light in weight*. When I say that my bicycle is light, I can mean either that it has a light color like tan or white or that it weighs very little. Note that my bicycle can be light in both senses, or that it can be light in one sense but not in the other. Indeed, the philosopher W. V. O. Quine proposed the existence of an object to which a term both does and doesn't apply as a test for ambiguity (Quine 1960, Sect. 27). Again, ambiguity is not a source of borderline cases, although an ambiguous term may also be vague in one or more of its several senses. There are objects that are borderline cases of being light in color, as well as objects that are borderline cases of being light in weight.

Finally, vagueness is also distinct from relativity. A term is **relative** if its applicability is determined relative to, and varies with, subclasses of objects in the term's field of application. Vague terms are frequently relative as well. When we say that a woman is tall, we may mean *tall for a woman*—in this case the application is relative to the class of women. In fact, we probably mean more specifically *tall for a certain race or ethnicity of women*. The applicability of the term *tall* thus varies relative to the class to which it is being applied.

1.3 The Problem of the Fringe

As we have seen, a term is vague if there exists a fringe in its field of applicability. Max Black noted another logical problem that arises from borderline cases to which the term neither applies nor fails to apply. Consider the statement that there are objects in a term's fringe:

There are objects that are neither tall nor not tall,

or equivalently

There are objects that are both not tall and not not tall.

The **Principle of Double Negation** states that a doubly negated expression is equivalent to the expression with both of the negations removed—double negations cancel out. So *not not tall* is equivalent to *tall*, and so the statement that an object is not not tall is equivalent to the statement that it is tall. But then, we can equivalently assert that there are objects in the term *tall*'s fringe as

There are objects that are not tall and also tall.

But this is a contradiction and its truth would violate the **Law of Noncontradiction**, which says that no proposition is both true and false, and specifically in this case, that no single object can both have and not have a property.[9] It looks as if the assertion that a term satisfies the criterion for vagueness, that is, the assertion that there are borderline cases, lands us in contradiction! We will call this the ***Problem of the Fringe***; it is another issue that needs to be addressed in an adequate logic for vagueness.

1.4 Preview of the Rest of the Book

This is a text in logic and in the philosophy of logic. We will study a series of logical systems, culminating in fuzzy logic. But we will also discuss ways to assess systems of logic, which lands us squarely in the philosophy of logic. Students who have taken a first course in logic are sometimes surprised to learn that we can question and critically analyze systems of logic. I hope that the issues and problems that have been introduced in this chapter make it clear that we can and will do just that: we will need to analyze systems of logic critically if we are interested in developing a logic that can handle vague statements. (If, on the other hand, we *refuse* to develop such a logic we are also taking a philosophical stand on logical issues—perhaps by insisting that the purpose of logic is to deal only with reasoning about precise claims.)

Our first task, in Chapters 2 and 3, is to review *classical* (bivalent) propositional logic and classical first-order logic. This will set out a framework for what follows and will serve to introduce notation and terminology that will be used in subsequent chapters. In Chapter 4 we introduce Boolean algebras, systems that capture the "algebraic" structure that classical logic imposes on truth-values. Boolean algebras are not usually covered in introductory symbolic logic courses, so we do not presume that the material in this chapter is a review. We include the topic because, as we will see, algebraic analyses feature prominently in the study of formal fuzzy logic systems.

9 Actually, the earlier assertion *There are objects that are not tall and also not not tall* already violates the Law of Noncontradiction, but we follow Black in removing the double negation in order to make the point.

In Chapters 5 and 6 we will present several well-known systems of three-valued propositional logic, systems in which the Principle of Bivalence is dropped. Chapters 7 and 8 present first-order versions of the three-valued systems. In Chapter 9 we explore algebraic structures for the three-valued systems. We consider three-valued logical systems as candidates for a logic of vagueness. Some readers may feel satisfied that three-valued systems are adequate to this purpose, while others will not. Whichever is the case, the study of three-valued systems will uncover many principles that generalize very nicely as we turn to fuzzy logic.

Our very brief Chapter 10 introduces two new problems concerning vagueness that arise in three-valued logical systems. These problems will motivate the move from three-valued logic to fuzzy logic, in which formulas can have any one of an infinite number of truth-values.

Finally, Chapters 11 and 12 present fuzzy propositional logic—semantics and derivation systems; Chapter 13 introduces algebras for fuzzy logics; and Chapters 14 and 15 present fuzzy first-order logic. Chapter 16 examines augmenting fuzzy logic to include fuzzy qualifiers (like *very*: *how tall is very tall?*) and fuzzy "linguistic" truth-values (*when is a statement more-or-less true?*), and Chapter 17 addresses issues about defining membership functions (used in fuzzy logic) for vague concepts.

1.5 History and Scope of Fuzzy Logic

Formal infinite-valued logics, which form the basis for formal fuzzy logic, were first studied by the Polish logician Jan Łukasiewicz in the 1920s. Łukasiewicz developed a series of many-valued logical systems, from three-valued to infinite-valued, each generalizing the earlier ones for a greater number of truth-values. Although some of the most widely studied fuzzy logics are based on Łukasiewicz's infinite-valued system, Łukasiewicz's philosophical interest in his systems was not based on vagueness but on indeterminism—we will discuss this in Chapter 5.

In 1965 Lotfi Zadeh published a paper (Zadeh (1965)) outlining a theory of fuzzy sets, sets in which members have varying degrees of membership. Fuzzy sets contrast with classical sets, to which something either (fully) belongs or (fully) doesn't belong. One of Zadeh's examples of a fuzzy set is the set of tall men, so the relationship between vague terms and fuzzy sets was clearly established. We'll talk more about fuzzy sets as we introduce fuzzy logic. Two years after Zadeh's paper on fuzzy sets, Joseph Goguen (1967) generalized Zadeh's concept of fuzzy set, relating it to more general algebraic structures, and Goguen (1968–1969) connected fuzzy sets with infinite-valued logic and presented a formal fuzzy logical analysis of the Sorites arguments. Goguen's second article was the beginning of formal fuzzy logic, also known as *fuzzy logic in the narrow sense.*

In 1979 Jan Pavelka published a three-part article (Pavelka 1979) that provides the full framework for fuzzy logic in the narrow sense. Acknowledging his debt to Goguen, Pavelka developed a (fuzzy) complete and consistent axiomatic system for propositional fuzzy logic with "graded" rules of inference: two-part rules that state that one formula can be derived from others and that define the (minimal) degree of truth for the derived formula based on the degrees of truth of the formulas from which it has been derived. Pavelka's paper contains several important metatheoretic results as well. In 1990 Vilém Novák (1990) extended this work to first-order fuzzy logic.[10] In 1995–1997 Petr Hájek made significant simplications to these systems (Hájek 1995a, 1995b), and in 1998 he introduced an axiomatic system BL (for *basic logic*) that captures the commonalities among the major formal fuzzy logics along with a corresponding type of algebra, the BL-algebra (Hájek 1998a). Since the 1990s, Novák and Hájek have dominated the field of fuzzy logic (in the narrow sense) with several texts and numerous articles, more of which will be cited later.

In this text we are strictly concerned with fuzzy logic in the narrow sense. But when many speak of fuzzy logic they often have in mind either fuzzy set theory or fuzzy logic *in the broad sense.* Needless to say, although fuzzy set theory is *used* in fuzzy logic, it is a distinct discipline. Fuzzy logic *in the broad sense* originated in a 1975 article in which Zadeh proposed to develop *fuzzy logic* as "a logic whose distinguishing features are (i) fuzzy truth-values expressed in linguistic terms, e.g., *true, very true, more or less true, rather true, not true, false, not very true and not very false,* etc.; (ii) imprecise truth tables; and (iii) rules of inference whose validity is approximate rather than exact" (Zadeh 1975, p. 407). It is a stretch to call what has developed here a *logic,* at least in the sense in which logicians use that word.

We'll take a brief look at Zadeh's linguistic truth-values at the end of this text, since they may be used to answer at least one philosophical objection to fuzzy logic. The approximate rules to which Zadeh alludes generate reasoning such as the following (Zadeh's example):

a is small
a and b are approximately equal
Therefore, b is more or less small.

As is evident, the logic behind these rules allows us to conclude that if two objects are "approximately" equal and one has a certain property, then the other object "more or less" has that property. The rules used in computational systems based on Zadeh's fuzzy logic *in the broad sense* are like rules of thumb, are stated in English, and are quite useful in contexts such as expert systems. A typical rule for a fuzzy expert system looks like

IF temperature is high AND humidity is low THEN garden is dry

where *temperature* and *humidity* are given as data and *high, low,* and *dry* are measures based on fuzzy sets. Zadeh has also called his version of fuzzy logic *linguistic logic,* and perhaps that would be a more appropriate name for this general area of

[10] This and further work of Novák's appears in Novák, Perfilieva, and Močkoř (1999).

research.[11] Ruspini, Bonissone, and Predrycz (1998) is a good introduction to fuzzy logic *in the broad sense.*

Finally, we note that certain technologies advertise the use of "fuzzy logic." Fuzzy logic rice cookers have been around for a decade or so, cookers that "[do] what a real cook does, using [their] senses and intuition when [they are] cooking rice, watching and intervening when necessary to turn heat up or down, and reacting to the kind of rice in the pot, the volume and the time needed" (Wu 2003, p. E1). And there are fuzzy logic washing machines, fuzzy logic blood pressure monitors, fuzzy logic automatic transmission systems in automobiles, and so forth. The "fuzzy logic" in these cases is the circuit logic built into microchips designed to handle fuzzy measurements. For more on fuzzy technologies see Hirota (1993).

1.6 Tall People

Visit the Web site http://members.shaw.ca/harbord/heights.html. This is fun and will get you thinking about what *tall* means.

1.7 Exercises

SECTION 1.2

1 In his article "Vagueness," Max Black claimed that all terms whose application involves use of the senses are vague. For example, we use color words like *green* and shape words like *round* to describe what we see—and both of these terms are vague. The sea sometimes appears greenish, and this is typically a borderline case of *green*—not really green, but not really not green. While the moon is round when full and not round when in one of its quarters, phases close to full are borderline cases of *round* for the moon—it's not really round, but also not clearly not round.

Give examples of vague terms whose application involves each of the other senses: one for hearing, one for smell, one for taste, and one for touch. Show that your terms are vague by describing one or more borderline cases—cases of things to which the term does not clearly apply or clearly fail to apply.

2 Show that each of the following terms is vague by giving an example of a borderline case: *young, fun, husband, sport, stale, chair, many, flat, book, sleepy.*

3 Are any of the terms in question 2 also ambiguous? General? Relative? Give examples to support your claims.

[11] Not only would such a term make clear the distinction between formal fuzzy logic originating from Goguen's work and Zadeh's version of fuzzy logic; its use would also make it clear when attacks on "fuzzy logic" by logicians (such as Susan Haack [1979]) are targeting the claim that fuzzy logic "in the broad sense" is logic, rather than work done in formal fuzzy logic.

SECTION 1.3

4 Produce a version of the Sorites paradox using the term *rich*.

5 Can Sorites arguments always be constructed for terms that exhibit multidimensional vagueness (defined in footnote 6), or do they arise mainly in the case of unidimensional vagueness? Defend your position.

2 Review of Classical Propositional Logic

2.1 The Language of Classical Propositional Logic

The basic linguistic units symbolized in **propositional logic** are (simple) sentences along with logical connectives that combine them. We'll use uppercase roman letters (with integer subscripts, if more than twenty-six are needed) as **atomic formulas** standing for simple sentences, and we'll symbolize English connectives as follows:[1]

English Connective	*Logical Operation*	*Symbol*
not	negation	$\neg$
and	conjunction	$\wedge$
or	disjunction	$\vee$
if... then	conditional	$\rightarrow$
if and only if	biconditional	$\leftrightarrow$

We say that the negation connective is a ***unary connective*** since it applies to a single formula, and the rest are ***binary connectives*** since they combine *pairs* of formulas. Here are some examples of symbolized English sentences using these connectives, where J stands for *John is a mathematician*, C stands for *Christy is a mathematician*, and P stands for *Christy is a philosopher*:

John is not a mathematician.	$\neg J$
John is a mathematician and so is Christy.	$J \wedge C$
Christy is either a mathematician or a philosopher.	$C \vee P$
If John is a mathematician, then Christy's a philosopher.	$J \rightarrow P$
John is a mathematician if and only if Christy is as well.	$J \leftrightarrow C$

The negation connective $\neg$ has the highest **binding priority**, with the other connectives being of equal priority. Giving $\neg$ the highest binding priority means that in the absence of parentheses, the negation in the formula $\neg J \vee C$ applies to the single formula J, not to $J \vee C$. The formula $\neg J \vee C$ symbolizes *Either John isn't a mathematician or Christy is.* Parentheses are required to override the default priority

[1] Some common alternative symbols are $\sim$, $-$, ! for negation; &, $\cdot$ for conjunction; $|$, $+$ for disjunction; $\supset$, $\Rightarrow$ for the conditional operation; and $=$, $\equiv$, $\Leftrightarrow$ for the biconditional operation.

and to indicate grouping among connectives of the same priority. So, for example, $\neg (J \vee C)$ symbolizes *It's not true that John or Christy is a mathematician* (i.e., *Neither John nor Christy is a mathematician*), and the parentheses are necessary in $(J \wedge C) \vee P$ to indicate that the formula means: *Either John and Christy are both mathematicians or Christy is a philosopher* rather than *John is a mathematician and Christy is either a mathematician or a philosopher.* The latter sentence would be symbolized as $J \wedge (C \vee P)$. We will always require parentheses to indicate the order of evaluation for the binary connectives. For example, a set of parentheses is required to indicate the order of evaluation in $J \vee C \vee P$ even though $(J \vee C) \vee P$ (one way of placing the parentheses) is equivalent to $J \vee (C \vee P)$.

The rules for forming **formulas** in the language of classical propositional logic are as follows:

1. Every uppercase roman letter, with or without an integer subscript, is a formula.
2. If **P** is a formula, so is $\neg\mathbf{P}$.
3. If **P** and **Q** are formulas, so are $(\mathbf{P} \wedge \mathbf{Q})$, $(\mathbf{P} \vee \mathbf{Q})$, $(\mathbf{P} \rightarrow \mathbf{Q})$, and $(\mathbf{P} \leftrightarrow \mathbf{Q})$.[2]

We call single (possibly subscripted) roman letters *A*, *B*, and so on, ***atomic formulas***, while formulas formed using one or more connectives are called ***compound formulas***. The connectives introduced in clauses 2 and 3 are the **main connectives** of the formulas so formed. For example, because the formula $((A \vee B) \vee C)$ is formed from the formulas $(A \vee B)$ and C by clause 3, the main connective of $((A \vee B) \vee C)$ is the second disjunction. By convention, we will drop outermost parentheses in compound formulas—thus $((A \vee B) \vee C)$ may be written as $(A \vee B) \vee C$.

A compound formula is named after the operation symbolized by its main connective: a compound formula whose main connective is $\neg$ is called a ***negation***, and so on. Context will always make it clear whether we are talking about the operation itself or a formula. The immediate subformulas **P** and **Q** of a conjunction $\mathbf{P} \wedge \mathbf{Q}$ are called its ***conjuncts***, and the immediate subformulas **P** and **Q** of a disjunction $\mathbf{P} \vee \mathbf{Q}$ are called its ***disjuncts***. **P** is the **antecedent** of the conditional $\mathbf{P} \rightarrow \mathbf{Q}$ and **Q** is its **consequent**. (At times we will also refer to connectives by the name of the operation they symbolize; e.g., we will call $\neg$ a *negation*.)

2.2 Semantics of Classical Propositional Logic

The five logical connectives that we have introduced are **truth-functional connectives**: the truth-values of formulas formed with these connectives are determined by (i.e., are a *function of*) the truth-values of the constituent formulas.

2 We use *boldface* letters **P**, **Q**, . . . to stand for arbitrary formulas of the language. This will make it clear when we are talking about a particular formula (we will use nonboldface letters) and when we are talking about formulas generally (we will use boldface letters). In this context the boldface letters are called *metavariables*.

The truth-functional operations in classical propositional logic are captured by the following **truth-tables**:

P	¬P
T	F
F	T

P Q	P ∧ Q
T T	T
T F	F
F T	F
F F	F

P Q	P ∨ Q
T T	T
T F	T
F T	T
F F	F

P Q	P → Q
T T	T
T F	F
F T	T
F F	T

P Q	P ↔ Q
T T	T
T F	F
F T	F
F F	T

where *T* and *F* stand for *true* and *false*, respectively. The information in these tables can be used to determine the truth-values of arbitrary formulas on *truth-value assignments*. A **truth-value assignment** is an assignment of truth-values to atomic formulas of the language, and in classical logic *true* and *false* are the only truth-values. If we have a truth-value assignment on which *J* is false and *C* is true, then $\neg J$ is true, $\neg C$ is false, $J \wedge C$ is false, $J \vee C$ is true, $J \rightarrow C$ is true, $C \rightarrow J$ is false, and $C \leftrightarrow J$ is false.

More generally, a truth-table can be used to display the values that a formula will have on all truth-value assignments. Here's a truth-table for the formula $\neg J \wedge (C \vee R)$:

C J R	¬ J ∧ (C ∨ R)
T T T	F T F T T T
T T F	F T F T T F
T F T	T F T T T T
T F F	T F T T T F
F T T	F T F F T T
F T F	F T F F F F
F F T	T F T F T T
F F F	T F F F F F

All combinations of truth-values that *C*, *J*, and *R* can have are listed to the left of the vertical bar. For each of these combinations we list the value of $\neg J \wedge (C \vee R)$ and of each of its subformulas to the right of the vertical bar. The truth-value for an atomic subformula is written immediately below the atomic subformula, and the truth-value for a compound subformula (as well as for the formula as a whole) is written under its main connective. Thus, the first **F** in the first row—under the negation—tells us that the subformula $\neg J$ is false on any truth-value assignment that assigns **T** to *C*, *J*, and *R*, the **F** under the conjunction tells us that the entire formula is false on such assignment, while the **T** under the disjunction tells us that

$C \vee R$ is true. The **semantics** of a language consists in its meaning (or interpretation), and in the case of classical propositional logic the semantics consists of bivalent truth-value assignments and the definitions of the truth-functional operations that can be used to construct a truth-table for any formula.

Logicians sometimes single out some connectives as **primitive** and introduce the other connectives as **defined** ones. One common way of doing this is to take $\neg$ and $\wedge$ as primitive and then to define the others as follows:

$$\mathbf{P} \vee \mathbf{Q} =_{\text{def}} \neg(\neg \mathbf{P} \wedge \neg \mathbf{Q})$$
$$\mathbf{P} \rightarrow \mathbf{Q} =_{\text{def}} \neg(\mathbf{P} \wedge \neg \mathbf{Q})$$
$$\mathbf{P} \leftrightarrow \mathbf{Q} =_{\text{def}} \neg(\mathbf{P} \wedge \neg \mathbf{Q}) \wedge \neg(\neg \mathbf{P} \wedge \mathbf{Q}).$$

The symbol $=_{def}$ means: *is defined as.* Simple reasoning confirms that these definitions are correct. For example, a formula $\neg(\neg \boldsymbol{P} \wedge \neg \boldsymbol{Q})$ is true when $\neg \boldsymbol{P} \wedge \neg \boldsymbol{Q}$ is false, and $\neg \boldsymbol{P} \wedge \neg \boldsymbol{Q}$ is false when either when one or both of $\neg \boldsymbol{P}$, $\neg \boldsymbol{Q}$ are false, that is, when one or both of $\boldsymbol{P}$, $\boldsymbol{Q}$ are true—and that is exactly when a disjunction $\boldsymbol{P} \vee \boldsymbol{Q}$ is true. So the preceding definition for disjunction is correct. Alternatively, we can verify the correctness with a truth-table:

P	Q	P	∨	Q	¬	(¬	P	∧	¬	Q)
T	T	T	T	T	T	F	T	F	F	T
T	F	T	T	F	T	F	T	F	T	F
F	T	F	T	T	T	T	F	F	F	T
F	F	F	F	F	F	T	F	T	T	F

The column of truth-values under the disjunction is identical to the column under the second formula's main connective—the first negation—so the formulas are equivalent. The correctness of the definitions for the conditional and biconditional can be verified similarly.

Another common (and similar) way of dividing the connectives into primitive and defined ones is to take $\neg$ and $\vee$ as primitive and then to introduce the others as:

$$\mathbf{P} \wedge \mathbf{Q} =_{\text{def}} \neg(\neg \mathbf{P} \vee \neg \mathbf{Q})$$
$$\mathbf{P} \rightarrow \mathbf{Q} =_{\text{def}} \neg \mathbf{P} \vee \mathbf{Q}$$
$$\mathbf{P} \leftrightarrow \mathbf{Q} =_{\text{def}} \neg(\neg \mathbf{P} \vee \neg \mathbf{Q}) \vee \neg (\mathbf{P} \vee \mathbf{Q}).$$

The two connectives $\neg$ and $\rightarrow$ can also be taken as primitive, as will be confirmed in an exercise.

Formulas in the language of classical propositional logic that are true on all truth-value assignments are called ***tautologies***, and formulas that are false on all truth-value assignments are called ***contradictions***. The truth-values appearing in the truth-table in the column under a formula's main connective will indicate whether that formula is a tautology, a contradiction, or neither. A symbolic version of the Law of Excluded Middle is $A \vee \neg A$, and this is a tautology in classical logic:

A	A ∨ ¬ A
T	T T F T
F	F T T F

The column of truth-values under the ∨ contains exclusively **T**s, confirming that the formula $A \vee \neg A$ is a tautology. Now for a technical point: a truth-value assignment always assigns truth-values to all atomic formulas of the language. So the truth-table we have just examined doesn't show complete truth-value assignments. However, since each truth-value assignment will have to assign one of the two values **T** and **F** to A, the truth-table shows us that the formula $A \vee \neg A$ is true on *all* truth-value assignments.

The formula $A \wedge \neg A$ is a contradiction, while its negation, which is often called the ***Law of Noncontradiction***, is a tautology:

A	A ∧ ¬ A
T	T F F T
F	F F T F

A	¬(A ∧ ¬ A)
T	T T F F T
F	T F F T F

(The main connective of the second formula is the *initial* negation connective, and underneath it we see only **T**s.) None of our previous formulas symbolizing claims about John and Christy are tautologies or contradictions. For example, we have

J	¬ J
T	F T
F	T F

C J	J ∧ C
T T	T T T
T F	F F T
F T	T F F
F F	F F F

C J	C → J
T T	T T T
T F	T F F
F T	F T T
F F	F T F

and in each case the column under the main connective contains both **T**s and **F**s.[3]

Two formulas of classical propositional logic are **equivalent** if they have the same truth-value on each truth-value assignment. The formulas $C \rightarrow J$ and $\neg J \rightarrow \neg C$ are equivalent:

C J	C → J	¬ J → ¬ C
T T	T T T	F T T F T
T F	T F F	T F F F T
F T	F T T	F T T T F
F F	F T F	T F T T F

3 For uniformity we always list the atomic constituents of formulas in alphabetical order, even when that is not the order in which they appear in compound formulas.

The columns under the main connectives of the two formulas, the conditional connective in each case, are identical. On the other hand, the formulas $C \to J$ and $J \to C$ are not equivalent:

C J	C → J	J → C
T T	T T T	T T T
T F	T F F	F T T
F T	F T T	T F T
F F	F T F	F T F

The formulas have different truth-values on truth-value assignments represented by the second and third rows—truth-value assignments on which C and J have different values. We note as a special case that all tautologies are equivalent to one another because they are all true on every truth-value assignment, and all contradictions are equivalent to one another for a similar reason.

A set Γ (pronounced *gamma*) of formulas **entails** a formula **P** if, whenever all of the formulas in the set Γ are true, **P** is true as well (that is, there is no truth-value assignment on which all the formulas in Γ are true and **P** is false). An **argument** consists of one or more formulas, the **premises**, and an additional formula, the **conclusion**. We say that an argument is ***valid*** if the set consisting of its premises entails the argument's conclusion. We also say that the conclusion of a valid argument ***follows*** from the premises.

The components of an argument are traditionally displayed by writing the premises, one per line, followed by a separator line and then the conclusion, for example,

$$\begin{array}{l} J \to C \\ \underline{J} \\ C \end{array}$$

This argument is valid, as is shown by the fact that every row in the following truth-table in which both premises are true also has the conclusion true (there is only one such row in this case, the first one):

C J	J → C	J	C
T T	T T T	T	T
T F	F T T	F	T
F T	T F F	T	F
F F	F T F	F	F

On the other hand, the argument

$$\begin{array}{l} J \to C \\ \underline{\neg J} \\ \neg C \end{array}$$

is *not* valid. In the following table, the premises are both true in the second row while the conclusion is false, and that is enough to establish **invalidity**:

C J	J → C	¬ J	¬ C
T T	T T T	F T	F T
T F	F T T	T F	F T
F T	T F F	F T	T F
F F	F T F	T F	T F

2.3 Normal Forms

In this section we introduce *disjunctive* and *conjunctive normal forms* for formulas. Each formula of propositional logic has an equivalent formula in disjunctive normal form, and an equivalent formula in conjunctive normal form. The normal forms are used to standardize (*normalize*) the forms of logical formulas for various reasons, such as allowing the use of a computational proof technique known as *resolution* (Robinson 1965). We will use normal forms to prove *functional completeness* in Section 2.5. More importantly, the normal forms will allow us to make some semantic connections among classical logic, three-valued logics, and fuzzy logics.

We'll begin with disjunctive normal form. First we define ***literals*** to include all atomic formulas and their negations: $A, \neg A, B, \neg B, \ldots$. Next we define what a ***phrase*** is:

1. A literal is a phrase
2. If **P** and **Q** are phrases, so is **(P ∧ Q)**

A phrase is either a single literal, or a conjunction of literals: A, $\neg A$, B, $\neg B$, $A \wedge A$, $A \wedge \neg B$, $(D \wedge \neg E) \wedge F$, and so forth. Finally, we define ***disjunctive normal form***:

1. Every phrase is in disjunctive normal form
2. If **P** and **Q** are in disjunctive normal form, so is **(P ∨ Q)**

So a formula is in disjunctive normal form if it contains at most the three connectives ¬, ∧, and ∨, such that negations only appear in front of atomic formulas and no subformula that is a conjunct contains a disjunction. Examples are A, $A \vee \neg B$, $A \wedge \neg B$, $(A \wedge \neg B) \vee B$, $(A \wedge \neg B) \wedge B$, and $(C \wedge \neg E) \vee (D \vee (E \wedge F))$.

We claimed that each formula of classical propositional logic is equivalent to a formula (at least one) that is in disjunctive normal form. To prove this, we show how to transform any formula of classical propositional logic into a formula in disjunctive normal form, where each step of the transformation produces an equivalent formula. The transformation uses the following equivalences:

$\mathbf{P} \rightarrow \mathbf{Q}$	is equivalent to	$\neg\mathbf{P} \vee \mathbf{Q}$	(*Implication*)
$\mathbf{P} \leftrightarrow \mathbf{Q}$	is equivalent to	$(\neg\mathbf{P} \vee \mathbf{Q}) \wedge (\neg\mathbf{Q} \vee \mathbf{P})$	(*Implication*)
$\neg(\mathbf{P} \wedge \mathbf{Q})$	is equivalent to	$\neg\mathbf{P} \vee \neg\mathbf{Q}$	(*DeMorgan's Law*)
$\neg(\mathbf{P} \vee \mathbf{Q})$	is equivalent to	$\neg\mathbf{P} \wedge \neg\mathbf{Q}$	(*DeMorgan's Law*)[4]
$\neg\neg\mathbf{P}$	is equivalent to	$\mathbf{P}$	(*Double Negation*)
$(\mathbf{P} \vee \mathbf{Q}) \wedge \mathbf{R}$	is equivalent to	$(\mathbf{P} \wedge \mathbf{R}) \vee (\mathbf{Q} \wedge \mathbf{R})$	(*Distribution*)
$\mathbf{P} \wedge (\mathbf{Q} \vee \mathbf{R})$	is equivalent to	$(\mathbf{P} \wedge \mathbf{Q}) \vee (\mathbf{P} \wedge \mathbf{R})$	(*Distribution*)

(Proof that these forms are equivalent is left as an exercise.) We'll explain the transformation process using the formula

$$\neg(\neg(P \rightarrow Q) \vee (\neg R \rightarrow (S \leftrightarrow T)))$$

First we use the Implication equivalences to eliminate all conditionals and biconditionals, producing the formula

$$\neg(\neg(\neg P \vee Q) \vee (\neg\neg R \vee ((\neg S \vee T) \wedge (\neg T \vee S))))$$

Next we use DeMorgan's Laws to move negations deeper into the formula until all negations appear in front of atomic formulas. In our example we use the second DeMorgan Law first to obtain

$$\neg\neg(\neg P \vee Q) \wedge \neg(\neg\neg R \vee ((\neg S \vee T) \wedge (\neg T \vee S)))$$

and then

$$\neg(\neg\neg P \wedge \neg Q) \wedge (\neg\neg\neg R \wedge \neg((\neg S \vee T) \wedge (\neg T \vee S))$$

Next we can use the first DeMorgan Law twice to obtain

$$(\neg\neg\neg P \vee \neg\neg Q) \wedge (\neg\neg\neg R \wedge (\neg(\neg S \vee T) \vee \neg(\neg T \vee S)))$$

and then the second Law twice more to obtain

$$(\neg\neg\neg P \vee \neg\neg Q) \wedge (\neg\neg\neg R \wedge ((\neg\neg S \wedge \neg T) \vee (\neg\neg T \wedge \neg S)))$$

Now Double Negation eliminates all double negations:

$$(\neg P \vee Q) \wedge (\neg R \wedge ((S \wedge \neg T) \vee (T \wedge \neg S))).$$

We're almost done, but note that this is a conjunction, and some of the conjuncts are *dis*junctions, so the formula is not yet in disjunctive normal form (if all of the conjuncts were themselves conjunctions, it *would* be in disjunctive normal form). We can use the first Distribution equivalence to convert the overall formula into a disjunction, with $\neg P$ in place of $\mathbf{P}$, Q in place of $\mathbf{Q}$, and $(\neg R \wedge ((S \wedge \neg T) \vee (T \wedge \neg S)))$ in place of $\mathbf{R}$:

$$(\neg P \wedge (\neg R \wedge ((S \wedge \neg T) \vee (T \wedge \neg S)))) \vee (Q \wedge (\neg R \wedge ((S \wedge \neg T) \vee (T \wedge \neg S)))).$$

4 These two laws are named after the nineteenth-century mathematician Augustus DeMorgan, who stated these laws set-theoretically. According to one historian, these laws did not actually originate with DeMorgan but "were known in scholastic times and probably even in antiquity" (Delong 1970, p. 106).

Next we can apply the second Distribution equivalence to both occurrences of the subformula $(\neg R \wedge ((S \wedge \neg T) \vee (T \wedge \neg S))$, with $\neg R$ in place of **P**, $(S \wedge \neg T)$ in place of **Q**, and $(T \wedge \neg S)$ in place of **R**, to obtain

$$(\neg P \wedge ((\neg R \wedge (S \wedge \neg T)) \vee (\neg R \wedge (T \wedge \neg S)))) \vee$$
$$(Q \wedge ((\neg R \wedge (S \wedge \neg T)) \vee (\neg R \wedge (T \wedge \neg S))))$$

This formula is in disjunctive normal form, and it is equivalent to the original formula

$$\neg(\neg(P \to Q) \vee (\neg R \to (S \leftrightarrow T)))$$

because each step of the transformation produces an equivalent formula.

This series of steps will convert *any* formula to an equivalent formula in disjunctive normal form—we first get rid of conditional and biconditional connectives, then we use the DeMorgan Laws to move negations in as far as they will go, and we use Double Negation to eliminate all double negations. At this point all negations occur in front of atomic formulas, and the binary connectives are either conjunctions or disjunctions. The Distribution laws enable us to convert the result to disjunctive normal form by replacing each conjunction with a disjunctive conjunct into a disjunction of conjunctive disjuncts. In Section 2.5 we will see another way to produce formulas in disjunctive normal form.

In contrast, formulas in conjunctive normal form are conjunctions of disjunctions, rather than disjunctions of conjunctions. We define a ***clause*** as:

1. A literal is a clause
2. If **P** and **Q** are clauses, so is **(P** $\vee$ **Q)**

and ***conjunctive normal form*** as:

1. Every clause is in conjunctive normal form
2. If **P** and **Q** are in conjunctive normal form, so is **(P** $\wedge$ **Q)**

Any formula of classical propositional logic can be converted to an equivalent formula in conjunctive normal form by following the same steps as we did for disjunctive normal form, but using the following Distribution equivalences at the end:

A formula **(P** $\wedge$ **Q)** $\vee$ **R** is equivalent to **(P** $\vee$ **R)** $\wedge$ **(Q** $\vee$ **R)** (*Distribution*)
A formula **P** $\vee$ **(Q** $\wedge$ **R)** is equivalent to **(P** $\vee$ **Q)** $\wedge$ **(P** $\vee$ **R)** (*Distribution*)

Let us call a pair of literals ***complementary*** if one is the negation of the other; for instance, S and $\neg S$ are a complementary pair of literals. We now state two important general results:

Result 2.1: A clause **C** of classical propositional logic is a tautology if and only if **C** contains a complementary pair of literals.

Proof: If a clause **C** is a tautology then **C** must be a disjunction (since no literal is a tautology), and on each truth-value assignment at least one disjunct is true. But then **C** must contain a complementary pair of literals, because

ls are independent of one
signment on which they are
T) is false when P, Q, and T
ains a complementary pair
of these literals will be true
e formula **C** will be true as

gic is contradictory if and

onjunctive normal form is
s a complementary pair of

ther a clause or a conjunc-
gy if and only if **P** contains
on, then **P** is a tautology if
this is the case if and only
e built are tautologies. By
Result 2.1, each of these clauses is a tautology if and only if each clause contains a complementary pair of literals.

Result 2.4: A formula **P** of classical logic that is in disjunctive normal form is a contradiction if and only if each phrase in **P** contains a complementary pair of literals.

Proof: Left as an exercise.

2.4 An Axiomatic Derivation System for Classical Propositional Logic

Alongside semantic means of assessing formulas and evaluating arguments, derivation systems may be used. There are several types of derivation systems, including axiomatic systems and natural deduction systems. Axiomatic systems are the norm in the field of fuzzy logic, and so we shall use axiomatic derivation systems throughout this text.[5] In an **axiomatic derivation system** we have a set of formulas proclaimed to be axioms along with a set of rules to derive new formulas from previous ones. There are many axiomatic systems that have been studied for classical propositional logic. The axiomatic system that we present here was developed by the Polish logician Jan Łukasiewicz (1930, 1934) to simplify a system proposed by the

[5] For examples of natural deduction derivation systems see Bergmann, Moor, and Nelson (2004).

German logician Gottlob Frege (Frege 1879). We have chosen the Frege-Łukasiewicz system—which we will designate as *CLA* (for *classical propositional logic axiomatic system*)—because some simple modifications will serve to axiomatize three-valued and fuzzy logics that were also developed by Łukasiewicz. CLA contains three *axiom schemata*:

CL1. $\mathbf{P} \rightarrow (\mathbf{Q} \rightarrow \mathbf{P})$
CL2. $(\mathbf{P} \rightarrow (\mathbf{Q} \rightarrow \mathbf{R})) \rightarrow ((\mathbf{P} \rightarrow \mathbf{Q}) \rightarrow (\mathbf{P} \rightarrow \mathbf{R}))$
CL3. $(\neg\mathbf{P} \rightarrow \neg\mathbf{Q}) \rightarrow (\mathbf{Q} \rightarrow \mathbf{P})$

and the single inference rule MP, which is short for the rule's traditional name, Modus Ponens:

MP (Modus Ponens). From **P** and **P**→ **Q**, infer **Q**.

An **axiom schema** stands for infinitely many axioms, namely, all formulas that have the overall form exemplified by the schema. We call such formulas **instances** of the axiom schema. We can define an ***instance of an axiom schema*** to be any formula that results from uniform substitution of formulas of the language (not necessarily distinct) for each of the letters **P**, **Q**, and **R**. By *uniform substitution* we mean that in a given instance, the same formula must be substituted for *every* occurrence of **P**, and similarly for **Q** and **R**. So, for example, the following formulas are all instances of the axiom schema CL1:

$P \rightarrow (Q \rightarrow P)$	Substituting P for **P** and Q for **Q**
$Q \rightarrow (P \rightarrow Q)$	Substituting Q for **P** and P for **Q**
$P \rightarrow (P \rightarrow P)$	Substituting P for **P** and P for **Q**
$(A \wedge B) \rightarrow ((F \leftrightarrow (G \vee H)) \rightarrow (A \wedge B))$	Substituting $(A \wedge B)$ for **P** and $(F \leftrightarrow (G \vee H))$ for **Q**

Note that each of the axiom schemata has the form of a tautology. For example, here is the truth-table for CL2:

P	Q	R	(P→	(Q→	R))	→	((P→	Q)	→	(P→	R))
T	T	T	T T	T T	T	T	T T	T	T	T T	T
T	T	F	T F	T F	F	T	T T	T	F	T F	F
T	F	T	T T	F T	T	T	T F	F	T	T T	T
T	F	F	T T	F T	F	T	T F	F	T	T F	F
F	T	T	F T	T T	T	T	F T	T	T	F T	T
F	T	F	F T	T F	F	T	F T	T	T	F T	F
F	F	T	F T	F T	T	T	F T	F	T	F T	T
F	F	F	F T	F T	F	T	F T	F	T	F T	F

It is left as an exercise to verify that the other two axiom schemata also have tautologous forms. In addition, the rule Modus Ponens is a **truth-preserving** rule—a rule that when applied to true formulas results in a true formula: if **P** and **P** → **Q** are both true, then **Q** must be true as well.

A **derivation** is a sequence of formulas each of which is designated as an assumption, or is an instance of an axiom schema, or can be derived from earlier formulas in the sequence using the derivation rule MP. In addition, the formulas designated as assumptions must begin the derivation. (This last stipulation is not theoretically necessary but will make rules in later chapters easy to state, with no loss of derivational power.) Here is an example of a derivation annotated in a standard way:

1	A	Assumption
2	$(B \rightarrow A) \rightarrow (A \rightarrow M)$	Assumption
3	$A \rightarrow (B \rightarrow A)$	CL1, with A / **P**, B / **Q**
4	$B \rightarrow A$	1,3 MP
5	$A \rightarrow M$	2,4 MP
6	M	1,5 MP

We have *derived* the conclusion M from the assumptions A and $(B \rightarrow A) \rightarrow (A \rightarrow M)$. We have numbered each of the formulas in the sequence constituting the derivation and have added annotations. Each line of the displayed derivation is annotated to indicate the justification for entering that formula in the derivation: either as an assumption (lines 1 and 2), as an instance of an axiom schema (line 3; the annotation indicates that axiom schema CL1 has been used, with A substituted for **P** and B for **Q**), or by virtue of being derived from earlier formulas (lines 4–6; each annotation indicates to which formulas the rule MP was applied). When a formula appears in a derivation that begins with a set of assumptions, we say that the formula is ***derived from*** those assumptions. So each of the formulas on lines 1–6 is derived from the assumptions on lines 1 and 2. If a formula can be derived from a set of assumptions, we also say that the formula is ***derivable from*** any set containing those assumptions. So the preceding derivation shows that the formula M is derivable from *any* set that contains both A and $(B \rightarrow A) \rightarrow (A \rightarrow M)$. As another example, the following derivation shows that $A \rightarrow B$ is derivable from $\neg A$:

1	$\neg A$	Assumption
2	$\neg A \rightarrow (\neg B \rightarrow \neg A)$	CL1, with A / **P**, B / **Q**
3	$\neg B \rightarrow \neg A$	1,2 MP
4	$(\neg B \rightarrow \neg A) \rightarrow (A \rightarrow B)$	CL3, with B / **P**, A / **Q**
5	$A \rightarrow B$	3,4 MP

Some formulas, namely, the tautologies of classical logic, can be derived without using any assumptions. For example, $A \rightarrow A$ is a tautology and we can derive it without any assumptions as follows:

1	$A \rightarrow ((A \rightarrow A) \rightarrow A)$	CL1, with A / **P**, A → A / **Q**
2	$A \rightarrow ((A \rightarrow A) \rightarrow A)) \rightarrow ((A \rightarrow (A \rightarrow A)) \rightarrow (A \rightarrow A))$	CL2, with A / **P**, A → A / **Q**, A / **R**
3	$(A \rightarrow (A \rightarrow A)) \rightarrow (A \rightarrow A)$	1,2 MP
4	$A \rightarrow (A \rightarrow A)$	CL1, with A / **P**, A / **Q**
5	$A \rightarrow A$	3,4 MP

A derivation that does not contain any assumptions is called a ***proof***, and a formula is called a ***theorem*** if there is a proof ending with that formula. The proof just given establishes that $A \to A$ is a theorem. We may call the proof a ***proof of*** that theorem.

Another theorem is $\neg A \to (A \to B)$. Before presenting a proof, we draw the reader's attention to the relation between this theorem and the fact that (as we have shown) $A \to B$ is derivable from $\neg A$, namely: the consequent and the antecedent of the theorem are the two formulas in question. It turns out that whenever a formula $\mathbf{Q}$ is derivable from a formula $\mathbf{P}$ in CLA there is a corresponding theorem $\mathbf{P} \to \mathbf{Q}$ in CLA, and vice versa. This general fact is known as the ***Deduction Theorem:***[6]

Result 2.5 (The Deduction Theorem for Classical Propositional Logic): $\mathbf{Q}$ is derivable from $\mathbf{P}$ in CLA if and only if $\mathbf{P} \to \mathbf{Q}$ is a theorem in CLA.

Proof: The *if* part—if $\mathbf{P} \to \mathbf{Q}$ is a theorem then $\mathbf{Q}$ is derivable from $\mathbf{P}$—is easy to establish. If $\mathbf{P} \to \mathbf{Q}$ is a theorem then there is a proof of $\mathbf{P} \to \mathbf{Q}$ in CLA. We can add $\mathbf{P}$ to this proof as an assumption and then use Modus Ponens to derive $\mathbf{Q}$ from $\mathbf{P}$ and $\mathbf{P} \to \mathbf{Q}$.

For the *only if* part we will explain how to convert any derivation of $\mathbf{Q}$ from the assumption $\mathbf{P}$ into a proof of $\mathbf{P} \to \mathbf{Q}$, thus establishing the theoremhood of the latter. (We'll then illustrate the method by converting our derivation of $A \to B$ from $\neg A$ into a proof of $\neg A \to (A \to B)$.) The strategy is to show how, given a derivation of $\mathbf{Q}$ from the assumption P—a derivation consisting of the sequence $\mathbf{P}, \mathbf{R}_1, \mathbf{R}_2, \ldots, \mathbf{R}_{n-1}, \mathbf{R}_n$ where $\mathbf{R}_n$ is $\mathbf{Q}$—we can produce a derivation in which each of the formulas $\mathbf{P} \to \mathbf{P}, \mathbf{P} \to \mathbf{R}_1, \mathbf{P} \to \mathbf{R}_2, \ldots, \mathbf{P} \to \mathbf{R}_{n-1}, \mathbf{P} \to \mathbf{R}_n$ occurs as a theorem. The last theorem in this sequence is the desired theorem for this result. The formulas will all be theorems because the new derivation will not include the assumption $\mathbf{P}$ or any other assumption.

We begin the derivation by deriving the first new formula $\mathbf{P} \to \mathbf{P}$ exactly as we derived $A \to A$ on the previous page, using $\mathbf{P}$ in place of A. Now, each of $\mathbf{R}_1, \mathbf{R}_2, \ldots, \mathbf{R}_{n-1}, \mathbf{R}_n$ either was an instance of an axiom schema or followed from previous formulas in the derivation by MP. For each one of these formulas $\mathbf{R}_i$, if $\mathbf{R}_i$ is an instance of an axiom schema we can add the following lines to derive $\mathbf{P} \to \mathbf{R}_i$ in the new derivation:

m	$\mathbf{R}_i$	{by relevant axiom schema}
$m+1$	$\mathbf{R}_i \to (\mathbf{P} \to \mathbf{R}_i)$	CL1, with $\mathbf{R}_i$ / $\mathbf{P}$, $\mathbf{P}$ / $\mathbf{Q}$
$m+2$	$\mathbf{P} \to \mathbf{R}_i$	$m+1$, $m+2$ MP

If $\mathbf{R}_i$ followed by MP from earlier formulas $\mathbf{R}_k$ and $\mathbf{R}_k \to \mathbf{R}_i$ in the sequence $\mathbf{R}_1, \mathbf{R}_2, \ldots, \mathbf{R}_{i-1}, \mathbf{R}_n$, then we already have $\mathbf{P} \to \mathbf{R}_k$ and $\mathbf{P} \to (\mathbf{R}_k \to \mathbf{R}_i)$ in the new derivation, say, on lines m and n, and three additional lines will derive $\mathbf{P} \to \mathbf{R}_i$:

[6] There is also a *semantic* version of the Deduction Theorem: the set $\{\mathbf{P}\}$ entails $\mathbf{Q}$ if and only if $\mathbf{P} \to \mathbf{Q}$ is a tautology. This is easy to prove directly, but we also note that the semantic version follows from Result 2.5 and the fact that CLA is sound and complete for classical propositional logic (see p. 26).

m	$\mathbf{P} \to \mathbf{R}_k$	
...		
n	$\mathbf{P} \to (\mathbf{R}_k \to \mathbf{R}_i)$	
...		
p	$(\mathbf{P} \to (\mathbf{R}_k \to \mathbf{R}_i)) \to ((\mathbf{P} \to \mathbf{R}_k) \to (\mathbf{P} \to \mathbf{R}_i))$	CL2, with **P** / **P**, $\mathbf{R}_k$ / **Q**, $\mathbf{R}_i$ / **R**
p+1	$(\mathbf{P} \to \mathbf{R}_k) \to (\mathbf{P} \to \mathbf{R}_i)$	n, p MP
p+2	$\mathbf{P} \to \mathbf{R}_i$	m, p+1 MP

Using the method described in the proof of the Deduction Theorem and the earlier derivation

1	$\neg A$	Assumption
2	$\neg A \to (\neg B \to \neg A)$	CL1, with A / **P**, B / **Q**
3	$\neg B \to \neg A$	1,2 MP
4	$(\neg B \to \neg A) \to (A \to B)$	CL3, with B / **P**, A / **Q**
5	$A \to B$	3,4 MP

we construct the following derivation establishing the theoremhood of $\neg A \to (A \to B)$ (to the left of the lines containing the conditionals whose consequents are formulas from the earlier derivation we write the line numbers from that derivation):

	1	$\neg A \to ((\neg A \to \neg A) \to \neg A)$	CL1, with ¬A / **P**, ¬A → ¬A / **Q**
	2	$\neg A \to ((\neg A \to \neg A) \to \neg A)) \to$ $((\neg A \to (\neg A \to \neg A)) \to (\neg A \to \neg A))$	CL2, with ¬A / **P**, ¬A → ¬A / **Q**, ¬A / **R**
	3	$(\neg A \to (\neg A \to \neg A)) \to (\neg A \to \neg A)$	1,2 MP
	4	$\neg A \to (\neg A \to \neg A)$	CL1, with ¬A / **P**, A /**Q**
1.	5	$\neg A \to \neg A$	3,4 MP
	6	$\neg A \to (\neg B \to \neg A)$	CL1, with ¬A / **P**, ¬B /**Q**
	7	$(\neg A \to (\neg B \to \neg A)) \to (\neg A \to (\neg A \to (\neg B \to \neg A)))$	CL1, with ¬A → (¬B → ¬A) / **P**, ¬A / **Q**
2.	8	$\neg A \to (\neg A \to (\neg B \to \neg A))$	6,7 MP
	9	$\neg A \to (\neg A \to (\neg B \to \neg A)) \to$ $((\neg A \to \neg A) \to (\neg A \to (\neg B \to \neg A)))$	CL2, with ¬A / **P**, ¬A / **Q**, ¬ A → ¬B / **R**
	10	$(\neg A \to \neg A) \to (\neg A \to (\neg B \to \neg A))$	8,9 MP
3.	11	$\neg A \to (\neg B \to \neg A)$	5,10 MP
	12	$(\neg B \to \neg A) \to (A \to B)$	CL3, with B / **P**, A / **Q**
	13	$((\neg B \to \neg A) \to (A \to B)) \to$ $(\neg A \to ((\neg B \to \neg A) \to (A \to B)))$	CL1, with (¬B → ¬A) → (A → B) / **P**, ¬A / **Q**
4.	14	$\neg A \to ((\neg B \to \neg A) \to (A \to B))$	12,13 MP
	15	$(\neg A \to ((\neg B \to \neg A) \to (A \to B))) \to$ $((\neg A \to (\neg B \to \neg A)) \to (\neg A \to (A \to B)))$	CL2, with ¬A / **P**, ¬B → ¬A / **Q**, A → B / **R**
	16	$(\neg A \to (\neg B \to \neg A)) \to (\neg A \to (A \to B))$	14,15 MP
5.	17	$\neg A \to (A \to B)$	11,16 MP

Shorter derivations are certainly possible—for example, in this case we did not need to derive the formula on line 11 since it already appears on line 6; our point here was to illustrate the mechanical conversion procedure introduced in the proof of the Deduction Theorem, a procedure that shows that we can always convert a derivation of **Q** from **P** into a proof of **P** $\rightarrow$ **Q**.

A derivation system is said to be ***sound*** for classical propositional logic if (a) all theorems are tautologies and (b) whenever a formula **P** is derivable from a set Γ of formulas, then **P** is also entailed by the set Γ. The system CLA is sound. For example, we have already noted that the theorem $A \rightarrow A$ is a tautology of classical logic, and so is the theorem $\neg A \rightarrow (A \rightarrow B)$. The arguments corresponding to our other two derivations:

$$\frac{\begin{array}{l}\mathrm{A}\\ (\mathrm{B} \rightarrow \mathrm{A}) \rightarrow (\mathrm{A} \rightarrow \mathrm{M})\end{array}}{\mathrm{M}}$$

and

$$\frac{\mathrm{A}}{\neg \mathrm{A} \rightarrow \neg \mathrm{B}}$$

are both valid (as can be confirmed by truth-tables). CLA is a sound derivation system precisely because the axioms are tautologies and the single rule is truth-preserving.

A derivation system is said to be ***weakly complete*** for classical propositional logic if every tautology of classical logic is a theorem in the system (the converse of what goes into soundness) and it is said to be ***strongly complete***, or just ***complete***, if in addition whenever a set Γ of formulas entails a formula **P**, **P** is also derivable from Γ within the system (again, the converse of soundness). CLA is complete as well as sound for classical propositional logic, and when a system has both of these properties we say that it is ***adequate*** for classical propositional logic.[7] Simply put, soundness and completeness mean that you can derive everything you should be able to derive (the system is complete), and nothing you shouldn't (the system is sound).

We point out that there is prima facie ("at first sight") something fishy about the claim that CLA is complete for classical propositional logic: completeness means that for every entailment or tautology there is a corresponding derivation in the system, but it is unclear that there is any derivation for entailments or tautologies whose validity depends on connectives other than $\neg$ and $\rightarrow$, because no other connectives appear in either the axioms or the derivation rule MP. As an example, $M \vee (M \rightarrow S)$ is a tautology, but how can it be derived in CLA? The answer is that when we want to use CLA to produce derivations for formulas containing the connectives $\wedge$, $\vee$, and $\leftrightarrow$,

[7] See Hunter (1971) for several completeness proofs for the system CLA (called *PS* in Hunter), along with a soundness proof.

we first rewrite the formulas to contain only $\neg$ and $\to$, using some standard set of definitions. For this purpose we'll use the definitions

$$\mathbf{P} \vee \mathbf{Q} =_{\text{def}} \neg\mathbf{P} \to \mathbf{Q}$$
$$\mathbf{P} \wedge \mathbf{Q} =_{\text{def}} \neg(\mathbf{P} \to \neg\mathbf{Q})$$
$$\mathbf{P} \leftrightarrow \mathbf{Q} =_{\text{def}} (\mathbf{P} \to \mathbf{Q}) \wedge (\mathbf{Q} \to \mathbf{P}), \text{ which turns into } \neg((\mathbf{P} \to \mathbf{Q}) \to \neg(\mathbf{Q} \to \mathbf{P}))$$

and then produce derivations for the rewritten formulas. To show that $M \vee (M \to S)$ is a theorem we first convert the formula to $\neg M \to (M \to S)$ using the preceding definition for disjunction, then we construct a derivation for the latter formula (*caution*: do not spend too much time reading through this derivation; we will shorten it momentarily):

1	$\neg M \to (\neg S \to \neg M)$	CL1, with $\neg M$ / $\mathbf{P}$, $\neg S$ / $\mathbf{Q}$
2	$(\neg S \to \neg M) \to (M \to S)$	CL3, with S / $\mathbf{P}$, M / $\mathbf{Q}$
3	$((\neg M \to ((\neg S \to \neg M) \to (M \to S))) \to ((\neg M \to (\neg S \to \neg M)) \to (\neg M \to (M \to S)))) \to (((\neg S \to \neg M) \to (M \to S)) \to ((\neg M \to ((\neg S \to \neg M) \to (M \to S))) \to ((\neg M \to (\neg S \to \neg M)) \to (\neg M \to (M \to S)))))$	CL1, with $(\neg M \to ((\neg S \to \neg M) \to (M \to S))) \to ((\neg M \to (\neg S \to \neg M)) \to (\neg M \to (M \to S)))$ / $\mathbf{P}$, $(\neg S \to \neg M) \to (M \to S)$ / $\mathbf{Q}$
4	$(\neg M \to ((\neg S \to \neg M) \to (M \to S)) \to ((\neg M \to (\neg S \to \neg M)) \to (\neg M \to (M \to S)))$	CL2, with $\neg M$ / $\mathbf{P}$, $\neg S \to \neg M$ / $\mathbf{Q}$, $M \to S$ / $\mathbf{R}$
5	$((\neg S \to \neg M) \to (M \to S)) \to ((\neg M \to ((\neg S \to \neg M) \to (M \to S))) \to ((\neg M \to (\neg S \to \neg M)) \to (\neg M \to (M \to S))))$	3,4 MP
6	$(((\neg S \to \neg M) \to (M \to S)) \to ((\neg M \to ((\neg S \to \neg M) \to (M \to S))) \to ((\neg M \to (\neg S \to \neg M)) \to (\neg M \to (M \to S))))) \to ((((\neg S \to \neg M) \to (M \to S)) \to (\neg M \to ((\neg S \to \neg M) \to (M \to S)))) \to (((\neg S \to \neg M) \to (M \to S)) \to ((\neg M \to (\neg S \to \neg M)) \to (\neg M \to (M \to S)))))$	CL2, with $(\neg S \to \neg M) \to (M \to S)$ / $\mathbf{P}$, $\neg M \to ((\neg S \to \neg M) \to (M \to S))$ / $\mathbf{Q}$, $(\neg M \to (\neg S \to \neg M)) \to (\neg M \to (M \to S))$ / $\mathbf{R}$
7	$(((\neg S \to \neg M) \to (M \to S)) \to (\neg M \to ((\neg S \to \neg M) \to (M \to S)))) \to (((\neg S \to \neg M) \to (M \to S)) \to ((\neg M \to (\neg S \to \neg M)) \to (\neg M \to (M \to S))))$	5,6 MP
8	$((\neg S \to \neg M) \to (M \to S)) \to (\neg M \to ((\neg S \to \neg M) \to (M \to S)))$	CL1, with $(\neg S \to \neg M) \to (M \to S)$ / $\mathbf{P}$, $\neg M$ / $\mathbf{Q}$
9	$((\neg S \to \neg M) \to (M \to S)) \to ((\neg M \to (\neg S \to \neg M)) \to (\neg M \to (M \to S)))$	7,8 MP
10	$(\neg M \to (\neg S \to \neg M)) \to (\neg M \to (M \to S))$	2,9 MP
11	$\neg M \to (M \to S)$	1,10 MP

We may conclude that $M \vee (M \to S)$ is a theorem of CLA.

At this point it is probably (frighteningly) clear that it can be difficult to design derivations in CLA – and to follow them; that is why we added the previous caution. Why, you may ask, would anyone propose such a terse system, with merely three axioms and a single rule to choose from? Wouldn't it be better to have more axioms and rules? The reason for terse axiomatic systems is that authors of these systems strive for elegance, usually in the form of a very small set of axioms and rules, which in turn can make it easier to prove things *about* the system.[8]

But when we put axiomatic systems to action, terseness ceases to look like a virtue. In order to make derivations less difficult to construct or to follow, it is common to introduce **derived axiom schemata** and/or **derived rules**. For example, we can introduce $\mathbf{P} \rightarrow \mathbf{P}$ as a derived axiom schema:

CLD1. $\mathbf{P} \rightarrow \mathbf{P}$

The *D* is short for *derived*. We are justified in introducing $\mathbf{P} \rightarrow \mathbf{P}$ as a derived axiom schema because we have already presented a proof of $A \rightarrow A$, and it is clear that this proof can be converted into a proof of any instance of the schema $\mathbf{P} \rightarrow \mathbf{P}$ simply by substituting any other formula for A. In general, once we have proved a formula to be a theorem we may present the formula, with metavariables uniformly substituted for the formula letters, as a derived axiom schema. We can then use the derived schema in subsequent derivations, as we do in the following derivation of $B \rightarrow (C \rightarrow C)$:

1	$C \rightarrow C$	CLD1, with C / **P**
2	$(C \rightarrow C) \rightarrow (B \rightarrow (C \rightarrow C))$	CL1, with $C \rightarrow C$ / **P**, B / **Q**
3	$B \rightarrow (C \rightarrow C)$	1,2 MP

Using CLD1 we can also construct a short simple derivation showing that $A \vee \neg A$ is a theorem. First, we rewrite the formula as $\neg A \rightarrow \neg A$ so that it contains only the negation and conditional symbols. Here's the derivation!

1	$\neg A \rightarrow \neg A$	CLD1, with ¬A / **P**

Now let's return to the derivation of the theorem $\neg M \rightarrow (M \rightarrow S)$. The sole purpose of lines 3–11 of the derivation is to derive $\neg M \rightarrow (M \rightarrow S)$ from $\neg M \rightarrow (\neg S \rightarrow \neg M)$ and $(\neg S \rightarrow \neg M) \rightarrow (M \rightarrow S)$. Examining these formulas we see that they are instances of a general inference pattern that derives a formula of the form $\mathbf{P} \rightarrow \mathbf{R}$ from formulas having the form $\mathbf{P} \rightarrow \mathbf{Q}$ and $\mathbf{Q} \rightarrow \mathbf{R}$. We may introduce this inference pattern, commonly called *hypothetical syllogism*, as a derived rule:

HS (Hypothetical Syllogism). From $\mathbf{P} \rightarrow \mathbf{Q}$ and $\mathbf{Q} \rightarrow \mathbf{R}$, infer $\mathbf{P} \rightarrow \mathbf{R}$.

[8] Claims *about* a derivation system are part of what is called the ***metatheory*** of logic. The claim that an axiomatic system is sound, or that it is complete, is a metatheoretic claim.

To justify HS we first introduce the derived axiom

CLD2. $(\mathbf{Q} \to \mathbf{R}) \to ((\mathbf{P} \to \mathbf{Q}) \to (\mathbf{P} \to \mathbf{R}))$

This axiom is justified as follows (compare with lines 3–9 of the previous derivation of $M \vee (M \to S)$:

1	$((\mathbf{P} \to (\mathbf{Q} \to \mathbf{R})) \to ((\mathbf{P} \to \mathbf{Q}) \to (\mathbf{P} \to \mathbf{R}))) \to$ $((\mathbf{Q} \to \mathbf{R}) \to ((\mathbf{P} \to (\mathbf{Q} \to \mathbf{R})) \to ((\mathbf{P} \to \mathbf{Q}) \to (\mathbf{P} \to \mathbf{R}))))$	CL1, with $(\mathbf{P} \to (\mathbf{Q} \to \mathbf{R})) \to ((\mathbf{P} \to \mathbf{Q}) \to (\mathbf{P} \to \mathbf{R}))$ / **P**, $\mathbf{Q} \to \mathbf{R}$ / **Q**
2	$(\mathbf{P} \to (\mathbf{Q} \to \mathbf{R})) \to ((\mathbf{P} \to \mathbf{Q}) \to (\mathbf{P} \to \mathbf{R}))$	CL2, with **P** / **P**, **Q** / **Q**, **R** / **R**
3	$(\mathbf{Q} \to \mathbf{R}) \to ((\mathbf{P} \to (\mathbf{Q} \to \mathbf{R})) \to ((\mathbf{P} \to \mathbf{Q}) \to (\mathbf{P} \to \mathbf{R})))$	1,2 MP
4	$((\mathbf{Q} \to \mathbf{R}) \to ((\mathbf{P} \to (\mathbf{Q} \to \mathbf{R})) \to ((\mathbf{P} \to \mathbf{Q}) \to (\mathbf{P} \to \mathbf{R})))) \to$ $(((\mathbf{Q} \to \mathbf{R}) \to (\mathbf{P} \to (\mathbf{Q} \to \mathbf{R}))) \to$ $((\mathbf{Q} \to \mathbf{R}) \to ((\mathbf{P} \to \mathbf{Q}) \to (\mathbf{P} \to \mathbf{R}))))$	CL2, with $\mathbf{Q} \to \mathbf{R}$ / **P**, $\mathbf{P} \to (\mathbf{Q} \to \mathbf{R})$ / **Q**, $(\mathbf{P} \to \mathbf{Q}) \to (\mathbf{P} \to \mathbf{R})$ / **R**
5	$((\mathbf{Q} \to \mathbf{R}) \to (\mathbf{P} \to (\mathbf{Q} \to \mathbf{R}))) \to$ $((\mathbf{Q} \to \mathbf{R}) \to ((\mathbf{P} \to \mathbf{Q}) \to (\mathbf{P} \to \mathbf{R})))$	3,4 MP
6	$(\mathbf{Q} \to \mathbf{R}) \to (\mathbf{P} \to (\mathbf{Q} \to \mathbf{R}))$	CL1, with $\mathbf{Q} \to \mathbf{R}$ / **P**, **P** / **Q**
7	$(\mathbf{Q} \to \mathbf{R}) \to ((\mathbf{P} \to \mathbf{Q}) \to (\mathbf{P} \to \mathbf{R}))$	5,6 MP

and the following derivation justifies HS (compare lines 3–5 with lines 9–11 of the longer dervivation):

1	$\mathbf{P} \to \mathbf{Q}$	Assumption
2	$\mathbf{Q} \to \mathbf{R}$	Assumption
3	$(\mathbf{Q} \to \mathbf{R}) \to ((\mathbf{P} \to \mathbf{Q}) \to (\mathbf{P} \to \mathbf{R}))$	CLD2, with **P** / **P**, **Q** / **Q**, **R** / **R**
4	$(\mathbf{P} \to \mathbf{Q}) \to (\mathbf{P} \to \mathbf{R})$	2,3 MP
5	$\mathbf{P} \to \mathbf{R}$	1,4 MP

This derivation shows that if we already have $\mathbf{P} \to \mathbf{Q}$ and $\mathbf{Q} \to \mathbf{R}$, whether or not they are assumptions, we can derive $\mathbf{P} \to \mathbf{R}$. Using HS, we can seriously shorten the previous derivation of $\neg M \to (M \to S)$:

1	$\neg \mathrm{M} \to (\neg \mathrm{S} \to \neg \mathrm{M})$	CL1, with ¬M / **P**, ¬S / **Q**
2	$(\neg \mathrm{S} \to \neg \mathrm{M}) \to (\mathrm{M} \to \mathrm{S})$	CL3, with S / **P**, M / **Q**
3	$\neg \mathrm{M} \to (\mathrm{M} \to \mathrm{S})$	1,2 HS

Another useful derived rule is

TRAN (Transposition). From $\mathbf{P} \to (\mathbf{Q} \to \mathbf{R})$ infer $\mathbf{Q} \to (\mathbf{P} \to \mathbf{R})$.

The rule is called *Transposition* because it transposes antecedents. The following derivation, which uses both CLD2 and HS, justifies TRAN:

1	$\mathbf{P \to (Q \to R)}$	Assumption
2	$\mathbf{((P \to Q) \to (P \to R)) \to ((Q \to (P \to Q)) \to (Q \to (P \to R)))}$	CLD2, with $\mathbf{Q / P, P \to Q / Q, P \to R / R}$
3	$\mathbf{(((P \to Q) \to (P \to R)) \to ((Q \to (P \to Q)) \to (Q \to (P \to R)))) \to}$ $\mathbf{(((P \to Q) \to (P \to R)) \to (Q \to (P \to Q))) \to}$ $\mathbf{(((P \to Q) \to (P \to R)) \to (Q \to (P \to R)))}$	CL2, with $\mathbf{(P \to Q) \to (P \to R) / P,}$ $\mathbf{Q \to (P \to Q) / Q, Q \to (P \to R) / R}$
4	$\mathbf{((P \to Q) \to (P \to R)) \to (Q \to (P \to Q))) \to}$ $\mathbf{(((P \to Q) \to (P \to R)) \to (Q \to (P \to R))}$	2,3 MP
5	$\mathbf{Q \to (P \to Q)}$	CL1, with $\mathbf{Q / P, P / Q}$
6	$\mathbf{(Q \to (P \to Q)) \to (((P \to Q) \to (P \to R)) \to (Q \to (P \to Q)))}$	CL1, with $\mathbf{Q \to (P \to Q) / P,}$ $\mathbf{(P \to Q) \to (P \to R) / Q}$
7	$\mathbf{((P \to Q) \to (P \to R)) \to (Q \to (P \to Q))}$	5,6 MP
8	$\mathbf{((P \to Q) \to (P \to R)) \to (Q \to (P \to R))}$	4,7 MP
9	$\mathbf{(P \to (Q \to R)) \to ((P \to Q) \to (P \to R))}$	CL2, with $\mathbf{P / P, Q / Q, R / R}$
10	$\mathbf{(P \to (Q \to R)) \to (Q \to (P \to R))}$	8,9 HS
11	$\mathbf{Q \to (P \to R)}$	1,10 MP

Two additional useful derived axiom schemata are

CLD3. $\mathbf{\neg\neg P \to P}$
CLD4. $\mathbf{P \to \neg\neg P}$

Here is a derivation justifying CLD3:

1	$\mathbf{\neg\neg P \to (\neg\neg\neg\neg P \to \neg\neg P)}$	CL1, with $\mathbf{\neg\neg P / P, \neg\neg\neg\neg P / Q}$
2	$\mathbf{(\neg\neg\neg\neg P \to \neg\neg P) \to (\neg P \to \neg\neg\neg P)}$	CL3, with $\mathbf{\neg\neg\neg P / P, \neg P / Q}$
3	$\mathbf{\neg\neg P \to (\neg P \to \neg\neg\neg P)}$	1,2 HS
4	$\mathbf{(\neg P \to \neg\neg\neg P) \to (\neg\neg P \to P)}$	CL3, with $\mathbf{P / P, \neg\neg P / Q}$
5	$\mathbf{\neg\neg P \to (\neg\neg P \to P)}$	3,4 HS
6	$\mathbf{(\neg\neg P \to (\neg\neg P \to P)) \to ((\neg\neg P \to \neg\neg P) \to (\neg\neg P \to P))}$	CL2, with $\mathbf{\neg\neg P / P, \neg\neg P / Q, P / R}$
7	$\mathbf{(\neg\neg P \to \neg\neg P) \to (\neg\neg P \to P)}$	5,6 MP
8	$\mathbf{\neg\neg P \to \neg\neg P}$	CLD1, with $\mathbf{\neg\neg P / P}$
9	$\mathbf{\neg\neg P \to P}$	7,8 MP

and we may use CLD3 to justify CLD4:

1	$\mathbf{\neg\neg\neg P \to \neg P}$	CLD3, with $\mathbf{\neg P / P}$
2	$\mathbf{(\neg\neg\neg P \to \neg P) \to (P \to \neg\neg P)}$	CL3, with $\mathbf{\neg\neg P / P, P / Q}$
3	$\mathbf{P \to \neg\neg P}$	1,2 MP

With the help of these derived axioms and rule, we can also derive

CLD5. $\mathbf{(P \to Q) \to (\neg Q \to \neg P)}$

which is the *converse* of CL3:

1	$(Q \to \neg\neg Q) \to ((P \to Q) \to (P \to \neg\neg Q))$	CLD2, with $P / P, Q / Q, \neg\neg Q / R$
2	$Q \to \neg\neg Q$	CLD4, with Q / P
3	$(P \to Q) \to (P \to \neg\neg Q)$	1,2 MP
4	$(P \to \neg\neg Q) \to ((\neg\neg P \to P) \to (\neg\neg P \to \neg\neg Q))$	CLD2, with $\neg\neg P / P, P / Q, \neg\neg Q / R$
5	$(\neg\neg P \to P) \to ((P \to \neg\neg Q) \to (\neg\neg P \to \neg\neg Q))$	4, TRAN
6	$\neg\neg P \to P$	CLD3, with P / P
7	$(P \to \neg\neg Q) \to (\neg\neg P \to \neg\neg Q)$	5,6 MP
8	$(P \to Q) \to (\neg\neg P \to \neg\neg Q)$	3,7 HS
9	$(\neg\neg P \to \neg\neg Q) \to (\neg Q \to \neg P)$	CL3, with $\neg P / P, \neg Q / Q$
10	$(P \to Q) \to (\neg Q \to \neg P)$	8,9 HS

Another useful rule:

MT (**Modus Tollens**). From $P \to Q$ and $\neg Q$, infer $\neg P$

is readily derived using CLD5:

1	$P \to Q$	Assumption
2	$\neg Q$	Assumption
3	$(P \to Q) \to (\neg Q \to \neg P)$	CLD5, with $P / P, Q / Q$
4	$\neg Q \to \neg P$	1,3 MP
5	$\neg P$	2,4 MP

As a final example, with the help of our derived axioms and rules we can show that the argument

If the economy is sound, then either the unemployment rate is low or spending is high.
If the unemployment rate is low, most people are well off.
If spending is high, most people are well off.
It's not true that most people are well off.
____
The economy isn't sound.

is valid. First we symbolize the argument:

$E \to (U \vee S)$
$U \to W$
$S \to W$
$\neg W$
_____
$\neg E$

and we rewrite the first premise as $E \rightarrow (\neg U \rightarrow S)$. Here is a derivation:

1	$E \rightarrow (\neg U \rightarrow S)$	Assumption
2	$U \rightarrow W$	Assumption
3	$S \rightarrow W$	Assumption
4	$\neg W$	Assumption
5	$\neg U$	2,4 MT
6	$\neg S$	3,4 MT
7	$(\neg U \rightarrow S) \rightarrow (\neg S \rightarrow \neg\neg U)$	CLD5, with $\neg U$ / **P**, S / **Q**
8	$\neg S \rightarrow ((\neg U \rightarrow S) \rightarrow \neg\neg U)$	7, TRAN
9	$(\neg U \rightarrow S) \rightarrow \neg\neg U$	6,8 MP
10	$((\neg U \rightarrow S) \rightarrow \neg\neg U) \rightarrow (\neg\neg\neg U \rightarrow \neg (\neg U \rightarrow S))$	CLD5, with $\neg U \rightarrow S$ / **P**, $\neg\neg U$ / **Q**
11	$\neg\neg\neg U \rightarrow \neg (\neg U \rightarrow S)$	9,10 MP
12	$\neg U \rightarrow \neg\neg\neg U$	CLD4, with $\neg U$ / **P**
13	$\neg\neg\neg U$	5,12 MP
14	$\neg(\neg U \rightarrow S)$	11,13 MP
15	$\neg E$	1,14 MT

We stress that these derived axiom schemata and rules are a *convenience* for constructing derivations; the set of axiom schemata CL1–CL3 alone with the single rule MP form a complete derivation system for classical propositional logic, and so the additional axioms and rules do not add to the power of the system. Nor do they affect its soundness, since they are all derivable within a system that was sound to begin with.

2.5 Functional Completeness

In Section 2.2 we pointed out that we could take one of several pairs of connectives as primitive and define the rest in terms of these. In Section 2.4 we took advantage of this fact: the axiomatic system CLA uses only two connectives, since formulas containing the other connectives can be rewritten using the two connectives $\neg$ and $\rightarrow$ that appear in CLA.

There is an important related theoretical issue to which we now turn. First, we formally define the concept of a truth-function: a **truth-function** is a function that maps truth-values to truth-values. More specifically, a truth-function is always a function of a given finite number of arguments: one, two, three, whatever; it is a function that maps each sequence of the appropriate number of truth-values to a truth-value. The negation truth-function is a truth-function of one argument as specified by the following truth-table template:

T	**F**
F	**T**

This function maps the single truth-value **T** (more precisely, the single-membered sequence <T>) to the truth-value **F**, and it maps the single truth-value **F** (<F>) to the truth-value **T**. The conditional truth-function is a truth-function of two arguments:

T	T	T
T	F	F
F	T	T
F	F	T

It maps the sequence <T, F> to the truth-value **F**, and all other sequences of two truth-values to the truth-value **T**.

We say that a formula $\mathbf{P}$ of propositional logic *expresses* a truth-function of n arguments if the truth-table for $\mathbf{P}$ specifies that truth-function; that is, the values under $\mathbf{P}$'s main connective are the values to which the function maps each sequence of n truth-values listed to the left of the vertical line. So, for example, and by design, $\neg P$ expresses the negation truth-function, and $P \rightarrow Q$ expresses the conditional truth-function (other formula letters may be used). Other truth-functions may require more complicated formulas. For example, the **neither-nor** truth-function, captured in the following truth-table template,

T	T	F
T	F	F
F	T	F
F	F	T

is expressed by the formula $\neg(P \vee Q)$ or equivalently by $\neg P \wedge \neg Q$.

Here is the theoretical issue: can *every* classical truth-function be expressed by a formula of classical propositional logic using only the five connectives $\neg$, $\wedge$, $\vee$, $\rightarrow$, and/or $\leftrightarrow$? The answer is *yes*. We shall show how, given *any* truth-function, to construct a formula that expresses exactly that truth-function. To facilitate the proof we shall assume that the function in question has been laid out in a truth-table template as previously, and that n is the number of arguments that the truth-function operates on.

First, we choose n atomic formulas $\mathbf{P}_1, \ldots, \mathbf{P}_n$, one corresponding to each argument place. These will head the columns to the left of the vertical line in the truth-table template. Next, for each row of the truth-function template we form a corresponding conjunction conjoining the atomic formulas that have the value **T** in that row along with the negations of the atomic formulas that have the value **F**—in the terminology of Section 2.3, each such conjunction is a *phrase*. So, for example, phrases corresponding to the four rows of the *neither-nor* function template are, respectively, $P \wedge Q$, $P \wedge \neg Q$, $\neg P \wedge Q$, and $\neg P \wedge \neg Q$. Note that each of these phrases is true exactly when P and Q have the truth-values in its corresponding row. Next we form a disjunction of the phrases corresponding to the rows that have **T** to the right of the vertical line, thus producing a formula in disjunctive normal form. In the case of the *neither-nor* function there is one such row, the fourth, so we form the "disjunction" of the single phrase for that row: $\neg P \wedge \neg Q$. (A "disjunction" of a single formula is simply the formula.) This formula expresses the function captured in the *neither-nor* truth-table template.

As a more complicated example consider the function of three arguments:

T	T	T	T
T	T	F	T
T	F	T	T
T	F	F	F
F	T	T	F
F	T	F	F
F	F	T	T
F	F	F	F

Assuming that we have chosen the atomic formulas P, Q, and R, a disjunction of phrases corresponding to the rows with **T** to the right of the vertical line is $((((P \wedge Q) \wedge R) \vee ((P \wedge Q) \wedge \neg R)) \vee ((P \wedge \neg Q) \wedge R)) \vee ((\neg P \wedge \neg Q) \wedge R)$. This formula expresses the truth-function specified in the truth-table template, as the reader may easily confirm.

We must add two special cases. In one we have a truth-function of one argument, such as,

T	T
F	T

In this case the phrase corresponding to a row is a single atomic formula or its negation: P for the first row and $\neg P$ for the second. Since both rows contain **T** to the right of the vertical line, the disjunction $P \vee \neg P$ of these two phrases expresses the truth-function. In the other case there are no **T**s to the right of the vertical line. In this case we may simply conjoin the phrase corresponding to the first (or any other) row with its negation. So for the truth-function

T	F
F	F

we have the formula $P \wedge \neg P$, and for the truth-function

T	T	F
T	F	F
F	T	F
F	F	F

we have the formula $(P \wedge Q) \wedge \neg(P \wedge Q)$.

Note that there are other formulas—in fact infinitely many other formulas—that express these same truth-functions, so it is important to keep in mind that we are only showing that, given any truth-function, there is at least one formula using the five connectives that expresses it. Now, we've claimed that our procedure will always work—but how do we *know* this? It's rather simple. Each phrase corresponding to a row of a truth-function template is true on the truth-value assignments represented by that row that is false on *all other* truth-value assignments. Thus a disjunction of the phrases corresponding to rows that have a **T** to the right of the vertical line will be true on the truth-value assignments represented by those rows and false on all other truth-value assignments. In the case where there are no **T**s to the right of the vertical line we produce a contradictory formula of the general form **P** $\wedge$ $\neg$**P**, which is always false. Conclusion: we've specified a way to construct, for any classical truth-function, a formula that exactly expresses that truth-function.

When a set of connectives is sufficient to express every truth-function, we say that the set of connectives is ***functionally complete***. Thus, we have proved

Result 2.6: The set of connectives $\{\neg, \wedge, \vee\}$ is functionally complete,

since these are the only connectives we have used in formulas to express any truth-function. From this it follows that the full set $\{\neg, \wedge, \vee, \rightarrow, \leftrightarrow\}$ is also functionally complete—since we are not required to use *all* of the connectives in the candidate formulas expressing the various truth-functions. But since we know that there are three subsets consisting of only two of our connectives that are sufficient to define the others, we may also conclude that those three subsets, $\{\neg, \wedge\}$, $\{\neg, \vee\}$, and $\{\neg, \rightarrow\}$, are truth-functionally complete.[9]

2.6 Decidability

Classical propositional logic has a desirable property that isn't shared by all logical systems: its set of tautologies is *decidable*. A set Γ of formulas is **decidable** if there is a *decision procedure* for membership in the set, that is, a mechanical procedure that will, given any formula, correctly decide after a finite number of steps whether that formula is a member of Γ.[10]

The set of tautologies of classical logic is decidable because there exist mechanical procedures for testing whether a formula is a tautology. We've already seen one such procedure: given any formula we can construct a truth-table for that formula and examine the column of truth-values under the formula's main connective. If that column consists solely of **T**s then the formula is a tautology; otherwise it is not. Clearly truth-tables can be constructed mechanically, and just as clearly the construction and examination of the relevant column of truth-values take only a finite number of steps. Similarly, the set of contradictions of classical propositional logic

9 These are the only sets consisting of two of the five connectives that are truth-functionally complete. First, we note that we need the negation connective, for without it we can never produce a formula that is false when all of its atomic components are true. Second, we note that negation and the biconditional won't suffice because, for example, every formula constructed from two atomic formulas using only these two connectives will have an even number of **T**s and an even number of **F**s in its truth-table (proof is left as an exercise).

10 Sets that contain things other than formulas can also be said to be decidable, but that is not our interest here.

is decidable. We can use an analogous decision procedure, except that here we are looking for a column of truth-values containing only **F**s. There are other mechanical procedures to test whether a formula is a tautology in classical propositional logic, for example, resolution and semantic tableaux.[11]

The set of theorems in our axiomatic system for classical propositional logic, CLA, is also decidable. To be sure, we haven't shown how to construct proofs as a mechanical method to test for theoremhood, but we don't need to. Because CLA is sound and complete for classical propositional logic, truth-tables afford a mechanical procedure for testing for theoremhood! Given a formula we use the truth-table procedure for determining whether or not it is a tautology. If it is, then it is also a theorem of CLA because CLA is complete. If the formula isn't a tautology, then it isn't a theorem either because CLA is sound. It *is* possible to produce a mechanical method for proving theorems in CLA: given a formula that is a theorem such a method can be used to generate a proof. But a mechanical method for constructing proofs of theorems does not constitute a decision procedure if it does not also yield negative results telling us of nontheorems *that* they are nontheorems. Truth-tables can correctly test for negative as well as positive results.

2.7 Exercises

SECTION 2.2

1 Produce truth-tables for the following formulas, and state whether each formula is a tautology, a contradiction, or neither.

a. $P \leftrightarrow \neg P$

b. $(A \wedge B) \rightarrow (A \vee B)$

c. $\neg (A \wedge B) \rightarrow \neg (A \vee B)$

d. $(A \rightarrow B) \vee (B \rightarrow A)$

e. $(A \rightarrow B) \vee (\neg A \rightarrow \neg B)$

f. $(A \rightarrow B) \rightarrow \neg (B \rightarrow A)$

g. $((P \rightarrow Q) \rightarrow R) \leftrightarrow ((P \wedge Q) \rightarrow R)$

h. $(P \rightarrow Q) \wedge (P \rightarrow \neg(Q \vee R))$

2 Produce truth-tables for each of the following pairs of formulas, to confirm that they are equivalent:

a. $P \vee Q, \neg P \rightarrow Q$

b. $P \wedge Q, \neg(P \rightarrow \neg Q)$

c. $P \leftrightarrow Q, (P \rightarrow Q) \wedge (Q \rightarrow P)$

3 Produce truth-tables for each of the following arguments, and state whether each of the arguments is valid or invalid:

a. $\dfrac{(P \rightarrow \neg P) \rightarrow \neg P}{\neg P}$

11 Smullyan (1968) is an excellent reference for semantic tableaux.

b. $(P \leftrightarrow \neg Q) \rightarrow R$
P
$\neg R$
$\neg Q$

c. $(A \vee B) \wedge (A \vee \neg B)$
A

d. $P \rightarrow (Q \rightarrow R)$
Q
$P \rightarrow R$

e. $P \vee Q$
$P \rightarrow R$
$Q \rightarrow R$
R

SECTION 2.3

4 Produce truth-tables to verify the following:

a.	$\mathbf{P \rightarrow Q}$ is equivalent to $\mathbf{\neg P \vee Q}$	(*Implication*)
b.	$\mathbf{P \leftrightarrow Q}$ is equivalent to $\mathbf{(\neg P \vee Q) \wedge (\neg Q \vee P)}$	(*Implication*)
c.	$\mathbf{\neg(P \wedge Q)}$ is equivalent to $\mathbf{\neg P \vee \neg Q}$	(*DeMorgan's Law*)
d.	$\mathbf{\neg(P \vee Q)}$ is equivalent to $\mathbf{\neg P \wedge \neg Q}$	(*DeMorgan's Law*)
e.	$\mathbf{P}$ is equivalent to $\mathbf{\neg\neg P}$	(*Double Negation*)
f.	$\mathbf{(P \vee Q) \wedge R}$ is equivalent to $\mathbf{(P \wedge R) \vee (Q \wedge R)}$	(*Distribution*)
g.	$\mathbf{P \wedge Q) \vee R}$ is equivalent to $\mathbf{(P \vee R) \wedge (Q \vee R)}$	(*Distribution*)

(The other two distribution laws are trivial variations of f and g.)

5 Convert each of the formulas in problem 1 to disjunctive normal form.

6 Convert each of the formulas in problem 1 to conjunctive normal form.

7 a. Prove Result 2.2.
b. Prove Result 2.4.

SECTION 2.4

8 Produce truth-tables to verify that the forms of axiom schemata CL1 and CL3 of system CLA are both tautologies of classical propositional logic.

9 Produce derivations in CLA showing that

a. M is derivable from $M \wedge S$

b. S is derivable from $M \wedge S$

c. $(P \rightarrow Q) \rightarrow ((Q \rightarrow R) \rightarrow (P \rightarrow R))$ is a theorem

d. The conclusion of the following argument is derivable from its premises:
$P \rightarrow (Q \rightarrow R)$
Q
$P \rightarrow R$

In each case you may use derived axioms and rules and derive your own axioms and/or rules if convenient.

SECTION 2.5

10 Prove that there are infinitely many formulas of propositional logic that express any given truth-function.

11 Explain why every formula constructed from two atomic formulas using only negation and the biconditional will have an even number of **T**s and an even number of **F**s in its truth-table. *Hint*: construct a variety of such truth-tables and look for this pattern. What is causing it?

3 Review of Classical First-Order Logic

3.1 The Language of Classical First-Order Logic

First-order logic (sometimes called *predicate logic*) includes all of the connectives of propositional logic. Unlike propositional logic, however, first-order logic analyzes simple sentences into terms and predicates. We use uppercase roman letters as **predicates**, lowercase roman letters *a* through *t* as **(individual) constants**, and lowercase roman letters *u* through *z* as **(individual) variables.** Predicates, constants, and variables may be augmented with subscripts if necessary, thus guaranteeing an infinite supply of each.

Constants function like names in English, and variables function like pronouns. Together constants and variables count as **terms**. Predicates have *arities*, where an **arity** is the number of terms to which a predicate applies. In English, for example, the arity of the predicate *runs* in *John runs* is 1—it combines with a single term, *John* in this case—while the arity of the predicate *loves* in *John loves Sue* is 2—it combines with two terms. **Atomic formulas** are formed by writing predicates in initial position followed by an appropriate number of terms (determined by the predicate's arity). *John runs* and *John loves Sue* might thus be symbolized as *Rj* and *Ljs*.

There are two standard quantifiers in first-order logic, the universal and the existential quantifiers. We'll use ∀ as the **universal quantifier** symbol and ∃ as the **existential quantifier** symbol.[1] The universal quantifier is used to symbolize claims made about *everything* (usually *everything of such-and-such a type*), while the existential quantifier symbolizes claims about *some* things (again, usually *some things of such-and-such a type*). Variables are used along with quantifiers to mark quantified positions with respect to predicates. We might symbolize *Everything runs* as *(∀x)Rx—every x is such that x runs*, and *Some things run* as *(∃x)Rx*. We can use any other variable in place of *x* as long as we are consistent, for example, *(∀y)Ry* or *(∀z)Rz* but not *(∀y)Rz*. Position-marking has an important function in formulas whose predicates have arity greater than 1. For example, if *j* symbolizes *John* and *Lxy*

1 Alternative symbols are a large ∧ for the universal quantifier and a large ∨ for the existential quantifier.

symbolizes *x loves y*, then *(∀x)Ljx* symbolizes *John loves everything* (*every x is such that John loves x*) while *(∀x)Lxj* symbolizes *Everything loves John* (*every x is such that x loves John*). We can read *(∀x)(∀y)Lxy* as *Everything (every x) loves everything (every y)* and *(∀x)(∀y)Lyx* as *Everything (every x) is loved by everything (every y)*. The sentences *Something runs*, *John loves something*, *Everything loves something*, and *Something loves everything* are similarly symbolized as *(∃x)Rx*, *(∃x)Ljx*, *(∀x)(∃y)Lxy*, and *(∃x)(∀y)Lxy*. Note that the existential quantifier *(∃x)*, which we informally read as *some x*, more specifically will mean: *at least one x*, so a formula like *(∃x)Rx* will signify that *at least one thing* runs.

Here are the rules for forming formulas of classical first-order logic:

1. Every predicate of arity *n* followed by *n* terms is a formula.
2. If **P** is a formula, so is ¬**P**.
3. If **P** and **Q** are formulas, so are (**P** ∧ **Q**), (**P** ∨ **Q**), (**P** → **Q**), and (**P** ↔ **Q**).
4. If **P** is a formula, so are (∀**x**)**P** and (∃**x**)**P**.

Formulas formed in accordance with clause 1 are **atomic formulas**, and the others are **compound formulas**. Formulas formed in accordance with 2 and 3 are, respectively, called (as they are in propositional logic) *negations*, *conjunctions*, *disjunctions*, *conditionals*, and *biconditionals*. (∀**x**) is called a ***universal quantifier*** and (∀**x**)**P** is called a *universally quantified formula* or a *universal quantification*. Similarly, (∃**x**) is called an ***existential quantifier***, and a compound formula (∃**x**)**P** is called an *existentially quantified formula* or an *existential quantification*.

In what follows we'll need the concepts of **free** and **bound occurrences** of **variables**. Each separate appearance of a variable in a formula counts as a separate occurrence: the variable *x* has one occurrence in *Faxy* and in *Bx* → *By* but two occurrences in both *Fxax* and *Bx* → *Bx*. If **P** is an atomic formula (like *Faxy* or *By*), every occurrence of every variable in **P** is free. So the single occurrences of *x* and *y* in *Faxy* are both free, and both occurrences of *x* in *Dxx* are free. If an occurrence of a variable **x** is free in **P** and **Q**, then it is also free in ¬**P**, (**P** ∧ **Q**), (**P** ∨ **Q**), (**P** → **Q**), and (**P** ↔ **Q**); and all free occurrences of variables in a formula **P** other than the variable **x** are also free in the quantified formulas (∀**x**)**P** and (∃**x**)**P**. Thus, for example, all occurrences of *x* and *y* are free in *Faxy*, *By* → *By*, ¬*Faxy*, and *Faxy* ∧ (*By* → *By*); and the single occurrence of *x* is free in both *(∀y)Faxy* and *(∀y)Faxy* ∧ *(∀y)* (*By* → *By*).

On the other hand, all occurrences of *y* in each of the formulas *(∀y)Faxy*, *(∀y)(By* → *By)*, and *(∀y)Faxy* ∧ *(∀y)(By* → *By)* are *bound*. Every free occurrence of a variable **x** in **P** is bound by the quantifier (∀**x**) in (∀**x**)**P** and by the quantifier (∃**x**) in (∃**x**)**P**, as is the occurrence of **x** *within* the quantifier, and these occurrences of **x** are also bound in any formula of which (∀**x**)**P** or (∃**x**)**P** is a subformula. A variable can occur both free and bound in a single formula: the first occurrence of *y* in *Faxy* ∧ *(∀y)(By* → *By)* is free while the remaining three occurrences are bound.

We note two unusual but allowable cases. First, *(∃y)Fxaxa* is a legal formula although the quantifier *(∃y)* doesn't bind any variable other than the *y* occurring within the quantifier, because there is no *y* in the subformula *Fxaxa*. Second, *(∃y)(∀y)Fxaxy* is also a legal formula although no occurrence of *y* in the subformula *(∀y)Fxaxy* is free in that subformula—that means that in this case as well the quantifier *(∃y)* doesn't bind any variable other than the *y* occurring in the quantifier. We usually say that the quantifier *(∃y)* is *trivial* in both of these cases; it's not doing any useful work—each formula with a trivial occurrence of a quantifier will turn out to be equivalent to the formula with the trivial occurrence removed.

We can now define a **closed formula** to be a formula in which every occurrence of a variable is bound. *(∀x)(∀y)Lxy* is a closed formula, since both *x* and *y* are bound in all of their occurrences, but *(∀y)Lxy* is not a closed formula, since the single occurrence of *x* is not bound. We will be interested in closed formulas when we symbolize English and when we define the first-order versions of semantic concepts such as *tautology* and *entailment* as well as in our axiomatic systems.

In our earlier examples we read the quantifiers as referring to every*thing* and some*thing* rather than, say, every*one* and some*one*. Of course, we frequently want to talk about every thing or some thing *of such-and-such a kind*, for example, everything that is a person. In this case, we may do one of two things. First, we may state explicitly that we are only talking of people, and then do the symbolizations. In this case, we call the restricted group to which we refer the ***domain***. If we stipulate that the domain includes all and only people, then formulas like *(∀x)Rx* and *(∀x)(∀y)Lxy* may be read as *everyone runs* and *everyone loves everyone*. In the absence of an explicit stipulation, we assume that the domain includes everything. But even in this case, we can symbolize universal and existential claims about people by making explicit qualifications within our quantified formulas. For example, if we let *Px* symbolize *x is a person*, then *Everyone runs* might be symbolized as *(∀x)(Px → Rx)*: everything is such that *if* it is a person, *then* it runs; that is, *everyone runs*. We can symbolize *Someone runs* as *(∃x)(Px ∧ Rx)*—something both is a person and runs. *Everyone loves someone* can be symbolized as *(∀x)(Px → (∃y)(Py ∧ Lxy)*—every x is such that *if* x is a person then there is some y that is a person and that x loves.

Some other examples of symbolized English sentences (using an unrestricted domain) are:

If John runs, then everyone runs	Rj → (∀x)(Px → Rx)
If John runs, then everyone loves him	Rj → (∀x)(Px → Lxj)
Everyone who loves John runs	(∀x)((Px ∧ Lxj) → Rx)
If someone runs, everyone runs	(∃x)(Px ∧ Rx) → (∀x)(Px → Rx)
Everyone loves everyone who runs	(∀x)(Px → (∀y)((Py ∧ Ry) → Lxy))
Everyone loves someone who runs	(∀x)(Px → (∃y)((Py ∧ Ry) ∧ Lxy))

In each case, alternative (equivalent) symbolizations are possible. For example, the universal quantifiers in the first two formulas could be placed at the very beginning:

$(\forall x)(Rj \rightarrow (Px \rightarrow Rx))$
$(\forall x)(Rj \rightarrow (Px \rightarrow Lxj))$

There are two especially important equivalences involving quantifiers. *Everything is such-and-such* is equivalent to *nothing is not such-and-such*: that is, a universal quantifier $(\forall \mathbf{x})$ can always be replaced with $\neg(\exists \mathbf{x})\neg$, and vice versa. Similarly, an existential quantifier $(\exists \mathbf{x})$ is interchangeable with $\neg(\forall \mathbf{x})\neg$. Thus *Everyone runs*, which we have symbolized as $(\forall x)(Px \rightarrow Rx)$, can equivalently be symbolized as $\neg(\exists x)\neg(Px \rightarrow Rx)$ or, owing to the equivalence of $\neg(Px \rightarrow Rx)$ and $(Px \wedge \neg Rx)$, as $\neg(\exists x)\ (Px \wedge \neg Rx)$—there's no person who doesn't run. *Someone runs*, which is most naturally symbolized as $(\exists x)\ (Px \wedge Rx)$, can also be symbolized as $\neg(\forall x)\neg(Px \wedge Rx)$. Since $\neg(Px \wedge Rx)$ is equivalent to $(Px \rightarrow \neg Rx)$, the formula $(\exists x)\ (Px \wedge Rx)$ is equivalent to $\neg(\forall x)(Px \rightarrow \neg Rx)$—it's not true that every person doesn't run.

We can now symbolize the Sorites argument from Chapter 1. To simplify matters, we'll assume that the domain consists of heights and we'll restate the argument as:

6′ 7″ is tall.
Any height that is ⅛″ less than a tall height is also tall.
Therefore 4′ 7″ is tall.

We let *s* stand for 6′ 7″ and *f* for 4′ 7″, and we symbolize *x is tall* and *x is ⅛″ less than y* as *Tx* and *Exy*. The symbolized argument is:

Ts
$(\forall x)(\forall y)((Tx \wedge Eyx) \rightarrow Ty)$
Tf

(A common equivalent way of symbolizing the second premise is $(\forall x)(Tx \rightarrow (\forall y)(Eyx \rightarrow Ty))$.) In Chapter 1 we introduced Max Black's way of saying that the term *tall* has borderline cases: *There are heights that are neither tall nor not tall.* We can now symbolize this as

$(\exists x)\ (\neg Tx \wedge \neg\neg Tx)$

which, in the present context, we will read as: at least one height is not tall and also not *not* tall. To evaluate the Sorites argument and Black's Problem of the Fringe we must first present the semantics for first-order logic.

3.2 Semantics of Classical First-Order Logic[2]

The basis for the semantics for first-order logic, an *interpretation*, tells us what we are quantifying over, as well as what our constants and predicates stand for:

2 We present a version of so-called *satisfaction semantics*, which was first developed by Tarski (1936).

An **interpretation I** consists of

1. A nonempty set D, called the *domain*
2. An assignment of a (possibly empty) set of n-tuples of members of D to each predicate **P** of arity n:
   $\mathrm{I}(\mathbf{P}) \subseteq \mathrm{D}^n$
3. An assignment of a member of D to each individual constant **a**:
   $\mathrm{I}(\mathbf{a}) \in \mathrm{D}$

An ***n*-tuple** is an ordered set of n items. We use angle brackets when listing the members of an n-tuple. For example, the 2-tuple, or *ordered pair*, consisting of Gina Biggerly and Tina Littleton in that order is written as <*Gina Biggerly, Tina Littleton*>. The idea behind clause 2 is that the sets of entities that stand in the relation denoted by a predicate **P** (or that have the property denoted by the predicate, if its arity is 1) will constitute the n-tuples in I(**P**). For example, if R is to mean *runs* then I(R) consists of all 1-tuples of runners, and if L is to mean *loves* then I(L) consists of all ordered pairs of entities such that the first member of the pair loves the second one. If John runs and John loves Gina and himself, then <John> will be a member of I(R) and <John, John> and <John, Gina> will be members of I(L).

As an example of how the semantics works, we'll use the following interpretation I:

D: set of positive integers
I(P) = {<i>: i ∈ D and i is prime}
    (i.e., {<1>, <2>, <3>, <5>, <7>, <11>, . . .})
I(E) = {<i>: i ∈ D and i is even}
    (i.e., {<2>, <4>, <6>, <8>, <10>, <12>, . . .})
I(G) = {<i, j>: i ∈ D, j ∈ D, and i is greater than j}
    (i.e., {<2,1>, <3,1>, <3,2>, <4,1>, <4,2>, <4,3>, <5,1>, <5,2>, <5,3>, <5,4>, . . .})
I(a) = 1
I(b) = 2
I(c) = 3
I(d) = 4

Actually, this is only *part* of an interpretation, since an interpretation must assign an extension to every predicate and every constant in the language. But it will do, because the values assigned to other predicates and constants won't affect the truth-values of the formulas we'll look at.

We call I(**P**) and I(**a**) the *interpretations* of **P** and of **a** or, to use some well-entrenched philosophical jargon, the ***extensions*** of **P** and of **a**. Note that we do not assign extensions to variables. This is because a variable is like a pronoun. *She*, unlike *Gina Biggerly*, does not by itself denote any particular individual. But we may use a quantifier phrase to indicate which "she" we are talking about, for example, ***every woman** has the property that John loves **her*** (here, we are talking about *every* "she"). To specify how variables are used in conjunction with quantifiers, we first need to define a *variable assignment*: a **variable assignment v** is a function that assigns to each individual variable **x** a member of the domain D: v(**x**) ∈ D.

Our truth-conditions are given in terms of **satisfaction** by a variable assignment on an interpretation:

1. An atomic formula $\mathbf{Pt}_1 \ldots \mathbf{t}_n$ is satisfied by a variable assignment v on an interpretation I if $<I^*(\mathbf{t}_1),\ldots,I^*(\mathbf{t}_n)> \in I(\mathbf{P})$, where $I^*(\mathbf{t}_i)$ is $I(\mathbf{t}_i)$ if $\mathbf{t}_i$ is a constant and is $v(\mathbf{t}_i)$ if $\mathbf{t}_i$ is a variable.
2. A formula ¬**P** is satisfied by a variable assignment v on an interpretation I if **P** is not satisfied by v on I.
3. A formula **P** ∧ **Q** is satisfied by a variable assignment v on an interpretation I if both **P** and **Q** are satisfied by v on I.
4. A formula **P** ∨ **Q** is satisfied by a variable assignment v on an interpretation I if either **P** or **Q** (or both) is satisfied by v on I.
5. A formula **P** → **Q** is satisfied by a variable assignment v on an interpretation I if either **P** is *not* satisfied by v on I or **Q** *is* satisfied by v on I (or both).
6. A formula **P** ↔ **Q** is satisfied by a variable assignment v on an interpretation I if either both **P** and **Q** are satisfied by v on I, or neither **P** nor **Q** is satisfied by v on I.

For the following clauses we need one more definition: an **x-variant** v′ of a variable assignment v is an assignment v′ that assigns the same values as v, except that it may assign a different value to **x**—that is, for any variable **y** other than **x**, v′(**y**) = v(**y**) but v′(**x**) may be any member of the domain. Do note that any variable assignment v is itself included among its **x**-variants.

7. A formula (∀**x**)**P** is satisfied by a variable assignment v on I if **P** is satisfied by *every* **x**-variant of v on I.
8. A formula (∃**x**)**P** is satisfied by a variable assignment v on I if **P** is satisfied by *at least one* **x**-variant of v on I.

Finally, a formula **P** is **true on an interpretation I** if **P** is satisfied by *every* variable assignment v on I, and **P** is **false** on I if **P** is satisfied by *no* variable assignment v on I. It turns out that every closed formula is, according to our definitions, either true or false on any given interpretation because every closed formula will either be satisfied by all variable assignments or be satisfied by none. On the other hand, formulas that aren't closed may fail to be true or false by virtue of being satisfied by some, but not all, variable assignments.

Using our earlier interpretation I, the formulas *Pa*, *Pb*, *Pc*, and *Pd*, respectively, may be read as: *1 is prime*, *2 is prime*, *3 is prime*, and *4 is prime*. The formulas *Pb* and *Pc* are both true on this interpretation. Consider *Pb*. It is true if satisfied by every variable assignment on this interpretation. By clause 1, a variable assignment v will satisfy *Pb* if $<I(b)> \in I(P)$—that is, if $<2> \in I(P)$, since $I(b) = 2$. $<2>$ is a member of

I(*P*), and so every variable assignment will satisfy *Pb*, making the formula true on this interpretation. Note that in this case, and in the case of all other formulas that don't contain variables, we can bypass looking at variable assignments and simply check values assigned by the interpretation—because the assignments made by a variable assignment only matter when a formula contains variables. The formula *Pd* is false on this interpretation because I(*d*), which is <4>, is not a member of I(*P*). The formulas *Gab* and *Gba* may be read as *1 is greater than 2* and 2 *is greater than 1*. *Gab* is false on this interpretation because <I(*a*), I(*b*)>, that is, <1,2>, is not a member of I(*G*), while the formula *Gba* is true because <I(*b*), I(*a*)>, that is, <2,1>, *is* a member of I(*G*). *Gaa*, which may be read *as 1 is greater than 1*, is false because <I(*a*), I(*a*)>, that is, <1,1>, is not a member of I(*G*).

The compound formulas ¬*Pa*, ¬*Pb*, ¬*Pc*, and ¬*Gba* are all false on this interpretation, while ¬*Pd*, ¬*Gab*, and ¬*Gaa* are all true. The formula *Ec* ∧ *Pc (3 is an even prime)* is false because *Ec* is false (<I(*c*)>, that is, <3>, is not a member of I(*E*)); while *Ec* ∨ *Pc (3 is even or prime)* is true because *Pc* is true. The formula *Ec* ∨ ¬*Ec*—which may be read as *Either 3 is even or it isn't*—is true, as is every formula like this one but with a different constant in place of *c*. This is because the formula's truth requires only that each variable assignment satisfy either *Ec* or ¬*Ec*, and we know that that will always be the case no matter what I(*c*) and I(*E*) may be.

The formula *(∃x)Px*, which may be read as *At least one positive integer is prime*, is satisfied by every variable assignment and is therefore true. We may show this by beginning with an arbitrary variable assignment, say, the variable assignment v that assigns the integer 4 to every variable:

	u	v	x	y	z	u_1	v_1	x_1	y_1	z_1	...
v:	4	4	4	4	4	4	4	4	4	4	...

According to clause 8, this assignment v satisfies *(∃x)Px* if at least one *x*-variant of v satisfies *Px*. We may display these variant assignments schematically thus; they are the assignments that assign the value 4 to every variable besides *x* and that assign any value to *x* (including 4, as v itself does):

	u	v	x	y	z	u_1	v_1	x_1	y_1	z_1	...
v:	4	4	4	4	4	4	4	4	4	4	...
v_1:	4	4	1	4	4	4	4	4	4	4	...
v_2:	4	4	2	4	4	4	4	4	4	4	...
v_3:	4	4	3	4	4	4	4	4	4	4	...
v_4:	4	4	5	4	4	4	4	4	4	4	...
v_5:	4	4	6	4	4	4	4	4	4	4	...
					...						

Now, clearly v itself doesn't satisfy *Px* because v(*x*), which is 4, is not a member of I(*P*). But the variable assignment v_2 that assigns 2 to *x*, does satisfy *Px*—as do v_3 and v_4 and every other variable assignment that assigns a prime number to *x*. So because at least one of the *x*-variants of v satisfies *Px*, v satisfies the existential quantification

(∃x)Px. All variable assignments other than v satisfy *(∃x)Px* for the same reason: for each such assignment there is at least one *x*-variant that assigns a prime positive integer to *x*, thus satisfying *Px*. We can summarize the reason for the truth of *(∃x)Px* thus: the extension I(*P*) is nonempty; at least one integer is prime.

The formula *(∀x)Px*, which may be read as *Every positive integer is prime*, is false on I because it is satisfied by no variable assignment v on I. Consider again, for example, the assignment v displayed earlier. According to clause 7, *(∀x)Px* will be satisfied by v only if *Px* is satisfied by every *x*-variant of v. But assignment v itself, for example, is one of these variants and v(*x*) ∉ I(*P*)—4 is not prime—so v doesn't satisfy *Px* (nor does v_1 or v_5). The existence of nonsatisfying *x*-variants entails that v doesn't satisfy *(∀x)Px*. Nor will any other variable assignment satisfy *(∀x)Px*, for the same reason: some members of the domain, members which may be assigned to *x*, are not in the extension of *P*. We conclude that there is no variable assignment that satisfies the universally quantified *(∀x)Px* and therefore that the formula is false on interpretation I.

The reader may wonder why a variable assignment assigns values to every variable, since the values assigned to variables that do not occur free in a particular formula can be ignored when determining the truth-value of that formula. The answer is, at bottom, *perspicuity*: by requiring that every variable assignment assign a value to *every* variable, we make it easier to define the semantics. Having said that, we need consider only the values assigned to free variables when we evaluate formulas. In particular, when we ask whether a closed formula is true, that is, satisfied by every variable assignment v, we can speak complete generally about an arbitrary variable assignment v, ignoring altogether any actual values that v may assign. Specific values will matter only as quantifiers are stripped and alternative variable assignments are examined.

The formula *(∀x)(∃y)Gyx* is true on interpretation I. A variable assignment v will satisfy this formula if every *x*-variant v′ of v satisfies the subformula *(∃y)Gyx*, and *(∃y)Gyx* will be satisfied by an *x*-variant v′ if at last one assignment v″ that is a *y*-variant of v′ satisfies *Gxy*. So we have: v satisfies *(∀x)(∃y)Gyx* if for each value that can be assigned to *x* (by an *x*-variant v′) there is a value that can be assigned to *y* (by a *y*-variant v″) such that *Gyx* is satisfied. That is, the overall formula is satisfied if for each positive integer *i* there is at least one positive integer *j* such that $<j, i>\in$ I(*G*). This condition is met—for each positive integer *i* there is at least one positive integer *j* that is greater than *i*—and so *(∀x)(∃y)Gyx* is true.

On the other hand, the formula *(∃y)(∀x)Gyx*—where the order of the quantifiers has been reversed—is false on interpretation I. A variable assignment v can satisfy *(∃y)(∀x)Gyx* only if at least one *y*-variant v′ of x satisfies *(∀x)Gyx*. But if *(∀x)Gyx* is to be satisfied by a *y*-variant v′ it must be the case that *Gyx* is satisfied by every *x*-variant v of v′. In sum: an assignment v satisfies *(∃y)(∀x)Gyx* if there is at least one value *i* that can be assigned to *y* such that for every value *j* that can be assigned to *x*, *Gyx* is satisfied, that is, if there is at least positive integer *i* such that for every positive integer *j*, $<i, j> \in \mathrm{I}(G)$. But there is no single positive integer *i* that is greater than every positive integer *j*, so the formula $(\exists y)(\forall x)Gyx$ is false on I.

As a final example, the formula $(\forall x)(Ex \rightarrow (\exists y)(Py \wedge Gyx))$—which may be read as *Every even positive integer is smaller than at least one prime positive integer*—is true on interpretation I. A variable assignment v satisfies $(\forall x)(Ex \rightarrow (\exists y)(Py \wedge Gyx))$ if every *x*-variant satisfies $Ex \rightarrow (\exists y)(Py \wedge Gyx)$, and such an assignment v′ will satisfy $Ex \rightarrow (\exists y)(Py \wedge Gyx)$ on the condition that *if* v′ satisfies *Ex then* v′ also satisfies $(\exists y)(Py \wedge Gyx)$. Every *x*-variant v′ that does not satisfy *Ex* will thus trivially satisfy $Ex \rightarrow (\exists y)(Py \wedge Gyx)$. Now consider an *x*-variant v′ that *does* satisfy *Ex*; in this case $\mathrm{v}'(x) \in \mathrm{I}(E)$ and so $\mathrm{v}'(x)$ must be an even integer. Such an assignment v′ will also satisfy the consequent $(\exists y)(Py \wedge Gyx)$ if there is at least one *y*-variant v″ of v′ that satisfies $Py \wedge Gyx$, that is, if $\mathrm{v}''(y) \in \mathrm{I}(P)$ and $<\mathrm{v}''(y), \mathrm{v}''(x)> = <\mathrm{v}''(y), \mathrm{v}'(x)> \in \mathrm{I}(G)$. So the question now is, for each even positive integer *i* (v′(x)), is there at least one positive integer *j* (v″(y)) that both is prime and is greater than *i*? The answer is *yes*, so we may conclude that $Ex \rightarrow (\exists y)(Py \wedge Gxy)$ is satisfied no matter what value is assigned to *x* and therefore that $(\forall x)(Ex \rightarrow (\exists y)(Py \wedge Gxy))$ is true on interpretation I.

Earlier we explained that the formula $Ec \vee \neg Ec$, along with every formula that is like this one but has a different constant in place of *c*, is true on interpretation I. Our reasoning was in fact sufficient to show that the formula is true on *every* interpretation; it is a *tautology* in classical first-order logic. A closed formula is a **tautology** if it is true on every interpretation, and a closed formula is a **contradiction** if it is false on every interpretation.[3] (Note that these are the analogues in first-order logic of the concepts of tautologies and contradictions in propositional logic.) How do we know that $Ec \vee \neg Ec$ is true on every interpretation? We showed that its truth on an interpretation depended on *Ec* being either satisfied or not satisfied by any variable assignment on that interpretation. But this will always be the case—no matter what the interpretations of *E* and *c* are. We can extend the reasoning slightly to show that the universally quantified $(\forall x)(Ex \vee \neg Ex)$—an instance of the Law of Excluded Middle—is also logically true. For this formula is true on an interpretation if satisfied by all variable assignments on that interpretation, as will be the case

[3] Anticipating the first-order axiomatic systems in this text, we have chosen to designate only *closed* formulas as tautologies or contradictions—since only closed formulas will play a role in the axiomatic systems and we want, for example, all tautologies to turn out to be theorems. Other authors allow open formulas to be tautologies true and to be theorems—in such cases, the open formulas have implicit universal quantifiers binding all of their free variables.

The logics resulting from these different policies are "equivalent" in the sense that an open formula is true on every interpretation if and only if the closed formula obtained by prefixing the formula with universal quantifiers for its free variables is true on every interpretation. (The truth of an open formula on every interpretation requires satisfaction by every variable assignment on every interpretation, and the truth of the formula that results from prefixing the formula with universal quantifiers binding all of its variables requires the same thing, given the satisfaction clause for universally quantified formulas.) Given this equivalence it follows that if a universally quantified formula is a tautology by our account, then it as well as the formula(s) that results from removing the initial quantifier(s) will be a tautology on the alternative account, and vice versa. Similar comments apply to other semantic concepts.

if $Ex \vee \neg Ex$ is satisfied by all variable assignments. Clearly this is true, because for each variable assignment will either satisfy Ex or fail to satisfy Ex and in the latter case it will satisfy $\neg Ex$. We may also conclude that, since the extension of the predicate E doesn't figure in this reasoning, a formula expressing the Law of Excluded Middle using *any* unary predicate (as well as any variable) will turn out to be a tautology in classical first-order logic.

Black's formula asserting that there are borderline cases of *tall*, $(\exists x)(\neg Tx \wedge \neg\neg Tx)$, is a contradiction in classical first-order logic—it is false on every interpretation. No matter what $I(T)$ is, no variable assignment can satisfy $\neg Tx \wedge \neg\neg Tx$ because no variable assignment can satisfy both $\neg Tx$ and $\neg\neg Tx$.

A set Γ of closed formulas of classical first-order logic **entails** a closed formula **P** if, whenever all of the formulas in Γ are true **P** is true as well. An argument of first-order logic is **valid** if the set consisting of its premises entails the argument's conclusion. We repeat here the symbolized Sorites argument from Section 3.1:

$$\begin{array}{l} \text{Ts} \\ \underline{(\forall x)\ (\forall y)\ ((Tx \wedge Eyx) \rightarrow Ty)} \\ \text{Tf} \end{array}$$

This argument, as presented, is *not* valid in classical first-order logic because there is at least one interpretation on which the premises are true and the conclusion false. Here is one such interpretation:

D: set of positive integers[4]
$I(T) = \{<u>: u \in D \text{ and } u \text{ is even}\}$
$I(E) = \{<u_1,u_2>: u_1 \in D,\ u_2 \in D, \text{ and } u_1 \text{ evenly divides } u_2$
(i.e., u_1 goes into u_2 without remainder)$\}$
$I(s) = 2$
$I(f) = 3$

The formula *Ts* is true on this interpretation since $I(s) \in I(T)$—the number 2 is even. The formula $(\forall x)(\forall y)((Tx \wedge Eyx) \rightarrow Ty)$ is also true, since no matter what values a variable assignment v may assign to x and y, if $<v(x)> \in I(T)$ and $<v(x), v(y)> \in I(E)$, then it is also the case that $<v(y)> \in I(T)$. That is, if some positive integer is even and evenly divides a second positive integer, then the second integer must be even as well. But the conclusion *Tf* is false since $I(f) \notin I(T)$—the number 3 is *not* even.

Yet we have the feeling that the Sorites argument

6′ 7″ is a tall height.
<u>Any height that is $\frac{1}{8}$″ less than a tall height is also tall.</u>
4′ 7″ is a tall height.

4 The set of positive integers turns out to be a convenient domain for examples; that is why we are using it again. Other domains would also work here. In proving invalidity we always choose clear-cut cases, like those involving crisp predicates applied to positive integers, to make the point.

is valid. What has happened?!!? Well, our feeling is based on an implicit additional premise—that you can get from 6′ 7″ to 4′ 7″ by repeatedly subtracting $^1/_8$″. Without this premise, we can only conclude that 6′ 7″ is tall, since there might not be a height that is $^1/_8$″ less than 6′ 7″. *Logically might not,* that is, since we know that there is in fact such a height.[5] To expedite matters, we will add not one premise but 192 premises: *6′ 6$^7/_8$″ is $^1/_8$″ less than 6′ 7″; 6′ 6$^6/_8$″ is $^1/_8$″ less than 6′ 6$^7/_8$″; 6′ 6$^5/_8$″ is $^1/_8$″ less than 6″ 6$^6/_8$″;...; 4′ 7$^1/_8$″ is $^1/_8$″ less than 4′ 7$^2/_8$″*; and *4′ 7″ is $^1/_8$″ less than 4′ 7$^1/_8$″*. Using s_1 to represent 6′ 7″, s_2 to represent 6′ 6$^7/_8$″,..., down to s_{193} representing 4′ 7″, the Sorites argument *augmented to state the implicit premise explicitly* is

$$
\begin{array}{l}
Ts_1 \\
(\forall x)(\forall y)((Tx \wedge Eyx) \rightarrow Ty) \\
Es_2s_1 \\
Es_3s_2 \\
Es_4s_3 \\
\ldots \\
\underline{Es_{193}s_{192}} \\
Ts_{193}
\end{array}
$$

The augmented argument *is* valid in classical first-order logic. On any interpretation on which all the premises are true, so is the conclusion. Here's why: Assuming that all of the premises are true we will be able to infer that Ts_2 is true, then that Ts_3 is true, and so on, until we finally infer that the conclusion Ts_{193} is true. We'll just show the first step; the rest are similar. From the truth of the first and third premises it must be the case that $<I(s_1)> \in I(T)$ and that $<I(s_2), I(s_1)> \in I(E)$. From the truth of the second premise, we know that whatever v(x) and v(y) may be, if $Tx \wedge Eyx$ is satisfied by v then so is Ty. In particular, this holds when $v(x) = I(s_1)$ and $v(y) = I(s_2)$. Because $Tx \wedge Eyx$ *is* satisfied by v in this case, so is Ty—that is, $<I(s_2)>$ (*which is* $<v(y)>$) is a member of I(T). This means that Ts_2 must be true. Repeating this reasoning we'll conclude that Ts_{193} must be true as well.

3.3 An Axiomatic Derivation System for Classical First-Order Logic

By adding two axiom schemata and one rule to the axiomatic derivation system CLA for classical propositional logic we can produce a sound and complete axiomatic system for classical first-order logic.[6] We call the system *CL∀A*. We

5 OK, I guess we'd better quibble here. Some view mathematics as a branch of logic, and since it is a matter of mathematics that for any positive measure of height in feet and inches there is another measure that is ⅛″ less, those people would say that what we are entertaining is *not* a logical possibility. So to be more specific: it is not a matter of *classical first-order logic* that there is a height that is ⅛″ less than 6′ 7″.

6 The rule and axioms are from Stoll (1961, pp. 388–390). Stoll says that they are essentially the philosopher Bertrand Russell's rule and axioms. Chapter 9 of Stoll proves the soundness and completeness of the resulting system.

stipulate that only *closed* formulas may occur as assumptions or instances of axiom schemata in derivations in CL∀A (as suggested in footnote 3, some axiomatic systems allow open as well as closed formulas in derivations). Our rules preserve closure, so it follows that *every* formula in a derivation will be closed as long as assumptions and instances of axiom schemata that occur in derivations are all closed.

We will now refer to the axiom schemata CL1–CL3 from CLA as *CL∀1–CL∀3*:

CL∀1. $\mathbf{P} \rightarrow (\mathbf{Q} \rightarrow \mathbf{P})$
CL∀2. $(\mathbf{P} \rightarrow (\mathbf{Q} \rightarrow \mathbf{R})) \rightarrow ((\mathbf{P} \rightarrow \mathbf{Q}) \rightarrow (\mathbf{P} \rightarrow \mathbf{R}))$
CL∀3. $(\neg\mathbf{P} \rightarrow \neg\mathbf{Q}) \rightarrow (\mathbf{Q} \rightarrow \mathbf{P})$

while *Modus Ponens* retains its name:

MP. From **P** and $\mathbf{P} \rightarrow \mathbf{Q}$, infer **Q**.

We'll take the existential quantifier to be defined in terms of the universal quantifier:

$$(\exists x)P =_{\text{def}} \neg(\forall x)\neg P$$

so the additional axiom schemata and rules for first-order logic will mention only the universal quantifier. The first new axiom schema is

CL∀4. $(\forall \mathbf{x})(\mathbf{P} \rightarrow \mathbf{Q}) \rightarrow (\mathbf{P} \rightarrow (\forall \mathbf{x})\mathbf{Q})$
where **P** is a formula in which **x** does not occur free

That is, as long as the initial quantifier isn't quantifying over anything in the antecedent of the conditional $\mathbf{P} \rightarrow \mathbf{Q}$, the quantifier may be moved to the consequent. Here's an example of a derivation using this axiom schema:

1	$(\exists x)Hx$	Assumption
2	$(\forall y)((\exists x)Hx \rightarrow Py)$	Assumption
3	$(\forall y)((\exists x)Hx \rightarrow Py) \rightarrow ((\exists x)Hx \rightarrow (\forall y)Py)$	CL∀4, with $(\forall y)((\exists x)Hx \rightarrow Py)$ / $(\forall \mathbf{x})(\mathbf{P} \rightarrow \mathbf{Q})$
4	$(\exists x)Hx \rightarrow (\forall y)Py$	2,3 MP
5	$(\forall y)Py$	1,4 MP

The second axiom schema states that a universally quantified formula implies any one of its instances:

CL∀5. $(\forall \mathbf{x})\mathbf{P} \rightarrow \mathbf{P(a/x)}$
where **a** is any individual constant and the expression ***P(a/x)*** means: *the result of substituting the constant **a** for the variable **x** wherever **x** occurs free in **P***

We call **P(a/x)** a ***substitution instance of* P**. So, for example, if **P** is *Bby*, **a** is *c*, and **x** is *y*, then **P(a/x)** (or *Bby(c/y)*) is *Bbc*. Using CL∀5 we can continue the previous derivation to derive the formula *Pa* (or any other substitution instance of the formula on line 5):

6	(∀y)Py → Pa	CL∀5, with (∀y)Py / (∀**x**)**Px**, a / **a**
7	Pa	5,6 MP

Every instance of each of the two new axiom schemata is a tautology in classical first-order logic.

The new rule in CL∀A is

UG (Universal Generalization). From **P(a/x)** infer (∀**x**)**P**
where **x** is any individual variable, provided that no assumption contains the constant **a** and that **P** itself does not contain the constant **a**.

The first part of the condition ensures that we can derive a universally quantified formula from one of its substitution instances *only if* we have not made any assumptions involving the constant in that substitution instance. It rules out, for example, inferring *(∀x)Rx* from the assumption *Ra*—if we assume that Ann is Romanian it doesn't follow that everyone is Romanian! The second part of the condition rules out generalizations that are not truly general. Without it, we could have derivations like

1	(∀x)Lxx	Assumption
2	(∀x)Lxx → Laa	CL∀5, with (∀x)Lxx / (∀**x**)**P**, a / **a**
3	Laa	1,3 MP
4	(∀x)Lxa	3, UG **MISTAKE!**
5	(∀x) (∀y)Lxy	4, UG

This derivation is not truth-preserving: from the assumption that everything stands in the relation L to itself it does not follow that everything stands in the relation L to everything! Line 4 violates the second half of the condition for correct use of the rule UG, since the formula *Lxa* retains an occurrence of the variable *a* that was generalized on. *With* the condition that **P** does not contain the constant **a**, however, the new rule UG is truth-preserving. A correct use of UG would produce the formula *(∀x)Lxx* on line 4, and this is clearly acceptable since it's the assumption we started with! Of course, we cannot then go on to infer the formula on line 5 since there's no constant in *(∀x)Lxx* to generalize upon.

The following derivation illustrates the combined uses of CL∀5 and UG:

1	$(\forall x)(Fx \rightarrow Gx)$	Assumption
2	$(\forall x)Fx$	Assumption
3	$(\forall x)(Fx \rightarrow Gx) \rightarrow (Fa \rightarrow Ga)$	CL∀5, with $(\forall x)(Fx \rightarrow Gx)$ / $(\forall \mathbf{x})\mathbf{P}$, a / $\mathbf{a}$
4	$Fa \rightarrow Ga$	1,3 MP
5	$(\forall x)Fx \rightarrow Fa$	CL∀5, with $(\forall x)Fx$ / $(\forall \mathbf{x})\mathbf{P}$, a / $\mathbf{a}$
6	Fa	2,5 MP
7	Ga	4,6 MP
8	$(\forall x)Gx$	7, UG

UG has been correctly used on line 8 because *a* does not occur in either of the assumptions, nor does it occur in $(\forall x)Gx$.

We can use UG to derive a quantified version of the Law of the Excluded Middle, $(\forall x)(Ax \vee \neg Ax)$, which is converted to $(\forall x)(\neg Ax \rightarrow \neg Ax)$ when the disjunction is eliminated:

1	$\neg Aa \rightarrow \neg Aa$	CL∀D1, with $\neg Aa$ / $\mathbf{P}$
2	$(\forall x)\ (\neg Ax \rightarrow \neg Ax)$	1, UG

Although the constant *a* occurs in the formula on the first line, that formula is an instance of an axiom schema rather than an assumption, and so the use of UG on the second line is legitimate. Note that in our derivations we may use derived axioms and rules from Chapter 2 since the axioms and rules of CLA are also axioms and rules of CL∀A. We add ∀ to the names of the derived axioms, to make clear that we are now working within the first-order axiomatic system.

To show that $(\forall x)Gx \rightarrow (\exists x)Gx$, which is a tautology, is a theorem we first rewrite the existential quantifier using its definition to obtain $(\forall x)Gx \rightarrow \neg(\forall x)\neg Gx$. We can derive the rewritten formula with the help of derived axiom schemata and rules as follows:

1	$(\forall x)Gx \rightarrow Ga$	CL∀5, with $(\forall x)Gx$ / $(\forall \mathbf{x})\mathbf{P}$, a / $\mathbf{a}$
2	$((\forall x)Gx \rightarrow Ga) \rightarrow (\neg Ga \rightarrow \neg(\forall x)Gx)$	CL∀D5,with $(\forall x)Gx$ / $\mathbf{P}$, Ga / $\mathbf{Q}$
3	$\neg Ga \rightarrow \neg(\forall x)Gx$	1,2 MP
4	$(\forall x)\neg Gx \rightarrow \neg Ga$	CL∀5, with $(\forall x)\neg Gx$ / $(\forall \mathbf{x})\mathbf{P}$, a / $\mathbf{a}$
5	$(\forall x)\neg Gx \rightarrow \neg(\forall x)Gx$	3,4 HS
6	$\neg\neg(\forall x)\neg Gx \rightarrow (\forall x)\neg Gx$	CL∀D3, with $(\forall x)\neg Gx$ / $\mathbf{P}$
7	$\neg\neg(\forall x)\neg Gx \rightarrow \neg(\forall x)Gx$	5,6 HS
8	$(\neg\neg(\forall x)\neg Gx \rightarrow \neg(\forall x)Gx) \rightarrow ((\forall x)Gx \rightarrow \neg(\forall x)\neg Gx)$	CL∀3, with $\neg(\forall x)\neg Gx$ / $\mathbf{P}$, $(\forall x)Gx$ / $\mathbf{Q}$
9	$(\forall x)Gx \rightarrow \neg(\forall x)\neg Gx$	7,8 MP

We may also derive new axiom schemata and rules that are specific to first-order logic. The last formula in the preceding derivation is quite useful, so we will generalize to the derived axiom schema:

CL∀D6. $(\forall \mathbf{x})\mathbf{P} \rightarrow \neg(\forall \mathbf{x})\neg \mathbf{P}$

(We use *6* to number this derived axiom because we already have five derived axioms from Chapter 2.) The preceding derivation justifies this axiom schema since we can replace *(∀x)Gx* with any formula $(\forall \mathbf{x})\mathbf{P}$ and *Ga* with any substitution instance $\mathbf{P(a/x)}$ of $(\forall \mathbf{x})\mathbf{P}$, and the result (with corresponding changes made throughout) will still be a legal derivation. Another derived axiom schema, related to CL∀D6, is

CL∀D7. $\mathbf{P(a/x)} \rightarrow \neg(\forall \mathbf{x})\neg \mathbf{P}$

(The formula is $\mathbf{P(a/x)} \rightarrow (\exists \mathbf{x})\mathbf{P}$ when we substitute the defined existential quantifier.) Proof that this schema can be derived is left as an exercise.

Max Black's fringe formula $(\exists x)\ (\neg Tx \wedge \neg\neg Tx)$ is a contradiction in classical first-order logic, so we know that $\neg(\exists x)\ (\neg Tx \wedge \neg\neg Tx)$ is a tautology in classical first-order logic. It should therefore be a theorem of CL∀A, and indeed it is. We leave it as an exercise to construct a derivation that shows this.

Finally, we would like to derive the conclusion of the Sorites argument

Ts_1
$(\forall x)\ (\forall y)\ ((Tx \wedge Eyx) \rightarrow Ty)$
Es_2s_1
Es_3s_2
Es_4s_3
. . .
$\underline{Es_{193}s_{192}}$
Ts_{193}

from its premises. A useful derived rule for this purpose is

UI (Universal Instantiation). From $(\forall \mathbf{x})\mathbf{P}$ infer $\mathbf{P(a/x)}$.

This rule is justified as follows:

1	$(\forall \mathbf{x})\mathbf{P}$	Assumption
2	$(\forall \mathbf{x})\mathbf{P} \rightarrow \mathbf{P(a/x)}$	CL∀5, with $(\forall \mathbf{x})\mathbf{P}$ / $(\forall \mathbf{x})\mathbf{P}$, **a** / **a**
3	$\mathbf{P(a/x)}$	2,3 MP

This rule will shorten our next derivation considerably. We use the definition of ∧ to rewrite the second premise as the formula $(\forall x)\ (\forall y)\ (\neg(Tx \rightarrow \neg\ Eyx) \rightarrow Ty)$. Here is the derivation:

1	Ts_1	Assumption
2	$(\forall x)\ (\forall y)\ (\neg(Tx \rightarrow \neg\ Eyx) \rightarrow Ty)$	Assumption
3	Es_2s_1	Assumption
4	Es_3s_2	Assumption
5	Es_4s_3	Assumption
...	...	
194	$Es_{193}s_{192}$	Assumption
195	$(\forall y)(\neg(Ts_1 \rightarrow \neg\ Eys_1) \rightarrow Ty)$	2, UI
196	$\neg(Ts_1 \rightarrow \neg Es_2s_1) \rightarrow Ts_2$	195, UI
197	$(Ts_1 \rightarrow \neg Es_2s_1) \rightarrow (Ts_1 \rightarrow \neg Es_2s_1)$	CL∀D1, with $Ts_1 \rightarrow \neg Es_2s_1$ / **P**
198	$Ts_1 \rightarrow ((Ts_1 \rightarrow \neg Es_2s_1) \rightarrow \neg Es_2s_1)$	197, TRAN
199	$(Ts_1 \rightarrow \neg Es_2s_1) \rightarrow \neg Es_2s_1$	1,198 MP
200	$((Ts_1 \rightarrow \neg Es_2s_1) \rightarrow \neg Es_2s_1) \rightarrow$ $(\neg\neg Es_2s_1 \rightarrow \neg(Ts_1 \rightarrow \neg Es_2s_1))$	CL∀D5, with $Ts_1 \rightarrow \neg Es_2s_1$ / **P**, $\neg Es_2s_1$ / **Q**
201	$\neg\neg Es_2s_1 \rightarrow \neg(Ts_1 \rightarrow \neg Es_2s_1)$	199, 200 MP
202	$Es_2s_1 \rightarrow \neg\neg Es_2s_1$	CL∀D4, with Es_2s_1 / **P**
203	$\neg\neg Es_2s_1$	3,202 MP
204	$\neg(Ts_1 \rightarrow \neg Es_2s_1)$	201,203 MP
205	Ts_2	196,204 MP
206	$(\forall y)(\neg(Ts_2 \rightarrow \neg\ Eys_2) \rightarrow Ty)$	2, UI
207	$\neg(Ts_2 \rightarrow \neg Es_3s_2) \rightarrow Ts_3$	206, UI
208	$(Ts_2 \rightarrow \neg Es_3s_2) \rightarrow (Ts_2 \rightarrow \neg Es_3s_2)$	CL∀D1, with $Ts_2 \rightarrow \neg Es_3s_2$ / **P**
209	$Ts_2 \rightarrow ((Ts_2 \rightarrow \neg Es_3s_2) \rightarrow \neg Es_3s_2)$	208, TRAN
210	$(Ts_2 \rightarrow \neg Es_3s_2) \rightarrow \neg Es_3s_2$	205,209 MP
211	$((Ts_2 \rightarrow \neg Es_3s_2) \rightarrow \neg Es_3s_2) \rightarrow$ $(\neg\neg Es_3s_2 \rightarrow \neg(Ts_2 \rightarrow \neg Es_3s_2))$	CL∀D5, with $Ts_2 \rightarrow \neg Es_3s_2$ / **P**, $\neg Es_3s_2$ / **Q**
212	$\neg\neg Es_3s_2 \rightarrow \neg(Ts_2 \rightarrow \neg Es_3s_2)$	210,211 MP
213	$Es_3s_2 \rightarrow \neg\neg Es_3s_2$	CL∀D4, with Es_3s_2 / **P**
214	$\neg\neg Es_3s_2$	4,213 MP
215	$\neg(Ts_2 \rightarrow \neg Es_3s_2)$	212,214 MP
216	Ts_3	207,215 MP
...	...{*repeating 195–205 with appropriate substitutions we end with*}	
2090	Ts_{193}	2081,2089 MP

Given the soundness of the axiomatic system CL∀A, we have now demonstrated the validity of the Sorites argument in classical first-order logic in a second way (the first was the semantic argument in Section 3.2).

A final note about first-order classical logic: unlike the case in propositional logic, the set of theorems of CL∀A (or of any other adequate derivation system for first-order classical logic) is *undecidable*. Equivalently, given the soundness and completeness of CLA, the set of tautologies of classical predicate logic is undecidable. For neither theoremhood nor tautologousness in classical predicate logic is it possible to devise a mechanical test that is guaranteed to yield, for any formula, a correct yes-or-no classification in a finite number of steps. On the other hand, first-order tautologousness (and hence theoremhood) is ***semi-decidable***: there are for example mechanical tests based on resolution or semantic tableaux that will, after a finite number of steps, always yield a correct *yes* classification for any first-order formula that *is* a tautology. But such tests do not yield full decidability because they may fail to yield any answer within a finite number of steps for formulas that are *not* tautologies.[7]

3.4 Exercises

SECTION 3.2

1 Determine the truth-value of each of the following formulas on an interpretation that makes the following assignments:

D: set of positive integers
I(O) = {<u>: u ∈ D and u is odd}
I(S) = {<u_1, u_2>: u_1 ∈ D, u_2 ∈ D, and u_1 squared is u_2}
I(E) = {<u_1, u_2>: u_1 ∈ D, u_2 ∈ D, and u_1 evenly divides u_2}
I(a) = 1
I(b) = 2
I(c) = 3

a. Oa ∧ Oc
b. Oa → Ob
c. (∀x)Ox
d. (∀x)Exx
e. (∀x)¬Sxx
f. (∃x)Sxb
g. (∀x) (∀y)((Ox ∧ Oy) → Exy)
h. (∀x)(∀y)(Exy ∨ Eyx)
i. (∃x)(∀y)(¬Ox ∧ Exy)
j. (∀x)(∀y)(Sxy → Exy)
k. (∃x)(∃y)(Sxy ∧ Syx)
l. (∀x)(Ox → (∃y)(∃z)(Sxy ∧ Syz))
m. (∀x)(∃y)Sxy
n. (∃y)(∀x)Sxy

7 Church (1936) proved that theoremhood in first-order logic is undecidable. See Part 4 of Hunter (1971) for a general proof of undecidability in various first-order systems. A semi-decision procedure for first-order classical logic based on semantic tableaux is presented in Smullyan (1968, pp. 59–60).

SECTION 3.3

2 a. Prove that every instance of the axiom schemata CL∀4 is a tautology in classical first-order logic.

b. Prove that every instance of the axiom schemata CL∀5 is a tautology in classical first-order logic.

3 Construct a derivation that justifies derived axiom schema CL∀D7.

4 Construct a derivation that shows that $\neg(\exists x)\ (\neg Tx \wedge \neg\neg Tx)$ is a theorem of CL∀A. You will first need to use the definitions of the existential quantifier and conjunction to write the formula without those operators.

5 Show that the following rule can be derived in CL∀A:

EG (*Existential Generalization*). From **P(a/x)** infer (∃**x**)**P**

where **x** is any individual variable

6 Construct derivations to show that the following formulas are theorems of CL∀A (rewriting each formula to obtain a formula that contains only negation, the conditional, and the universal quantifier as operators):

a. $(\forall x)(\forall y)Lxy \rightarrow (\exists x)(\exists y)Lxy$

b. $(\exists x)(Fa \rightarrow Gx) \rightarrow (Fa \rightarrow (\exists x)Gx)$

c. $(\exists x)(\forall y)Lxy \rightarrow (\forall y)(\exists x)Lxy$

d. $(\exists x)(\forall y)Lxy \rightarrow (\exists x)Lxx$

4 Alternative Semantics for Truth-Values and Truth-Functions: Numeric Truth-Values and Abstract Algebras

4.1 Numeric Truth-Values for Classical Logic

We've been using the letters **T** and **F** to stand in for the truth-values *true* and *false*. We could just as well use the numerals *1* and *0* to stand for *true* and *false*, recasting truth-tables using these two numerals, for example,

P	**¬P**
1	0
0	1

More interesting, though, is using the *integers* 1 and 0, rather than the numerals that name these integers, *as the truth-values* of formulas in classical logic. A propositional truth-value assignment will consist in the assignment of one of these values to each atomic formula, and we can then numerically define the values for complex formulas rather than simply listing these values in truth-tables. Letting *V(**P**)* mean *the value of **P** on a (numeric) truth-value assignment V*, the following definitions will do the job:

1. V(¬**P**) = 1 – V(**P**)
2. V(**P** ∧ **Q**) = min (V(**P**), V(**Q**)) (i.e., the minimum of these two values)
3. V(**P** ∨ **Q**) = max (V(**P**), V(**Q**)) (i.e., the maximum of these two values)
4. V(**P** → **Q**) = max (1–V(**P**), V(**Q**))
5. V(**P** ↔ **Q**) = min (max (1–V(**P**), V(**Q**)), max (1–V(**Q**), V(**P**)))

Clause 1 "reverses" the value for a negated formula—since (looking at the right-hand side of the formula) 1 – 1 is 0 and 1 – 0 is 1. Clause 2 indicates that a conjunction is only as true as its least true conjunct. By contrast, clause 3 indicates that a disjunction is as true as its most true disjunct. Clause 4 is based on the equivalence of **P** → **Q** and ¬**P** ∨ **Q**, using clauses 1 and 3 to define the truth-conditions for the latter. Clause 5 is based on the equivalence of **P** ↔ **Q** and (**P** → **Q**) ∧ (**Q** → **P**). If we had instead used the equivalence of **P** ↔ **Q** and (**P** ∧ **Q**) ∨ (¬**P** ∧ ¬**Q**) to capture the truth-conditionals for biconditionals, the right-hand side of clause 5 would have been written as max (min (V(**P**), V(**Q**)), min (1–V(**P**), 1–V(**Q**))). This is equivalent to the right-hand side that we have chosen to use for clause 5 (and is left as an exercise to verify).

Clauses 1–5 produce truth-tables that look exactly like our previous classical truth-tables, except that we now have 1 in place of **T** and 0 in place of **F**:

P	¬P
1	0
0	1

P Q	P ∧ Q
1 1	1
1 0	0
0 1	0
0 0	0

P Q	P ∨ Q
1 1	1
1 0	1
0 1	1
0 0	0

P Q	P → Q
1 1	1
1 0	0
0 1	1
0 0	1

P Q	P ↔ Q
1 1	1
1 0	0
0 1	0
0 0	1

Note that conjunction could equivalently be defined in terms of multiplication as $V(\mathbf{P} \wedge \mathbf{Q}) = V(\mathbf{P}) \cdot V(\mathbf{Q})$. This will become important when we turn to fuzzy logic, where truth-values are always defined numerically. And disjunction is almost like addition, except in the case where both disjuncts have the value 1, so an alternative clause for disjunction is $V(\mathbf{P} \vee \mathbf{Q}) = \min(1, V(\mathbf{P}) + V(\mathbf{Q}))$.

Having substituted 1 and 0 for *true* and *false* in clauses defining the values of complex formulas, we need to make the same substitutions in other semantic definitions: a tautology of classical propositional logic is a formula that always has the value 1; a contradiction is a formula that always has the value 0; an argument is valid if its conclusion has the value 1 whenever its premises have the value 1; and so on. Thus we will arrive at the same set of tautologies as we had based on the values *true* and *false*, and other semantic results also remain the same. The difference is that now we can compute values for formulas numerically rather than simply referring to truth-tables. For example, we can show that $A \vee \neg A$ is a tautology of classical propositional logic as follows: $V(A \vee \neg A) = \max(V(A), 1–V(A))$, and when $V(A) = 1$ the maximum is $V(A)$, that is, 1, while when $V(A) = 0$ the maximum is $1 – V(A)$, that is, 1. Thus the formula always has the value 1 and is therefore a tautology.

We can similarly define numerical values for formulas of classical first-order logic. Instead of talking about a variable assignment satisfying or not satisfying a formula, we may talk instead about formulas having the value 1 or 0 on variable assignments. We'll designate the value that a formula **P** has on a variable assignment v on an interpretation I with the notation $I_v(\mathbf{P})$:

1. $I_v(\mathbf{Pt}_1 \ldots \mathbf{t}_n) = 1$ if $<I^*(\mathbf{t}_1), \ldots, I^*(\mathbf{t}_n)> \in I(\mathbf{P})$, where $I^*(\mathbf{t}_i)$ is $I(\mathbf{t}_i)$ if $\mathbf{t}_i$ is a constant and is $V(\mathbf{t}_i)$ if $\mathbf{t}_i$ is a variable, and $I_V(\mathbf{Pt}_1 \ldots \mathbf{t}_n) = 0$ otherwise.
2. $I_v(\neg\mathbf{P}) = 1 – I_v(\mathbf{P})$.
3. $I_v(\mathbf{P} \wedge \mathbf{Q}) = \min(I_v(\mathbf{P}), I_v(\mathbf{Q}))$
4. $I_v(\mathbf{P} \vee \mathbf{Q}) = \max(I_v(\mathbf{P}), I_v(\mathbf{Q}))$
5. $I_v(\mathbf{P} \rightarrow \mathbf{Q}) = \max(1 – I(\mathbf{P}), I_v(\mathbf{Q}))$
6. $I_v(\mathbf{P} \leftrightarrow \mathbf{Q}) = \min(\max(1 – I_v(\mathbf{P}), I_v(\mathbf{Q})), \max(1 – I_v(\mathbf{Q}), I_v(\mathbf{P})))$
7. $I_v((\forall\mathbf{x})\mathbf{P}) = \min\{I_{v'}(\mathbf{P})$: v′ is an **x**-variant of v$\}$
8. $I_v((\exists\mathbf{x})\mathbf{P}) = \max\{I_{v'}(\mathbf{P})$: v′ is an **x**-variant of v$\}$

The notation *min*$\{I_{v'}$*(P): v' is an* **x***-variant of v*$\}$ means: *the minimum value that P has on any* **x***-variant of v*, and similarly *for max*$\{I_{v'}$*(P): v' is an* **x***-variant of v*$\}$. Thus a universally quantified formula has the value 1 only if the formula being quantified over has the value 1 for *every* value of **x**, and an existentially quantified formula will have the value 1 if the formula being quantified over has the value 1 for *at least one* value of **x**. Note that the use of the maximum and minimum functions in clauses 7 and 8 reflects our understanding of the two quantifiers, respectively, in terms of conjunction (min) and disjunction (max).

4.2 Boolean Algebras and Classical Logic

We will be interested in numeric values in the following chapters, but we will also be interested in more abstract characterizations of the semantic structures for our logics, particularly in the case of fuzzy logics. The semantics for classical logic can be studied abstractly using *Boolean algebras.*

To motivate this abstraction, we point out it's possible to use yet other pairs of values for classical propositional logic in place of *true* and *false* or 1 and 0 as long as we define operations for the propositional connectives that preserve the structure of the classical values. As another example, we could use the set $\{5\}$ in place of *true* and the empty set $\varnothing$ in place of *false* and then define the values for complex formulas of classical propositional logic as follows, based an assignment V of one or the other of these values to each atomic formula:

1. $V(\neg \mathbf{P}) = \{5\} - V(P)$

where – is set-theoretic difference, that is, X–Y is the set consisting of all the members of X that are not members of Y; so that V(¬**P**) is $\varnothing$ if V(**P**) is $\{5\}$, and V(¬**P**) is $\{5\}$ when V(**P**) is $\varnothing$.

2. $V(\mathbf{P} \wedge \mathbf{Q}) = V(\mathbf{P}) \bigcap V(\mathbf{Q})$,

where $\bigcap$ is set-theoretic intersection: X $\bigcap$ Y is the set consisting of all items that are members of both X and Y. Thus V(**P** ∧ **Q**) = $\{5\}$ if both V(**P**) and V(**Q**) are $\{5\}$, and $\varnothing$ otherwise.

3. $V(\mathbf{P} \vee \mathbf{Q}) = V(\mathbf{P}) \bigcup V(\mathbf{Q}))$

where $\bigcup$ is set-theoretic union: X $\bigcup$ Y is the set consisting of all items that are members of X or of Y or both. So V(**P** ∨ **Q**) is $\{5\}$ if V(**P**) or V(**Q**) is, and $\varnothing$ otherwise.

4. $V(\mathbf{P} \rightarrow \mathbf{Q}) = \{5\}$ if $V(\mathbf{P}) \subseteq V(\mathbf{Q})$, and $\varnothing$ otherwise

where $\subseteq$ is set-theoretic inclusion: X $\subseteq$ Y if (and only if) every member of X is also a member of Y—so the only case in which V(**P** → **Q**) is $\varnothing$ occurs when V(**P**) is $\{5\}$ and V(**Q**) is $\varnothing$.

5. $V(\mathbf{P} \leftrightarrow \mathbf{Q}) =$ (left as an exercise).

The structure imposed on truth-values by the semantic operations of classical propositional logic—a structure mirrored in the interpretations based on 1 and 0 and on {5} and ∅—yields a *Boolean algebra*.[1] A **Boolean algebra** <B, ∪, ∩, ′, *unit*, *zero*> consists of a set B (the ***domain*** of the Boolean algebra) that contains at least two elements designated as the *unit* and *zero* elements, binary operations ∪ and ∩ (respectively called *join* and *meet*),[2] and a unary operation ′ (called *complementation*), such that the following conditions are satisfied for all members x, y, z of B:

i. $x \cup y = y \cup x$, and $x \cap y = y \cap x$ *(commutation)*
ii. $x \cup (y \cup z) = (x \cup y) \cup z$, and $x \cap (y \cap z) = (x \cap y) \cap z$ *(association)*
iii. $x \cup x = x$, and $x \cap x = x$ *(idempotence)*
iv. $x \cup (x \cap y) = x$, and $x \cap (x \cup y) = x$ *(absorption)*
v. $x \cup (y \cap z) = (x \cup y) \cap (x \cup z)$, and $x \cap (y \cup z) = (x \cap y) \cup (x \cap z)$ *(distribution)*
vi. $x \cup \mathit{zero} = x$, and $x \cap \mathit{unit} = x$ *(identity for join and meet)*
vii. $x \cup x' = \mathit{unit}$, and $x \cap x' = \mathit{zero}$. *(complementation)*

Condition i stipulates that the meet and join operations are **commutative**; condition ii stipulates that they are **associative**; and condition iii stipulates that they are **idempotent**. Condition iv specifies **absorption** laws, while condition v stipulates that each of the two binary operations **distributes** over the other. Condition vi stipulates that the *zero* element and *unit* elements are, respectively, **identity** elements for the join operation ∪ and the meet operation ∩, and condition vii specifies **complementation** laws.

We'll describe the Boolean algebra based on the numeric values 1 and 0. The domain B is the set {1, 0}, where 1 is the *unit* member and 0 is the *zero* member. The maximum and minimum operations—which define disjunction and conjunction—serve as the algebra's meet (∪) and join (∩) operations; and the negation operation 1−, which produces 1−x when applied to x, serves as complementation. It is straightforward to verify that the seven conditions on a Boolean algebra's operations are met for the structure <{1,0}, max, min, 1−, 1, 0>, that is, for all x, y, z ∈ {1, 0}:

i. *Commutation:* max (x, y) = max (y, x), and min (x, y) = min (y, x)
   Proof: Obviously true.
ii. *Association:* max (x, max (y, z)) = max (max (x, y), z), and min (x, min (y, z)) = min (min (x, y), z)
   Proof: Obviously true.

[1] Our definition of Boolean algebras and of lattices (introduced later) follows MacLane and Birkhoff (1999). Alternative (equivalent) definitions of Boolean algebras appear in various places in the literature. These algebras are named after the mathematician George Boole, who first developed them to study logic.

[2] The join and meet symbols are standardly used to denote set-theoretic union and intersection—but in the context of Boolean algebras they are used to denote any operations, set-theoretic or otherwise, that meet the specified conditions.

iii. *Idempotence:* max (x, x) = x, and min (x, x) = x.

Proof: Obviously true.

iv. *Absorption:* max (x, min (x, y)) = x, and min (x, max (x, y)) = x.

Proof of first equation: For any x, y ∈ {1, 0} there are two possibilities: either x ≥ y, or x < y. If x ≥ y, then max (x, min (x, y)) = max (x, y) = x. If x < y, then max (x, min (x, y)) = max (x, x) = x. Either way, the equation holds.

Proof of second equation: Left as an exercise.

v. *Distribution:* max (x, min (y, z)) = min (max (x, y), max (x, z)), and min (x, max (y, z)) = max (min (x, y), min (x, z)).

Proof of first equation: We'll consider the different orderings that x, y, and z can have, as three cases:

a. If x ≥ y and x ≥ z, then max (x, min (y, z)) = x = min (x, x) = min (max (x, y), max (x, z)) because max (x, y) = max (x, z) = x

b. If y ≥ x and z ≥ x, then max (x, min (y, z)) = min (y, z) = min (max (x, y), max (x, z)) because max (x, y) = y and max (x, z) = z

c. If either y ≥ x ≥ z or z ≥ x ≥ y, then max (x, min (y, z)) = x because x ≥ min (y, z), and

if y ≥ x ≥ z then min (max (x, y), max (x, z)) = min (y, x) = x while

if z ≥ x ≥ y then min (max (x, y), max (x, z)) = min (x, z) = x

Proof of second equation: Left as an exercise.

vi. *Identity for join and meet:* max (x, 0) = x, and min (x, 1) = x.

Proof: Obviously true.

vii. *Complementation:* max (x, 1 – x) = 1, and min (x, 1 – x) = 0.

Proof of first equation: If x = 1, then max (x, 1 – x) = max (1, 0) = 1. If x = 0, then max (x, 1 – x) = max (0, 1) = 1.

Proof of second equation: Left as an exercise.

We can similarly prove that either the values **T** and **F** or the values {5} and ∅ taken, respectively, as *unit* and *zero* elements, along with the corresponding operations defining disjunction, conjunction, and negation, form Boolean algebras. When we interpret formulas of propositional logic based on a two-valued Boolean algebra by assigning either *unit* or *zero* to each atomic formula and using the algebra's join, meet, and complement operations to define the, respective, values of disjunctions, conjunctions, and negations, we call this an ***algebraic interpretation*** based on that Boolean algebra and the set of all such interpretations a ***semantics*** based on the Boolean algebra. We will say that a formula of propositional logic is a ***tautology of a Boolean algebra*** (a **BA-tautology**) if the formula evaluates to *unit* under every algebraic interpretation based on that algebra. For every two-valued Boolean algebra we obtain the same set of tautologies for propositional logic (and the same entailments, etc.) owing to the following result:

Result 4.1: Every two-valued Boolean algebra BA = <{*unit*, *zero*}, ∪, ∩, ′, *unit*, *zero*> generates the following truth-tables for assignments of *unit* or *zero* to each atomic formula of propositional logic when $\cup$, $\cap$, and $'$, respectively, define the disjunction, conjunction, and negation operations:

P	**¬P**
unit	*zero*
zero	*unit*

P	**Q**	**P ∧ Q**
unit	*unit*	*unit*
unit	*zero*	*zero*
zero	*unit*	*zero*
zero	*zero*	*zero*

P	**Q**	**P ∨ Q**
unit	*unit*	*unit*
unit	*zero*	*unit*
zero	*unit*	*unit*
zero	*zero*	*zero*

Proof: This follows from the four Boolean algebra conditions

i. $x \cup y = y \cup x$, and $x \cap y = y \cap x$

iii. $x \cup x = x$, and $x \cap x = x$

vi. $x \cup \mathit{zero} = x$, and $x \cap \mathit{unit} = x$

vii. $x \cup x' = \mathit{unit}$, and $x \cap x' = \mathit{zero}$

Consider the table for conjunction. By condition iii, $\mathit{unit} \cap \mathit{unit} = \mathit{unit}$ and $\mathit{zero} \cap \mathit{zero} = \mathit{zero}$. That gives us the first and fourth rows. By condition vi, $\mathit{zero} \cap \mathit{unit} = \mathit{zero}$, which gives us the third row. By condition i, $\mathit{unit} \cap \mathit{zero} = \mathit{zero} \cap \mathit{unit}$ and so $\mathit{unit} \cap \mathit{zero} = \mathit{zero}$ as well, which gives us the second row. The reader will be asked in the exercises to verify that the four conditions also generate the displayed tables for disjunction and negation.

It turns out that conditions ii, iv, and v defining Boolean algebras (the conditions that were not used in the proof of Result 4.1) can be derived from the remaining four conditions when the algebra is two-valued. This will be considerably easier to show after we prove some additional equations that hold in Boolean logics in Section 4.3, so we will defer proof to the exercises for that section.

To round out the Boolean operations corresponding to connectives of propositional logic, we'll use the definitions $\mathbf{P} \to \mathbf{Q} =_{\mathrm{def}} \neg\mathbf{P} \vee \mathbf{Q}$ and $\mathbf{P} \leftrightarrow \mathbf{Q} =_{\mathrm{def}} (\mathbf{P} \to \mathbf{Q}) \wedge (\mathbf{Q} \to \mathbf{P})$. As a consequence the Boolean algebraic operations $\Rightarrow$ and $\Leftrightarrow$ corresponding to the conditional and biconditional satisfy the equations $x \Rightarrow y = x' \cup y$ and $x \Leftrightarrow y = (x \Rightarrow y) \cap (y \Rightarrow x)$. With these definitions the following is a direct consequence of Result 4.1:

Result 4.2: For any two-valued Boolean algebra BA, the set of formulas of propositional logic that are BA-tautologies is exactly the set of classical tautologies under the standard semantics based on **T** and **F**.

Analogously, entailments of propositional logic under any two-valued Boolean semantics—where a set of formulas Γ entails a formula **P** if **P** evaluates to the algebra's *unit* element on every algebraic interpretation such that all of the members of Γ evaluate to *unit*—coincide with entailments under the standard semantics.

Interestingly, we also have the following result:

Result 4.3: The formulas of propositional logic that are tautologies for *every* Boolean algebraic semantics are exactly the tautologies under the standard semantics based on **T** and **F**.

Proof: Because the standard semantics based on **T** and **F** is a Boolean algebraic semantics, every formula that is a tautology for every Boolean algebraic semantics must therefore be a tautology under the standard semantics.

For the converse, we can draw on the fact that the system CLA presented in Chapter 2 is a complete axiomatization for propositional logic under the standard semantics; that is, every tautology under the standard semantics is a theorem of CLA. Then we shall only need to prove that every theorem of CLA is a tautology under every Boolean algebraic semantics—and this we can do by establishing that the three axiom schemata of CLA are tautologous under every Boolean algebraic semantics and that Modus Ponens preserves this property. Specifically, we need to establish that $x \Rightarrow (y \Rightarrow x) = unit$ in every Boolean algebra (the left-hand side is the algebraic formula corresponding to CLA1), and that $(x \Rightarrow (y \Rightarrow z)) \Rightarrow ((x \Rightarrow y) \Rightarrow (x \Rightarrow z)) = unit$ (CLA2) and $(x' \Rightarrow y') \Rightarrow (y \Rightarrow x) = unit$ (CLA3)—again, the reader will be asked to prove these equations in the exercises for Section 4.3, as well as the claim that Modus Ponens preserves tautologousness in any Boolean algebra.

Apropos of Result 4.3, our only examples of Boolean algebras so far have been two-valued. As a final and more general example, we note that given any nonempty set S there is a standard way to generate a Boolean algebra using set-theoretic operations. Let $\cup$ and $\cap$ denote set-theoretic union and intersection, let $'$ denote complementation relative to the set S (i.e., for any subset X of S, X$'$ will be the set of elements of S that are *not* members of X), and let P(S) denote the **power set** of S, that is, the set of all subsets of S. The structure $<P(S), \cup, \cap, ', S, \varnothing>$ is then a Boolean algebra. To verify this we need to check that each of conditions i–vii holds for this structure. For any sets X and Y, $X \cup Y = Y \cup X$ and $X \cap Y = Y \cap X$, so condition i is met. Moving to condition vi, we note that for any set $X \subseteq S$, $X \cup \varnothing = X$ and $X \cap S = X$. The interested reader can verify that all of the other conditions hold as well. Thus it is not surprising that the two-valued semantics based on $\{5\}$ and $\varnothing$ that we presented for classical logic is a Boolean algebra.

4.3 More Results about Boolean Algebras

Not all of the seven conditions of Section 4.2 are needed to define Boolean algebras. Conditions ii, iii, and iv are derivable from the others (we explain later why we have nevertheless included them in the definition). Here is how the first idempotence condition (iii) can be derived:

$$
\begin{aligned}
x \cup x &= (x \cup x) \cap \mathit{unit} && \text{(identity for meet)}\\
&= (x \cup x) \cap (x \cup x') && \text{(complementation)}\\
&= x \cup (x \cap x') && \text{(distribution)}\\
&= x \cup \mathit{zero} && \text{(complementation)}\\
&= x && \text{(identity for join)}
\end{aligned}
$$

To derive the first absorption condition we first derive the law $\mathit{unit} \cup x = \mathit{unit}$, which we call *unit consumption*:

$$
\begin{aligned}
\mathit{unit} \cup x &= (\mathit{unit} \cup x) \cap \mathit{unit} && \text{(identity for meet)}\\
&= \mathrm{unit} \cap (x \cup \mathit{unit}) && \text{(commutation, twice)}\\
&= (x \cup x') \cap (x \cup \mathit{unit}) && \text{(complementation)}\\
&= x \cup (x' \cap \mathit{unit}) && \text{(distribution)}\\
&= x \cup x' && \text{(identity for meet)}\\
&= \mathit{unit} && \text{(complementation)}
\end{aligned}
$$

and then we use unit consumption to derive the first absorption condition (iv):

$$
\begin{aligned}
x \cup (x \cap y) &= (x \cap \mathit{unit}) \cup (x \cap y) && \text{(identity for meet)}\\
&= x \cap (\mathit{unit} \cup y) && \text{(distribution)}\\
&= x \cap \mathit{unit} && \text{(unit consumption)}\\
&= x && \text{(identity for meet)}
\end{aligned}
$$

To derive the first association condition we begin with the following, in which the right-hand side $((x \cup y) \cup z)$ of the formula is associated to $(x \cup (y \cup z))$:[3]

$$
\begin{aligned}
x \cap ((x \cup y) \cup z) &= (x \cap (x \cup y)) \cup (x \cap z) && \text{(distribution)}\\
&= x \cup (x \cap z) && \text{(absorption)}\\
&= x && \text{(absorption)}\\
&= x \cup (x \cap (y \cup z)) && \text{(absorption)}\\
&= (x \cap x) \cup (x \cap (y \cup z)) && \text{(idempotence)}\\
&= x \cap (x \cup (y \cup z)) && \text{(distribution)}
\end{aligned}
$$

(Note that we have already shown how to derive idempotence from conditions i and v–vii on Boolean algebras.) Next we show that the same result can be derived when we replace x on the left-hand side with its complement:

[3] This proof, which admittedly is tricky and not at all the obvious way to go, is based on an outline in Stoll (1961, p. 253).

$$
\begin{aligned}
x' \cap ((x \cup y) \cup z) &= (x' \cap (x \cup y)) \cup (x' \cap z) && \text{(distribution)}\\
&= ((x' \cap x) \cup (x' \cap y)) \cup (x' \cap z) && \text{(distribution)}\\
&= (\mathit{zero} \cup (x' \cap y)) \cup (x' \cap z) && \text{(complementation)}\\
&= (x' \cap y) \cup (x' \cap z) && \text{(identity for join)}\\
&= x' \cap (y \cup z) && \text{(distribution)}\\
&= \mathit{zero} \cup (x' \cap (y \cup z)) && \text{(identity for join)}\\
&= (x' \cap x) \cup (x' \cap (y \cup z)) && \text{(complementation)}\\
&= x' \cap (x \cup (y \cup z)) && \text{(distribution)}
\end{aligned}
$$

Because of these identities we can assert

$$(x \cap ((x \cup y) \cup z)) \cup (x' \cap ((x \cup y) \cup z)) = (x \cap (x \cup (y \cup z))) \cup (x' \cap (x \cup (y \cup z)))$$

and from this we can derive

$$(((x \cup y) \cup z) \cap x) \cup (((x \cup y) \cup z) \cap x') = ((x \cup (y \cup z)) \cap x) \cup ((x \cup (y \cup z)) \cap x')$$

by commutation on both sides, then

$$((x \cup y) \cup z) \cap (x \cup x') = (x \cup (y \cup z)) \cap (x \cup x')$$

by distribution, then

$$((x \cup y) \cup z) \cap \mathit{unit} = (x \cup (y \cup z)) \cap \mathit{unit}$$

by complementation, and finally

$$(x \cup y) \cup z = x \cup (y \cup z)$$

by identity for meet. This last formula is the first associative condition for Boolean algebras.

The second idempotence, absorption, and association conditions can be similarly derived from conditions i and v–vii and are left as an exercise.

We have included the derivable conditions ii–iv in the definition of Boolean algebras to make clear the connection between Boolean algebras and another type of algebraic structure called a *lattice*. A **lattice** is a structure $<L, \cup, \cap>$ for which conditions i–iv of Boolean algebras hold. If condition v also holds, the lattice is said to be ***distributed***. If the set L contains two elements that can fill the role of *zero* and *unit* in the identity condition vi, then those elements will be the *zero* and *unit* elements and the lattice is accordingly said to be ***bounded*** or, alternatively, to ***contain zero and unit elements***. If in addition corresponding to each member x of L there is a member y of L such that $x \cup y = \mathit{unit}$ and $x \cap y = \mathit{zero}$, then each such x and y are called complements (because y functions as x' in the complementation condition for Boolean algebras) and the lattice is said to be ***complemented***. So a

Boolean algebra is a special type of lattice, namely, a complemented distributive lattice with *zero* and *unit* elements.

Each of the conditions defining Boolean algebras is specified by a pair of equations, where the second equation results from the first by exchanging the meet and join operations in each formula and exchanging *zero* and *unit*. Such pairs are called ***duals***, and the fact that the conditions come as dual pairs entails that in a Boolean algebra the dual of *any* provable equation is also provable (by using the same proof, except that each formula is replaced by its dual). So for example, since we can show that $unit = zero'$ as follows:

$unit = zero \cup zero'$	(complementation)
$= zero' \cup zero$	(commutation)
$= zero'$	(identity for join)

we know that $zero = unit'$ is also provable using the dual formulas:

$zero = unit \cap unit'$	(complementation)
$= unit' \cap unit$	(commutation)
$= unit'$	(identity for meet)

The following laws also hold in every Boolean algebra:

Double Negation Law: $x'' = x$
DeMorgan's Laws: $(x \cup y)' = x' \cap y'$, and $(x \cap y)' = x' \cup y'$

We first derive Double Negation:

$x'' = x'' \cup zero$	(identity for join)
$= x'' \cup (x \cap x')$	(complementation)
$= (x'' \cup x) \cap (x'' \cup x')$	(distribution)
$= (x'' \cup x) \cap (x' \cup x'')$	(commutation)
$= (x'' \cup x) \cap unit$	(complementation)
$= (x'' \cup x) \cap (x \cup x')$	(complementation)
$= (x \cup x') \cap (x \cup x'')$	(commutation, twice)
$= x \cup (x' \cap x'')$	(distribution)
$= x \cup zero$	(complementation)
$= x$	(identity for join)

To establish the first DeMorgan Law, we will use the

Unique Complement Principle for Boolean Algebras:
If $x \cup y = unit$ and $x \cap y = zero$, then $y = x'$.

Proof: Assume that (a) $x \cup y = unit$ and (b) $x \cap y = zero$. Then

$$\begin{aligned}
y &= y \cap unit && \text{(identity for meet)}\\
&= y \cap (x \cup x') && \text{(complementation)}\\
&= (y \cap x) \cup (y \cap x') && \text{(distribution)}\\
&= (x \cap y) \cup (y \cap x') && \text{(commutation)}\\
&= zero \cup (y \cap x') && \text{(by assumption (b))}\\
&= (x \cap x') \cup (y \cap x') && \text{(complementation)}\\
&= (x' \cap x) \cup (x' \cap y) && \text{(commutation, twice)}\\
&= x' \cap (x \cup y) && \text{(distribution)}\\
&= x' \cap unit && \text{(by assumption (a))}\\
&= x' && \text{(identity for meet)}
\end{aligned}$$

The first DeMorgan Law can now be established as follows:

A. $$\begin{aligned}
(x \cup y) \cup (x' \cap y') &= ((x \cup y) \cup x') \cap ((x \cup y) \cup y') && \text{(distribution)}\\
&= ((y \cup x) \cup x') \cap ((x \cup y) \cup y') && \text{(commutation)}\\
&= (y \cup (x \cup x')) \cap (x \cup (y \cup y')) && \text{(association, twice)}\\
&= (y \cup unit) \cap (x \cup unit) && \text{(complementation, twice)}\\
&= unit \cup unit && \text{(unit consumption, twice)}\\
&= unit && \text{(idempotence)}
\end{aligned}$$

B. $(x \cup y) \cap (x' \cap y') = zero$—is left as an exercise.

C. By the Unique Complement Principle, it follows from A and B that

$$(x \cup y)' = x' \cap y'.$$

The second DeMorgan Law has a dual proof, which is left as an exercise.

Once we have characterized the algebra of truth-values corresponding to a logical system—which is Boolean algebra in the case of classical propositional logic—we can prove things about that system algebraically rather than by reference to truth-tables. For example, we know that $A \vee \neg A$ is a tautology of classical propositional logic. But now we can prove this fact algebraically. Under any Boolean algebraic interpretation, the value of $A \vee \neg A$ is $x \cup x'$ if the value of A is x, and complementation condition vii sets $x \cup x' = unit$. Thus $A \vee \neg A$ is a tautology under every Boolean algebraic interpretation and therefore a tautology of classical propositional logic. As another example, we have shown that the DeMorgan Laws hold true in every Boolean algebra, and on that basis we may conclude that the formula $\neg(A \vee B)$ of classical propositional logic is equivalent to $\neg A \wedge \neg B$ and that $\neg(A \wedge B)$ is equivalent to $\neg A \vee \neg B$. These two equivalences give us the DeMorgan Laws used in Section 2.3 of Chapter 2. Two other equivalences used there, the Distribution equivalences, are the propositional logic counterparts to distribution in Boolean algebras; and Double Negation in that section is the propositional logic version of the Double Negation Law that we have show to hold true in every Boolean algebra.

In Section 4.2 we used the definition $\mathbf{P} \rightarrow \mathbf{Q} =_{def} \neg\mathbf{P} \vee \mathbf{Q}$ from classical propositional logic to give us a Boolean algebraic conditional operation $\Rightarrow$ satisfying the equation $x \Rightarrow y = x' \cup y$. But there is another (equivalent) way to define Boolean algebraic conditional operations, based on the standard lattice-theoretic ordering relation. When considering a lattice as such (i.e., an algebra in which conditions i–iv for Boolean algebras hold), there is a natural **ordering relation** $\leq$ on elements of the lattice defined as

$$x \leq y =_{def} x \cap y = x.$$

We can also use this definition for Boolean algebras since, as we noted earlier, a Boolean algebra is a special type of lattice (namely, a complemented distributed lattice with zero and unit elements). In a Boolean algebra $<P(S), \cup, \cap, ', S, \varnothing>$ (where S is a nonempty set), for example, the ordering relation $\leq$ so defined turns out to be the subset relation $\subseteq$—that is, $X \subseteq Y$ if and only if $X \cap Y = X$. For the Boolean algebra $<\{1,0\}$, max, min, 1–, 1, 0$>$ the ordering relation $\leq$ is simply the numeric ordering relation because here x is less than or equal to y if and only if $\min(x, y) = x$.

In every lattice the relation $\leq$ is **reflexive** (for all x, $x \leq x$), **antisymmetric** (for all x and y, if $x \leq y$ and $y \leq x$, then $x = y$), and **transitive** (for all x, y, and z, if $x \leq y$ and $y \leq z$, then $x \leq z$). To show that the relation is reflexive, we note that for any x, $x \cap x = x$ because $\cap$ is defined to be idempotent in every lattice and so $x \leq x$ by the definition of the ordering relation. To show that $\leq$ is antisymmetric we must prove that for any x and y, if $x \cap y = x$ and $y \cap x = y$, then $x = y$. This is straightforward, since $x \cap y = y \cap x$ by the requirement of commutativity in every lattice. It is left as an exercise to show that the relation $\leq$ must be transitive in a lattice. We also note that because $x \cap y = x$ if and only if $x \cup y = y$ (see exercises), we also have

$$x \leq y \text{ if and only if } x \cup y = y.$$

We can use the lattice ordering $\leq$ to define conditional operations $\Rightarrow$ in Boolean algebras to be operations that satisfy

$$x \Rightarrow y = \textit{unit} \text{ if and only if } x \leq y.$$

In the algebra of classical truth-values, for example, this yields exactly the logical conditional operation. Consider the classical truth-values under their lattice ordering. Because the ordering is reflexive we have $\mathbf{T} \leq \mathbf{T}$ and $\mathbf{F} \leq \mathbf{F}$. Moreover, because *zero* $\cup$ *unit* = *unit*, we will always have *zero* $\leq$ *unit* and so $\mathbf{F} \leq \mathbf{T}$. However, the converse (*unit* $\leq$ *zero*) does not hold when *unit* and *zero* are distinct. Given this ordering, the truth-value of a conditional $\mathbf{P} \rightarrow \mathbf{Q}$ in classical logic is **T** if and only if $V(\mathbf{P}) \leq V(\mathbf{Q})$.

Before closing, we note that in addition to characterizating the semantics for classical propositional logic there is another important and related way that Boolean algebras are used to study classical propositional logic. For any logical system there is a special type of algebra called a Lindenbaum algebra, constructed from equivalence classes of formulas in the system, and the Lindenbaum algebras for classical propositional logics are all Boolean algebras. Lindenbaum algebras are beyond the scope of this text, so we refer the interested reader to Dunn and Hardegree (2001).[4]

4.4 Exercises

SECTION 4.1

1 Prove that min(max(1–V(**P**), V(**Q**)), max(1–V(**Q**), V(**P**))), which we used to define the truth-conditions for conditional formulas is equivalent to the alternative definition based on the equivalence of **P** ↔ **Q** and (**P** ∧ **Q**) ∨ (¬**P** ∧ ¬**Q**):

max(min(V(**P**), V(**Q**)), min(1–V(**P**), 1–V(**Q**))).

SECTION 4.2

2 Complete the definition of clause 5:

5. V(P ↔ Q) = . . .

for a semantics in which the values *true* and *false* are replaced by {5} and ∅, respectively, as suggested at the beginning of Section 4.2.

3 Complete the proof that <{1,0}, max, min, 1–, 1, 0> forms a Boolean algebra by proving that the second equations for absorption, distribution, and complemention hold in this structure.

4 Complete the proof of Result 4.1: show that conditions i, iii, vi, and vi of (two-valued) Boolean algebras generate the following tables when negation and disjunction are defined as algebraic complementation and join:

P	¬P
unit	*zero*
zero	*unit*

P	Q	P ∨ Q
unit	*unit*	*unit*
unit	*zero*	*unit*
zero	*unit*	*unit*
zero	*zero*	*zero*

SECTION 4.3

5 Prove the dual to *unit* consumption, which we may call *zero consumption*:

zero ∩ x = *zero*.

6 Show how to derive the second idempotence, absorption, and association conditions for Boolean algebras from conditions i and v–vii of Boolean algebras.

[4] Lindenbaum algebras are named after the logician Adolf Lindenbaum and are sometimes called *Lindenbaum-Tarski algebras* after Alfred Tarski as well. These algebras were studied by both logicians.

7 Prove that the equality

$$(x \cup y) \cap (x' \cap y') = zero$$

which was used to establish the first DeMorgan Law holds in every Boolean algebra.

8 Prove that the second DeMorgan Law holds in every Boolean algebra.

9 Prove that the following formulas are tautologies of classical propositional logic by showing that they must evaluate to *unit* under any Boolean algebraic semantics:

a. $\neg(P \wedge \neg P)$

b. $(P \wedge Q) \rightarrow (P \vee Q)$

c. $\neg P \rightarrow (P \rightarrow Q)$

10 In Section 4.2 we claimed that Boolean algebra conditions ii, iv, and v can be derived from the remaining four conditions for every Boolean algebra in which the domain B contains exactly two elements. Prove this.

Hint: You can derive the the first associativity condition $x \cup (y \cup z) = (x \cup y) \cup z$, for example, by looking at four cases that among them will cover all possible combinations of values for x, y, and z in a two-valued Boolean algebra:

a. $x = unit$ (y and z can each be either *unit* or *zero*)

b. $y = unit$

c. $z = unit$

d. $x = y = z = zero$.

For case a, we have $unit \cup (y \cup z) = unit$ by unit consumption, and $(unit \cup y) \cup z = unit \cup z = unit$ by the same law, thus establishing the first associativity condition for case a. The reader can pick up the proof from here.

11 Prove that the lattice ordering relation $\leq$, defined as

$$x \leq y \text{ if and only if } x \cap y = x,$$

is transitive.

12 Prove that in every lattice, $x \cap y = x$ if and only if $x \cup y = y$. *Hint:* The absorption condition will prove useful.

13 Complete the proof of Result 4.3 by showing that the following hold in every Boolean algebra:

a. $x \Rightarrow (y \Rightarrow x) = unit$

b. $(x \Rightarrow (y \Rightarrow z) \Rightarrow ((x \Rightarrow y) \Rightarrow (x \Rightarrow z)) = unit$

c. $(x' \Rightarrow y') \Rightarrow (y \Rightarrow x) = unit$

d. Modus Ponens preserves tautologousness in any Boolean algebra:

$$\text{if } x = unit \text{ and } x \Rightarrow y = unit \text{ then } y = unit.$$

5 Three-Valued Propositional Logics: Semantics

5.1 Kleene's "Strong" Three-Valued Logic

We began Chapter 1 by noting that sentences concerning borderline cases of vague predicates pose counterexamples to the Principle of Bivalence. For example, the sentence *Mary Middleford is tall* appears to be neither true nor false. We begin our exploration of logics for vagueness by dropping the Principle of Bivalence and allowing sentences to be either true (**T**), false (**F**), or neither true nor false (**N**—if you like, you may also say that **N** is neutral). This gives rise to three-valued (trivalent) systems of logic.[1] We use the same language as classical propositional logic. Truth-value assignments can now assign **N** (as well as **T** or **F**) to atomic formulas, and we'll use this value to signal the application of a vague predicate to a borderline case.

How are the truth-functions for the standard propositional connectives defined over the three values? There are several plausible choices, and the set of truth-functions we choose will define a specific system of three-valued logic. In this chapter we present four well-known systems of three-valued logic. Many others have been developed, but these four systems are sufficient to explore the flavor of three-valued logics and how they might be used to tackle problems associated with vagueness.[2]

We begin with a system developed by the mathematician Stephen Kleene (Kleene 1938). We'll call this first system $K^S{}_3$—the *S* stands for *strong*, a term that Kleene used to distinguish the connectives in this system from those of another three-valued system he developed (which is identical to Bochvar's internal

1 There are also systems that admit "truth-value gaps" rather than a third truth-value, for example, supervaluational logics (introduced in van Fraassen [1966]; for an application to vagueness see Fine [1975]). We shall ignore "gappy" logics because fuzzy logics are generalizations of trivalent (three-valued) logics rather than supervaluational ones, and our general objections to trivalent accounts of vagueness in Chapter 10 also apply to gappy accounts.

2 An excellent introduction to a wide variety of trivalent logics can be found in Rescher (1969). More advanced material is covered in the also excellent text Gottwald (2001).

system, to be introduced in Section 5.3). The negation truth-function in K^S_3 is defined as

P	$\neg_K$**P**
T	F
N	N
F	T

(We will subscript connectives within nonclassical logical systems to make clear which system we're working in.) Note that when **P** has one of the values **T** or **F** (we'll call these the *classical* truth-values), $\neg_K$**P** is defined as in classical logic. When **P** has the value **N**, reflecting a vague predicate's application to a borderline case, so does its negation. If *Mary Middleford is tall* is neither true nor false, *Mary Middleford is not tall* is also neither true nor false.

The truth-functions corresponding to the binary connectives in Kleene's system are

P $\wedge_K$ **Q**

P\Q	T	N	F
T	T	N	F
N	N	N	F
F	F	F	F

P $\vee_K$ **Q**

P\Q	T	N	F
T	T	T	T
N	T	N	N
F	T	N	F

P $\rightarrow_K$ **Q**

P\Q	T	N	F
T	T	N	F
N	T	N	N
F	T	T	T

P $\leftrightarrow_K$ **Q**

P\Q	T	N	F
T	T	N	F
N	N	N	N
F	F	N	T

In these tables the expression ***P**********Q*** means that **P** has the value in the column listed below while **Q** has the value in the row listed to the right. So, for example, each row beginning with **T** covers cases where **P** has the value **T**, and each column headed by **T** covers cases where **Q** has the value **T**. The intersection of the **T** row and the **T** column represents the case where both **P** and **Q** have the value **T**. In the table for conjunction **T** is listed at this intersecting point—meaning that the conjunction **P** $\wedge_K$ **Q** has the value **T** when both **P** and **Q** have the value **T**.

Each of these truth-functions agrees with classical logic when the arguments are both either **T** or **F**. For example, restricting attention to the four corners of the truth-table for conjunction:

P $\wedge_K$ **Q**

P\Q	T	N	F
T	T	N	F
N	N	N	F
F	F	F	F

we see that the truth-value of a conjunction in these cases is the same as it would be in classical logic. Connectives with this property are *normal*:[3] a propositional

[3] This terminology, along with the technical use of the term *uniform* a few paragraphs hence, is from Rescher (1969, pp. 54–57).

connective in a three-valued logical system is **normal** if, whenever the connective combines formulas with classical truth-values, the resulting formula has the same truth-value as it does in classical logic (dropping the subscripts on the connectives, of course). We will also say that the truth-function denoted by the connective is normal.

How are the remaining values in the truth-table determined? Let's continue to look at conjunction. A classical conjunction is false whenever at least one conjunct is false, no matter what the value of the other conjunct. The same is true of Kleene's conjunction—the row and column representing the falsity of one of the conjuncts **P** and **Q** uniformly have the value **F**:

$\mathbf{P} \wedge_K \mathbf{Q}$

P \ Q	T	N	F
T	T	N	**F**
N	N	N	**F**
F	**F**	**F**	**F**

We say that a propositional connective in a three-valued system is ***uniform*** if, whenever the truth-value of a formula formed with that connective is uniquely determined by the truth-value of one of its constituent formulas in classical logic, the truth-value of the formula formed with that connective is also uniquely so determined in the three-valued system. (We will also say that the truth-function denoted by the connective is uniform.) In classical logic a false conjunct guarantees the falsehood of a conjunction, and the fact that this is also the case in $K^S{}_3$ means that conjunction is uniform in this system. The other connectives are uniform as well.

Normality and uniformity account for all but three of the values—namely, the three **N**s—in the truth-table for conjunction. Generally, the value **N** appears in $K^S{}_3$ whenever neither normality nor uniformity requires a particular value for one of the connectives. Owing to normality, these cases always involve **N** as one or both of the arguments to the truth function. Thus a conjunction has the value **N** when at least one conjunct has this value and neither conjunct has the value **F**.

Disjunction in $K^S{}_3$ is normal and it is also uniform since the truth of one disjunct is sufficient for the truth of the whole, as is the case in classical logic. The $K^S{}_3$ conditional and biconditional are both normal and uniform. The conditional is uniform because it forms a true formula whenever the antecedent is false or the consequent is true. (Note that in the case of the biconditional, there is no case in classical logic where the truth-value of a compound formula can be determined by the truth-value of only one of its immediate components—so the biconditional is *trivially* uniform.) Finally, negation in $K^S{}_3$ is also normal and uniform (the latter in an uninteresting way since it is a unary rather than binary connective). Moreover, it is only when neither normality nor uniformity determines a truth-value that one of these functions will assign the value **N**.

Here are truth-tables for some formulas in $K^S{}_3$:

P	P	$\vee_K$	$\neg_K$	P
T	T	T	F	T
N	N	N	N	N
F	F	T	T	F

P	Q	P	$\rightarrow_K$	$(P$	$\rightarrow_K$	$Q)$
T	T	T	T	T	T	T
T	N	T	N	T	N	N
T	F	T	F	T	F	F
N	T	N	T	N	T	T
N	N	N	N	N	N	N
N	F	N	N	N	N	F
F	T	F	T	F	T	T
F	N	F	T	F	T	N
F	F	F	T	F	T	F

P	Q	$(P$	$\wedge_K$	$Q)$	$\rightarrow_K$	$(P$	$\vee_K$	$Q)$
T	T	T	T	T	T	T	T	T
T	N	T	N	N	T	T	T	N
T	F	T	F	F	T	T	T	F
N	T	N	N	T	T	N	T	T
N	N	N	N	N	N	N	N	N
N	F	N	F	F	T	N	N	F
F	T	F	F	T	T	F	T	T
F	N	F	F	N	T	F	N	N
F	F	F	F	F	T	F	F	F

The first formula, an instance of the Law of Excluded Middle, can have the value **N** as well as the value **T** in $K^S{}_3$. It shouldn't be surprising that this classical tautology can fail to have the value **T** in $K^S{}_3$, since we noted in Chapter 1 that the Principle of Bivalence (which three-valued logic rejects) and the Law of Excluded Middle are closely (although not necessarily!) related. The third formula is, like the Law of Excluded Middle, a tautology of classical logic, but it can also have the value **N** in $K^S{}_3$. The second formula, neither a tautology nor a contradiction of classical logic, can have any of the three values **T**, **F**, or **N** in $K^S{}_3$.

We have introduced $K^S{}_3$ as a possible three-valued logic for vagueness, so it is interesting to note that Kleene's motivation for presenting this three-valued system was altogether different. On Kleene's interpretation, the value **N** means *not defined* rather than simply *neither true nor false*. Kleene introduced his system in connection with mathematical functions that may be undefined for certain values (just as division by 0 is undefined), and of the atomic formulas, only those in which all functions *are* defined for their arguments would one of the values be **T** or **F**. Nevertheless, athough Kleene's motivation was different from ours, his system nevertheless turns out to be one reasonable three-valued logic for vagueness. For example, if **P** is true but **Q** is neither true nor false because it concerns a borderline case, then $\mathbf{P} \wedge_K \mathbf{Q}$ is also neither true nor false—it's not true, since that would require the truth of both conjuncts, but it's also not false, since neither conjunct is false.

We may choose to designate some connectives as primitive and introduce the others as defined. In fact, any choice of primitive connectives and accompanying definitions for the others that works in classical logic also works in $K^S{}_3$. So, for example, we can take $\neg_K$ and $\wedge_K$ as primitive connectives and introduce the other ones with the definitions

$$\mathbf{P} \vee_K \mathbf{Q} =_{\text{def}} \neg_K (\neg_K \mathbf{P} \wedge_K \neg_K \mathbf{Q})$$
$$\mathbf{P} \rightarrow_K \mathbf{Q} =_{\text{def}} \neg_K (\mathbf{P} \wedge_K \neg_K \mathbf{Q})$$
$$\mathbf{P} \leftrightarrow_K \mathbf{Q} =_{\text{def}} \neg_K (\mathbf{P} \wedge_K \neg_K \mathbf{Q}) \wedge_K \neg_K (\neg_K \mathbf{P} \wedge_K \mathbf{Q})$$

We leave it as an exercise to explore this claim.

As in classical logic, we define a ***tautology*** in a three-valued logical system to be a formula that always has the value **T**—there is no assignment on which it has either the value **F** or the value **N**. We define a ***contradiction*** in a three-valued logical system to be a formula that always has the value **F**—that is, it never has the value **T** or **N**. It turns out that there are no tautologies or contradictions in $K^S{}_3$! We'll prove this in a moment, but in case the reader is alarmed by this fact we reassure you that later in the chapter we'll offer variations on the concepts of tautologies and contradictions, variations that will not be trivial for $K^S{}_3$.

Result 5.1: There are no tautologies or contradictions in $K^S{}_3$.

Proof: Examination of the truth-tables shows that whenever all of the atomic components of a compound formula have the value **N**, so does the compound formula. This means that for any formula, there is at least one truth-value assignment on which it has the value **N**—so no formula can be either a tautology or a contradiction in $K^S{}_3$.

Thus, the Law of Excluded Middle (as we already knew) is not a tautology in $K^S{}_3$ (nor is its negation a contradiction), but neither are there any other classical tautologies that are tautologies in $K^S{}_3$.

Before turning to entailment and validity in Kleene's system, we introduce a lemma to which we shall often refer. Let us call a three-valued truth-value assignment ***classical*** if it assigns only the classical values **T** and/or **F** to atomic formulas—that is, it doesn't make any assignments of **N**.

Normality Lemma: In a *normal* three-valued system, a classical truth-value assignment behaves exactly as it does in classical logic—every formula that is true on that assignment in the three-valued system is also true on that assignment in classical logic, and every formula that is false on that assignment in the three-valued system is also false on that assignment in classical logic.

Proof: The lemma follows from the fact that the connectives in a normal system behave exactly as they do in classical logic whenever they operate on formulas with classical truth-values.

We will say that a set Γ of formulas ***entails*** a formula **P** in three-valued logic if, whenever all of the formulas in Γ are true **P** is true as well (there is no truth-value assignment on which all the formulas in Γ have the value **T** while **P** has the value **F** or **N**), and an argument is ***valid*** in three-valued logic if the set of premises of the argument entails its conclusion. We will use a standard notation for entailment: where Γ is a set of formulas, $\boldsymbol{\Gamma} \models \boldsymbol{P}$ means *the set of formulas* Γ *entails the sentence P*. Since entailment is within a system, we'll use unsubscripted $\models$ to indicate entailment in classical logic and $\models_K$ to indicate entailment in $K^S{}_3$.

Result 5.2: If $\boldsymbol{\Gamma} \models_K \mathbf{P}$ then $\boldsymbol{\Gamma} \models \mathbf{P}$ (i.e., every entailment in $K^S{}_3$ is also an entailment in classical propositional logic).

Proof: Assume that $\boldsymbol{\Gamma} \models_K \mathbf{P}$. It follows from the definition of entailment that on every classical (and nonclassical) truth-value assignment in $K^S{}_3$ on which the formulas in Γ are all true, **P** is also true. But then since $K^S{}_3$ is normal, the same is true in classical logic by the Normality Lemma. So $\boldsymbol{\Gamma} \models \mathbf{P}$ as well.

In the opposite direction some, but not all, classical entailments hold in K^S_3. An example of a classically valid argument that is also valid in K^S_3 is

$$\frac{\begin{array}{l} P \\ P \rightarrow Q \end{array}}{Q}$$

We leave it as an exercise to verify that this argument is indeed valid in K^S_3. But not all classical entailments carry over:

Result 5.3: Not all entailments of classical propositional logic hold in K^S_3.

Proof: An example of an argument that is classically valid but not valid in K^S_3 is

$$\frac{\neg (P \leftrightarrow Q)}{(P \leftrightarrow R) \vee (Q \leftrightarrow R)}$$

This is classically valid because, for the premise to be true in classical logic, *P* and *Q* must have different truth-values. But then no matter what truth-value *R* has, it will be equivalent to one or the other of *P* and *Q* since there are only two truth-values in classical logic—the validity depends crucially on the fact that classical logic has only two truth-values. So it is not surprising that this argument isn't valid in K^S_3, where the premise can have the value **T** while the conclusion has the value **N** if *P* and *Q* have "opposite" classical values (one has the value **T** and the other has the value **F**) but *R* has the value **N**.

5.2 Łukasiewicz's Three-Valued Logic

Now we'll look at a three-valued system originating with the Polish logician Jan Łukasiewicz (Łukasiewicz 1930). This system, which we will call L_3, defines three of the propositional connectives identically to Kleene's strong connectives, but the conditional and biconditional differ from Kleene's in one truth-table entry each. This difference, we will see, yields a system that contains both tautologies and contradictions (in the sense defined in the previous section). Here are Ł$_3$'s truth-tables:

P	$\neg_L$P
T	F
N	N
F	T

P $\wedge_L$ Q

P\Q	T	N	F
T	T	N	F
N	N	N	F
F	F	F	F

P $\vee_L$ Q

P\Q	T	N	F
T	T	T	T
N	T	N	N
F	T	N	F

P $\rightarrow_L$ Q

P\Q	T	N	F
T	T	N	F
N	T	T	N
F	T	T	T

P $\leftrightarrow_L$ Q

P\Q	T	N	F
T	T	N	F
N	N	T	N
F	F	N	T

The differences between Łukasiewicz's conditional and biconditional and Kleene's are in the center of the tables. Each of these two connectives forms a true formula in

Łukasiewicz's system when both of its immediate components have the value **N**. Why didn't Łukasiewicz assign the compound formula the value **N** as well in this case? It's because he reasoned that any conditional whose antecedent and consequent are identical, for instance, $A \rightarrow_L A$, should be a tautology—as it is in Ł3, as can be verified by examining the diagonal in the truth-table that travels from the upper left to the lower right:

P $\rightarrow_L$ **Q**

P / Q	T	N	F
T	**T**	N	F
N	T	**T**	N
F	T	T	**T**

So even though Mary Middleford is a borderline case of tallness, the sentence *If Mary Middleford is tall, then she's tall* turns out true in Ł3 as do the similar conditionals about Gina Biggerly and Tina Littleton. On the other hand, *If Mary Middleford is tall, then so is Tina Littleton* is neither true nor false, and *If Mary Middleford is tall, then so is Gina Biggerly* is true. In a similar vein, Łukasiewicz wanted biconditionals like $A \leftrightarrow_L A$ to be tautologies. Note that even though the truth-tables for the two connectives differ from Kleene's truth-tables, the Ł3 connectives are also both normal and uniform.

Here are Łukasiewicz's truth-tables for the formulas that we examined in Section 4.1:

P	P	$\vee_L$	$\neg_L$	P
T	T	T	F	T
N	N	N	N	N
F	F	T	T	F

P	Q	P	$\rightarrow_L$	(P	$\rightarrow_L$	Q)
T	T	T	T	T	T	T
T	N	T	N	T	N	N
T	F	T	F	T	F	F
N	T	N	T	N	T	T
N	N	N	T	N	T	N
N	F	N	T	N	N	F
F	T	F	T	F	T	T
F	N	F	T	F	T	N
F	F	F	T	F	T	F

P	Q	(P	$\wedge_L$	Q)	$\rightarrow_L$	(P	$\vee_L$	Q)
T	T	T	T	T	T	T	T	T
T	N	T	N	N	T	T	T	N
T	F	T	F	F	T	T	T	F
N	T	N	N	T	T	N	T	T
N	N	N	N	N	T	N	N	N
N	F	N	F	F	T	N	N	F
F	T	F	F	T	T	F	T	T
F	N	F	F	N	T	F	N	N
F	F	F	F	F	T	F	F	F

The Law of Excluded Middle behaves as it did in $K^S{}_3$—not surprising because the connectives in this formula are defined identically in the two systems. The truth-table for the second formula makes a conditional true when its antecedent and consequent both have the value **N**, unlike Kleene's table. And the third formula always has the value **T** in Ł3, again unlike the treatment of that formula in $K^S{}_3$.

Having explained Łukasiewicz's reason for assigning the value **T** to a conditional whose antecedent and consequent both have the value **N**, a reason that seems reasonable, we add that formulas that are not tautologies that exemplify this assignment strike some as odd. If the unrelated formulas P and Q both have the value **N** why should $P \rightarrow Q$ have the value **T** rather than the value **N** as it does in $K^S{}_3$? The assignment of **T** would seem to make sense when antecedent and consequent are identical, or related as they are in the formula $(P \wedge_L Q) \rightarrow_L (P \vee_L Q)$ but not when they are completely unrelated. Some of the oddity is dispelled, however, when we consider that in the case where P and Q both have the value **N**, as well as in the other cases where $P \rightarrow Q$ has the value **T**, Q is "at least as true" as P.

Łukasiewicz, like Kleene, was not motivated by vagueness in constructing his system, but his motivation differed from Kleene's. Łukasiewicz was interested in the truth-values of so-called *future contingent sentences* (a concern raised by the ancient Greek philosopher Aristotle). A future contingent sentence is a sentence about the future that might turn out to be true and also might turn out to be false—neither its truth nor its falsehood is necessary. Consider, for example, the sentence *The U.S. president in the year 3000 will be a woman.* Such a sentence, according to Łukasiewicz's reasoning, is neither true nor false today—for if it is true, then there will *have* to be a female U.S. president in 3000, and if it is false, then there *cannot* be a female U.S. president in 3000. But since there doesn't *have* to be a female U.S. president in 3000, although there *might* be, it follows that the sentence is neither true nor false today. Given the assumption that future contingent sentences are neither true nor false Ł3 presents a nice logic for such sentences. For example, a conjunction of a true sentence and a future contingent one such as *George Bush was the U.S. president in 2004 and the U.S. president in 3000 will be a female* is neither true nor false. Despite the contingency of the sentence about the future presidency, however, the sentence *If the U.S. president in 3000 will be a female then the U.S. president in 3000 will be a female* is true in Łukasiewicz's system.

Because the truth-tables for the conditional and the biconditional assign **T** to a formula whose immediate components both have the value **N**, these connectives cannot be defined in Ł3 in terms of (any combination of) the other three. The reason is fairly simple. If you construct a formula using only $\neg_L$, $\wedge_L$, and $\vee_L$ as connectives, then whenever the atomic formulas from which it is constructed all have the value **N** the compound formula will have the value **N** as well. But now consider $A \rightarrow_L A$ and $A \leftrightarrow_L A$. Both formulas have the value **T** when A has the value **N**. Since we can't form a compound formula that has this property from A, $\neg_L$, $\wedge_L$, and $\vee_L$, we cannot define either $\rightarrow_L$ or $\leftrightarrow_L$ in terms of the other three connectives.

Łukasiewicz in fact took $\neg_L$ and $\rightarrow_L$ as primitive and used them to define the other three connectives:

$$\mathbf{P} \vee_L \mathbf{Q} =_{\text{def}} (\mathbf{P} \rightarrow_L \mathbf{Q}) \rightarrow_L \mathbf{Q}$$
$$\mathbf{P} \wedge_L \mathbf{Q} =_{\text{def}} \neg_L(\neg_L \mathbf{P} \vee_L \neg_L \mathbf{Q})$$
$$\mathbf{P} \leftrightarrow_L \mathbf{Q} =_{\text{def}} (\mathbf{P} \rightarrow_L \mathbf{Q}) \wedge_L (\mathbf{Q} \rightarrow_L \mathbf{P})$$

Proof that these definitions produce the correct truth-functions is left as an exercise.[4]

4 It is also possible to define different conjunction and disjunction operations in Ł3 that do support the classical interdefinability of the five connectives; more on this in Section 5.7.

Let us now consider various semantic concepts in Ł3:

Result 5.4: Every formula that is a tautology in Ł3 is also a tautology in classical logic, and every formula that is a contradiction in Ł3 is also a contradiction in classical logic.

Proof: A formula that is a tautology in Ł3 is true in Ł3 on every classical truth-value assignment. Since Ł3 is normal, it follows from the Normality Lemma that the formula is true on every truth-value assignment in classical logic and is therefore a tautology in classical logic. Similar reasoning holds for contradictions.

Result 5.5: Not every formula that is a tautology in classical logic is also a tautology in Ł3, and not every formula that is a contradiction in classical logic is also a contradiction in Ł3.

Proof: Any instance of the Law of the Excluded Middle, for example, $A \vee_L \neg_L A$, is an example of a classical tautology that does not always have the value **T** in Ł3. Another example is the formula $(P \rightarrow_L (Q \rightarrow_L R)) \rightarrow_L ((P \rightarrow_L Q) \rightarrow_L (P \rightarrow_L R))$. This formula always has the value **T** in classical logic, but in Ł3 it has the value **N** when P and Q have the value **N** and R has the value **F**. The formula $A \wedge_L \neg_L A$, which is a classical contradiction, is not a contradiction in Ł3—it has the value **N** when A has the value **N**.

Note that Result 5.5 does not claim that *all* classical tautologies fail to be tautologies of Ł3 (nor that all classical contradictions fail to be contradictions of Ł3). For example, $A \rightarrow_L A$ is a tautology in both systems.

Result 5.6: If $\Gamma \models_L$ P then $\Gamma \models$ P.

Proof: This follows from the Normality Lemma since Ł3 is normal.

Result 5.7: Not every entailment in classical propositional logic holds in Ł3.

Proof: The argument and truth-value assignment in Result 5.3 will suffice here as well.

We note that other classically valid arguments *are* valid in Ł3. For example, the classically valid

P
P → Q
———
Q

is valid in Ł3 as well as in K^S_3.

5.3 Bochvar's Three-Valued Logics

The Russian mathematician Dmitri Bochvar proposed two very different systems of three-valued logic (Bochvar 1937)—different from Kleene's strong system and Łukasiewicz's system but also different from each other. Bochvar was concerned with paradoxical sentences like the *Liar Paradox*, which, in its simplest form, is

This sentence is false.

The paradox begins with the assumption that the sentence is true or false. But now consider: if the sentence is true, then what it says is the case, and so it is false. So it can't be true since that leads to a contradiction. Is it false, then? Well, if it is false then what it says is *not* the case and so it must be true. So the sentence can't be false either. There's your paradox. The Liar Paradox has been extensively studied,[5] so we will only note here that Bochvar's position was that the sentence is meaningless and hence neither true nor false, since only meaningful sentences can say true or false things. The third truth-value **N** represents meaninglessness for Bochvar.

Bochvar's "internal" three-valued system, which we will designate as B^I_3, has the following truth-tables:

P	$\neg_{BI}$**P**
T	**F**
N	**N**
F	**T**

P $\wedge_{BI}$ **Q**

P\Q	**T**	**N**	**F**
T	**T**	**N**	**F**
N	**N**	**N**	**N**
F	**F**	**N**	**F**

P $\vee_{BI}$ **Q**

P\Q	**T**	**N**	**F**
T	**T**	**N**	**T**
N	**N**	**N**	**N**
F	**T**	**N**	**F**

P $\rightarrow_{BI}$ **Q**

P\Q	**T**	**N**	**F**
T	**T**	**N**	**F**
N	**N**	**N**	**N**
F	**T**	**N**	**T**

P $\leftrightarrow_{BI}$ **Q**

P\Q	**T**	**N**	**F**
T	**T**	**N**	**F**
N	**N**	**N**	**N**
F	**F**	**N**	**T**

We mentioned earlier that Kleene had a second system of three-valued connectives, which he called the *weak* connectives. That system is identical to B^I_3. We shall nevertheless refer to the system as *Bochvar's*, as is customary.

Negation in B^I_3 is identical to negation in the previous systems, but the truth-functions for the other connectives are all different. We might say that the truth-value **N** is ***contagious*** in B^I_3—whenever a component of a compound formula has the value **N**, so does the compound formula as a whole—regardless of the value of any other component. If **N** represents meaninglessness, then it is quite sensible that this value should be contagious. Just as *Thiggledy piggledy* is meaningless, so is *Thiggledy piggledy and grass is green*. (The expression *as a whole* is meaningless, although part of it is meaningful.) Of course, our interest in this text is vagueness and so the "contagiousness" is that of the value **N** based on borderline cases. So, for

5 See, for example, Martin (1970, 1984).

example, even if a disjunction has a true disjunct it is nevertheless vague as a whole if the other disjunct is vague.

Here are the B^I_3 truth-tables for our earlier formulas:

P	P	$\vee_{BI}$	$\neg_{Ł}$	P
T	**T**	**T**	**F**	**T**
N	**N**	**N**	**N**	**N**
F	**F**	**T**	**T**	**F**

P	Q	P	$\rightarrow_{BI}$	(P	$\rightarrow_{BI}$	Q)
T	**T**	**T**	**T**	**T**	**T**	**T**
T	**N**	**T**	**N**	**T**	**N**	**N**
T	**F**	**T**	**F**	**T**	**F**	**F**
N	**T**	**N**	**N**	**N**	**N**	**T**
N	**N**	**N**	**N**	**N**	**N**	**N**
N	**F**	**N**	**N**	**N**	**N**	**F**
F	**T**	**F**	**T**	**F**	**T**	**T**
F	**N**	**F**	**N**	**F**	**N**	**N**
F	**F**	**F**	**T**	**F**	**T**	**F**

P	Q	(P	$\wedge_{BI}$	Q)	$\rightarrow_{BI}$	(P	$\vee_{BI}$	Q)
T	**T**	**T**	**T**	**T**	**T**	**T**	**T**	**T**
T	**N**	**T**	**N**	**N**	**N**	**T**	**N**	**N**
T	**F**	**T**	**F**	**F**	**T**	**T**	**T**	**F**
N	**T**	**N**	**N**	**T**	**N**	**N**	**N**	**T**
N	**N**	**N**	**N**	**N**	**N**	**N**	**N**	**N**
N	**F**	**N**	**N**	**F**	**N**	**N**	**N**	**F**
F	**T**	**F**	**F**	**T**	**T**	**F**	**T**	**T**
F	**N**	**F**	**N**	**N**	**N**	**F**	**N**	**N**
F	**F**	**F**	**F**	**F**	**T**	**F**	**F**	**F**

Neither of the classical tautologies is a tautology here, and the second formula receives the value **N** more often than it did in K^S_3 or $Ł_3$.

The B^I_3 connectives are all normal—they agree with the classical tables when their components are either **T** or **F**—but of the binary connectives only the biconditional is uniform, and that only trivially so. Uniformity of conjunction, for example, would require that a conjunction be false whenever one of the conjuncts is. But since the value **N** is contagious, this is not the case. Similar comments show that disjunction and the conditional are also not uniform in this system.

As with K^S_3, any way of interdefining connectives in classical logic will also work for B^I_3. This is because not only are the connectives normal—so we will get the desired results for truth-value assignments involving only **T** and **F**—but they all agree on what happens when a formula has a component with the value **N** (namely, the compound formula is also assigned the value **N**).

Also like K^S_3, B^I_3 has no tautologies. Because **N** is contagious, every formula has the value **N** on at least one truth-value assignment to its atomic components—namely, on any truth-value assignment that assigns **N** to at least one atomic component. So no formula is true on every truth-value assignment in B^I_3. We thus have

Result 5.8: No formula is a tautology in B^I_3, and no formula is a contradiction in B^I_3.

Concerning entailment:

Result 5.9: If $\Gamma \models_{BI} P$ then $\Gamma \models P$.

Proof: This follows from the Normality Lemma since B^I_3 is normal.

Result 5.10: Not every entailment that holds in classical propositional logic holds in B^I_3 as well.

Proof: The example argument and truth-value assignment in Result 5.3 suffice here as well.

There are also significant examples of classically valid arguments that are not valid in B^I_3 but that *are* valid in both K^S_3 and Ł3. One example is

$$\frac{Q}{P \rightarrow Q}$$

In B^I_3, the premise has the value **T** but the conclusion has the value **N** when *Q* has the value **T** and *P* has the value **N**.

Bochvar introduced a second system of connectives that together constitute his *external* system of three-valued logic, B^E_3. In B^E_3, the value **N** acts as if it is actually the value **F**:

P	$\neg_{BE}$P
T	F
N	T
F	T

P $\wedge_{BE}$ Q

P\Q	T	N	F
T	T	F	F
N	F	F	F
F	F	F	F

P $\vee_{BE}$ Q

P\Q	T	N	F
T	T	T	T
N	T	F	F
F	T	F	F

P $\rightarrow_{BE}$ Q

P\Q	T	N	F
T	T	F	F
N	T	T	T
F	T	T	T

P $\leftrightarrow_{BE}$ Q

P\Q	T	N	F
T	T	F	F
N	F	T	T
F	F	T	T

Bochvar introduced both the internal and external connectives within a single system; in that system the external connectives were *defined* connectives, using the internal connectives and a special *external assertion* operator *a*:

P	aP
T	T
N	F
F	F

To define the external version of a connective, we apply the internal version of the connective to externally asserted formulas. Thus, for example, if we apply the internal $\neg_{BI}$ to **aP** we get the table for external negation:

P	$\neg_{BI}$	aP
T	F	T
N	T	F
F	T	F

and if we apply the internal $\wedge_{BI}$ to **aP** and **aQ** we get the table for external conjunction:

P	Q	aP	$\wedge_{BI}$	aQ
T	T	T	T	T
T	N	T	F	F
T	F	T	F	F
N	T	F	F	T
N	N	F	F	F
N	F	F	F	F
F	T	F	F	T
F	N	F	F	F
F	F	F	F	F

We may read the external assertion connective as asserting truth: **aP** means ***P*** *is true*. This assertion is true if **P** is true, and is false otherwise. In particular, if **P** has the truth-value **N** then it is false that **P** is true. Rather than take the external assertion operator as primitive, however, we will define it using external negation:

$$\mathbf{aP} =_{\text{def}} \neg_{BE}\neg_{BE}\mathbf{P}$$

Henceforth, for simplicity in comparing systems, when we speak of Bochvar's external connectives we will mean the five connectives other than external assertion. The external connectives are both normal and uniform.

By introducing the external connectives Bochvar created a system with tautologies as well as contradictions; indeed, we have the following results:

Result 5.11: The set of formulas that are tautologies in $B^E{}_3$ is exactly the set of formulas that are tautologies in classical logic, and the set of formulas that are contradictions in $B^E{}_3$ is exactly the set of formulas that are contradictions in classical logic.

Proof: Since $B^E{}_3$ is normal, it follows from the Normality Lemma that every formula that is a tautology in $B^E{}_3$ is a classical tautology, and similarly for contradictions.

Conversely, we note that every classical tautology is a compound formula. Since the connectives in $B^E{}_3$ treat their **N** components as if they were false, $B^E{}_3$ treats the atomic components of any compound formula on a truth-value assignment where they are **N** as if they were false—and hence assigns the truth-value to the formula that classical logic would in that case. So a classical tautology must be a tautology in $B^E{}_3$ as well, and similar reasoning holds for contradictions.

Entailment also behaves classically in $B^E{}_3$:

Result 5.12: If $\Gamma \models_{BE} \mathbf{P}$ then $\Gamma \models \mathbf{P}$.

Proof: This follows from the Normality Lemma, since B^E_3 is normal.

Result 5.13: If $\Gamma \models \mathbf{P}$ then $\Gamma \models_{BE} \mathbf{P}$.

Proof: We shall show this by contraposition; that is, we'll show that if an entailment does *not* hold in B^E_3 then it doesn't hold in classical logic either. So consider a set Γ and formula **P** such that $\Gamma \not\models_{BE} \mathbf{P}$. Then there is some three-valued assignment on which all the formulas in Γ have the value **T** but on which **P** has either the value **F** or the value **N**. We can convert this to a classical truth-value assignment by keeping the **T** and **F** assignments to atomic formulas but turning the **N** assignments (if any) to atomic formulas into **F** assignments. This classical truth-value assignment will make the premises of the argument true in classical logic because compound formulas in B^E_3 behave as if their **N**-valued atomic components have the value **F**, and if any of the formulas in Γ are atomic, then, since they have the value **T** on the B^E_3 assignment, they will have the value **T** on the classical assignment as well. But **P** has the value **F** on the classical truth-value assignment for similar reasons.

Thus tautologousness, contradictoriness, and entailment all coincide for classical logic and B^E_3.

5.4 Evaluating Three-Valued Systems; Quasi-Tautologies and Quasi-Contradictions

There are several ways in which we can measure the adequacy of a three-valued system as a logic of vagueness. First, we note that the Principle of Bivalence fails for three-valued systems, by definition! On this count, all four systems that we have presented are good candidates for such a logic. However, we note that although B^E_3 rejects the Principle of Bivalence it does so only for atomic formulas, and compound formulas behave exactly as they do in classical logic. So if we believe, for example, that the Law of Excluded Middle fails for vague sentences, that would be a reason for rejecting B^E_3 as a logic for vagueness.

How else might we evaluate the three-valued systems, and in particular are there significant criteria that will distinguish among the remaining systems K^S_3, Ł$_3$, and B^I_3? One way to evaluate different logical systems is to compare their sets of tautologies and contradictions. We have seen that in each of the three systems, there are classical tautologies that fail to be tautologies in the three-valued system (and that a similar situation holds for classical contradictions). This is desirable, at least if we believe that classical tautologies such as instances of the Law of Excluded Middle should fail in the case of borderline attributions of vague predicates. But there is a difference in the sets of classical tautologies that fail for the three systems. K^S_3 and B^I_3 have no tautologies. On the other hand, Ł$_3$ does have tautologies. If we believe that the simple classical tautologies $A \rightarrow A$ and $A \leftrightarrow A$ should remain tautologies within a three-valued system, that would be a reason for preferring $Ł_3$ to $K^S{}_3$ and $B^I{}_3$.

But there is a second way to define a tautology-like concept in three-valued logic. We will say that a formula is a ***quasi-tautology*** if it is never false. Note that in classical logic the concepts of being a tautology and being a quasi-tautology coincide, since a formula that is never false in classical logic is always true, and vice versa. But the concepts do not coincide in three-valued systems. For example, although $K^S{}_3$ and $B^I{}_3$ have no tautologies they both have quasi-tautologies; the formula $A \vee \neg A$ is a quasi-tautology in each of the two systems (as well as in $Ł_3$).

It is common to talk of **designated truth-values** in connection with tautologies and their kin: the designated truth-values include **T** and any other truth-values that we wish to count as "good" or at least as "not bad." We can then define tautologies in terms of designated truth-values: a formula is a tautology if it has a designated truth-value on every truth-value assignment. If only the value **T** is designated, we end up with the definition of tautology that we've been using: a formula that always has the value **T**. If both the values **T** and **N** are designated, we end up with the definition of quasi-tautologies instead.

Why might we be interested in quasi-tautologies? For one thing, we might be interested in avoiding falsehood as much as we are interested in embracing truth. If the former is the case, the set of quasi-tautologies should be of interest. But practical interests aside, the concept of a quasi-tautology is a second way of generalizing the classical notion of a tautology—as a formula that is never false rather than as a formula that is always true—and the concept therefore also has purely theoretical interest.

As we just noted, there are quasi-tautologies in $K^S{}_3$ and $B^I{}_3$ even though there are no tautologies in either of the systems. In fact, every classical tautology is a quasi-tautology in both systems, and vice versa. We'll prove this first for Bochvar's system, since that is the simpler of the two proofs.

Result 5.14: The set of $B^I{}_3$ quasi-tautologies is exactly the set of classical tautologies.

Proof: Let **P** be a $B^I{}_3$ quasi-tautology. Then **P** does not have the value **F** on any truth-value assignment. Since $B^I{}_3$ is normal, it follows from the Normality Lemma that **P** does not have the value **F** on any classical truth-value assignment in classical logic and is therefore a tautology of classical logic. So every $B^I{}_3$ quasi-tautology is a classical tautology.

Conversely, assume that a formula **P** is *not* a $B^I{}_3$ quasi-tautology. Then **P** has the value **F** on some truth-value assignment in $B^I{}_3$. This truth-value assignment must be a classical assignment, since the value **N** is contagious in $B^I{}_3$. It follows from the Normality Lemma that **P** has the value **F** on this assignment in classical logic and therefore is not a classical tautology.

Although we have the same result for K^S_3, the proof in the second direction is somewhat different:

Result 5.15: The set of K^S_3 quasi-tautologies is exactly the set of classical tautologies.

Proof: The proof that every K^S_3 quasi-tautology is a classical tautology follows from the Normality Lemma.

The converse claim, that a formula **P** that is not a K^S_3 quasi-tautology is also not a classical tautology, is equivalent to saying that a formula **P** that has the value **F** on some truth-value assignment in K^S_3 will also have the value **F** on some classical assignment. The restated claim holds trivially if the assignment on which **P** has the value **F** in K^S_3 is a classical assignment. So we need to establish that if a formula **P** has the value **F** on some *non*classical truth-value assignment in K^S_3, **P** will also have the value **F** on some classical assignment in K^S_3.

In order for **P** to have the value **F** in K^S_3 on an assignment on which one or more of its atomic components have the value **N**, uniformity must have kicked in at some point to override the **N**s in favor of classical truth-values. And at each point where uniformity kicked in, the same classical value would have resulted if the **N** had been a **T** or an **F** instead. So if we replace all of the **N**s that the three-valued assignment assigns with either **T**s or **F**s (it doesn't matter which), **P** will end up having the same value on the resulting classical assignment as it did on the three-valued assignment.

On the other hand, the quasi-tautologies of Ł$_3$ do not coincide with the tautologies of classical logic:

Result 5.16: Every Ł$_3$ quasi-tautology is a classical tautology; every classical tautology that contains only negation, conjunction, and disjunction is an Ł$_3$ tautology; but some classical tautologies containing the conditional or the biconditional are not Ł$_3$ quasi-tautologies.

Proof: The proof that every Ł$_3$ quasi-tautology is a classical tautology follows from the Normality Lemma.

It follows from Result 5.15 that every classical tautology that contains only negation, conjunction, and disjunction is an Ł$_3$ tautology because Ł$_3$ negation, conjunction, and disjunction are defined the same as in K^S_3.

An example of a classical tautology containing a conditional that is not a quasi-tautology in Ł$_3$ is $\neg(A \rightarrow \neg A) \vee \neg(\neg A \rightarrow A)$. When A has the truth-value **N**, this formula is false in Ł$_3$ and therefore is not a quasi-tautology. An example of a classical tautology containing a biconditional that is not a quasi-tautology in Ł$_3$ is $\neg(A \leftrightarrow \neg A)$; this formula is also false when A has the value **N**. (It is easily verified that these formulas are both classical tautologies.)

For B^E_3, we have

Result 5.17: The set of B^E_3 quasi-tautologies coincides with the set of classical tautologies.

Proof: The only formulas that are not tautologies in B^E_3 but might be quasi-tautologies are atomic formulas, for no other formulas can ever have the truth-value N in this system. But these formulas are neither classical tautologies nor quasi-tautologies in B^E_3, since they can have the value F. So quasi-tautologies and tautologies coincide in B^E_3, and we have already established that the B^E_3 tautologies coincide with the set of classical tautologies.

As a dual to the concept of quasi-tautology, we introduce *quasi-contradictions*: a formula is a **quasi-contradiction** if it is never true; that means that in a three-valued system it always has the value **T** or the value **N**.[6] The results concerning quasi-tautologies in the three-valued systems also hold for quasi-contradictions: namely, in each of B^I_3, B^E_3, and K^S_3 the set of quasi-contradictions coincides with the set of classical contradictions; every Ł3 quasi-contradiction is a classical contradiction; and some classical contradictions are not Ł3 quasi-contradictions. Proofs of these claims are left as an exercise.

We also introduce the concept of a *quasi-entailment*: a set Γ of formulas **quasi-entails** a formula **P** if there is no truth-value assignment on which each of the formulas in Γ has the value **T** or **N** while **P** has the value **F**; that is, whenever each formula in Γ has one of the values **T** or **N** so does **P**. An argument is **quasi-valid** if the set consisting of its premises quasi-entails its conclusion. We have

Result 5.18: Every quasi-entailment in each of K^S_3, Ł3, B^I_3, and B^E_3 is a classical entailment.

Proof: If a set of formulas Γ quasi-entails a formula **P** in any of the four systems then there is no classical truth-value assignment in any of these systems on which all of the formulas in Γ have the value **T** (none will have the value **N** on a classical truth-value assignment) and **P** has the value **F** in that system. Since these systems are all normal, it follows from the Normality Lemma that the entailment holds in classical logic.

The converse does not generally hold:

Result 5.19: Not every classical entailment is a quasi-entailment in K^S_3, and ditto for the other three systems Ł3, B^I_3, and B^E_3.

Proof: The argument

$$\frac{A \wedge \neg A}{B}$$

[6] We may also define contradictions via *anti*-designated truth-values—truth-values including **F** and other values that we wish to single out. The sets of designated and anti-designated truth-values may overlap—so **N** can be included in both sets.

is valid in classical logic. But it fails to be quasi-valid in any of K^S_3, $Ł_3$, or B^I_3 (although it is valid in all three systems). This is because the premise has the value **N** and the conclusion has the value **F** in each of these three systems when *A* has the value **N** and *B* has the value **F**.

The classically valid argument

$$\frac{A}{A \vee A}$$

isn't valid in B^E_3. When the premise has the value **N**, the conclusion is false.

Of course, some classically valid arguments are also quasi-valid in more than one of the four systems. The argument

$$\frac{P \wedge Q}{P}$$

is quasi-valid in K^S_3, $Ł_3$, and B^E_3, and the argument

$$\frac{P}{P \vee Q}$$

is quasi-valid in K^S_3, $Ł_3$, and B^I_3.

There is a third interesting version of entailment (and hence of validity) in three-valued systems: rather than simply preserving truth (as in entailment proper) or preserving non-falsehood (as in quasi-entailment), we can rank the three truth-values and require that the value of **P** be at least as great as the value of the lowest-ranked formula in Γ. We rank the three truth-values as $\mathbf{T} \geq \mathbf{N} \geq \mathbf{F}$. We will say that a set of formulas Γ ***degree-entails***[7] a formula **P** if **P**'s value can never be less than the least value of the formulas in Γ. If all of the formulas in Γ have the value **T**, then **P** must also have the value **T**, and if each of the formulas in Γ has either the value **T** or the value **N**, then **P** must have either the value **T** or the value **N** as well. An argument is **degree-valid** in a three-valued system if the set of its premises degree-entails its conclusion. Not surprisingly, we have

Result 5.20: Every degree-entailment that holds in B^I_3, B^E_3, K^S_3, or $Ł_3$ is a classical entailment.

Proof: Left as an exercise.

In fact, degree-entailment is equivalent to entailment proper *plus* quasi-entailment (to be proved in the exercises). So if Γ fails to entail or to quasi-entail **P** in any three-valued system then Γ will also fail to degree-entail **P** in that system. Consequently, we have

7 The name ***degree-entailment*** anticipates the concept of ***degrees*** *of truth* to be introduced when we turn to fuzzy logic.

Result 5.21: Not every classical entailment is a degree-entailment in $K^S{}_3$, and ditto for the other three systems $Ł_3$, $B^I{}_3$, and $B^E{}_3$.

Proof: For the reason just given, this follows from earlier results.

5.5 Normal Forms

For each of the four systems, we define ***phrases, clauses, disjunctive normal form***, and ***conjunctive normal form*** the same as we did for classical propositional logic:

A literal is a phrase.
If **P** and **Q** are phrases, so is $(\mathbf{P} \wedge \mathbf{Q})$.
Every phrase is in disjunctive normal form.
If **P** and **Q** are in disjunctive normal form, so is $(\mathbf{P} \vee \mathbf{Q})$.
A literal is a clause.
If **P** and **Q** are clauses, so is $(\mathbf{P} \vee \mathbf{Q})$.
Every clause is in conjunctive normal form.
If **P** and **Q** are in conjunctive normal form, so is $(\mathbf{P} \wedge \mathbf{Q})$.

Recall the equivalences that we used to convert formulas to these normal forms:

$\mathbf{P} \rightarrow \mathbf{Q}$	is equivalent to	$\neg\mathbf{P} \vee \mathbf{Q}$	(*Implication*)
$\mathbf{P} \leftrightarrow \mathbf{Q}$	is equivalent to	$(\neg\mathbf{P} \vee \mathbf{Q}) \wedge (\neg\mathbf{Q} \vee \mathbf{P})$	(*Implication*)
$\neg(\mathbf{P} \wedge \mathbf{Q})$	is equivalent to	$\neg\mathbf{P} \vee \neg\mathbf{Q}$	(*DeMorgan's Law*)
$\neg(\mathbf{P} \vee \mathbf{Q})$	is equivalent to	$\neg\mathbf{P} \wedge \neg\mathbf{Q}$	(*DeMorgan's Law*)
$\mathbf{P}$	is equivalent to	$\neg\neg\mathbf{P}$	(*Double Negation*)
$(\mathbf{P} \vee \mathbf{Q}) \wedge \mathbf{R}$	is equivalent to	$(\mathbf{P} \wedge \mathbf{R}) \vee (\mathbf{Q} \wedge \mathbf{R})$	(*Distribution*)
$\mathbf{P} \wedge (\mathbf{Q} \vee \mathbf{R})$	is equivalent to	$(\mathbf{P} \wedge \mathbf{Q}) \vee (\mathbf{P} \wedge \mathbf{R})$	(*Distribution*)
$(\mathbf{P} \wedge \mathbf{Q}) \vee \mathbf{R}$	is equivalent to	$(\mathbf{P} \vee \mathbf{R}) \wedge (\mathbf{Q} \vee \mathbf{R})$	(*Distribution*)
$\mathbf{P} \vee (\mathbf{Q} \wedge \mathbf{R})$	is equivalent to	$(\mathbf{P} \vee \mathbf{Q}) \wedge (\mathbf{P} \vee \mathbf{R})$	(*Distribution*)

All of these equivalences hold in $K^S{}_3$ and $B^I{}_3$. The implication equivalences fail in $Ł_3$, and the Double Negation equivalence fails in $B^E{}_3$. Proof of these claims is left as an exercise.

Because all of the equivalences hold in $K^S{}_3$ and $B^I{}_3$, we can claim that every formula in these two systems is equivalent to a formula in disjunctive normal form and to a formula in conjunctive normal form. Formulas in $Ł_3$ that contain the conditional or the biconditional may not be equivalent to formulas in either normal form. We showed in Section 5.2 that we can't define either $\rightarrow_L$ or $\leftrightarrow_L$ in terms of $\neg_L$, $\wedge_L$, and $\vee_L$. It follows that neither $P \rightarrow_L Q$ nor $P \leftrightarrow_L Q$ can be equivalent to a formula in either normal form.

It turns out that each $B^E{}_3$ formula is equivalent to a formula in disjunctive normal form and to one in conjunctive normal form, but we can't claim that this follows from the previous equivalences since Double Negation fails for B^E_3. However, we note that the single case where Double Negation fails occurs when the double negation appears in front of an atomic formula, such as $\neg\neg S$. In this case we can eliminate the double negation by replacing $\neg\neg\mathbf{P}$ with $\mathbf{P} \wedge \mathbf{P}$, since these two formulas are equivalent for any atomic formula **P**. (This would also work where **P** is a complex formula but is unnecessary since we can simply eliminate the double negation in this case.) Then the Distribution equivalences can be applied to produce a formula in either of the normal forms.

In Chapter 2 we proved that a clause of classical propositional logic is a tautology if and only if it contains a complementary pair of literals, and that a phrase of classical propositional logic is contradictory if and only if it contains a complementary pair of literals. We have shown in this chapter, in Results 5.14, 5.15, and 5.17, that the quasi-tautologies of K^S_3, B^I_3, and B^E_3 coincide with the classical tautologies. It follows that

Result 5.22: A clause of K^S_3, Ł$_3$, B^I_3, or B^E_3 is a quasi-tautology in that system if and only if it contains a complementary pair of literals.

(The result holds of Ł$_3$ because clauses contain only negation and conjunction, and in Ł$_3$ these connectives are identical to the K^S_3 connectives.) In Chapter 2 we also proved that a phrase of classical propositional logic is contradictory if and only if it contains a complementary pair of literals. Because the quasi-contradictions of the three systems in question coincide with classical contradictions, we also have

Result 5.23: A phrase of K^S_3, Ł$_3$, B^I_3, or B^E_3 is quasi-contradictory in that system if and only if it contains a complementary pair of literals.

And, as a consequence of these two results:

Result 5.24: A formula **P** of K^S_3, Ł$_3$, B^I_3, or B^E_3 that is in conjunctive normal form is a quasi-tautology in that system if and only if each clause in **P** contains a complementary pair of literals.

Result 5.25: A formula **P** of K^S_3, Ł$_3$, B^I_3, or B^E_3 that is in disjunctive normal form is quasi-contradictory in that system if and only if each clause in **P** contains a complementary pair of literals.

5.6 Questions of Interdefinability between the Systems and Functional Completeness

Although we have characterized each of the four systems K^S_3, Ł$_3$, B^I_3, and B^E_3 independently of the others, there are obviously important connections. For example, negation is defined identically in K^S_3, Ł$_3$ and B^I_3, and disjunction and conjunction are defined identically in the former two systems. This raises the general question, Which connectives are definable within which systems? We first establish some very general negative results:

Result 5.26: The binary connectives of $K^S{}_3$, Ł_3, and $B^E{}_3$ are not definable in $B^I{}_3$.

Proof: None of the connectives in $B^I{}_3$ produces a formula with a classical truth-value when any of its immediate components have the value **N**. But the binary connectives of the other three systems can produce such, so none of these can be defined using only the connectives of $B^I{}_3$.

Result 5.27: None of the connectives of $K^S{}_3$, Ł_3, or $B^I{}_3$ are definable in $B^E{}_3$.

Proof: The connectives of $B^E{}_3$ never produce formulas with the value **N**. Since each of the connectives in the other systems can produce such formulas, the result follows.

As a consequence of these two results, we know that neither $B^I{}_3$ nor $B^E{}_3$ can express everything that Ł_3 can or everything that $K^S{}_3$ can, nor can either of $B^I{}_3$ or $B^E{}_3$ express everything that the other system can.

We turn now to the expressive powers of $K^S{}_3$ and Ł_3. Both systems can express everything that can be expressed in $B^I{}_3$:

Result 5.28: All of the connectives of $B^I{}_3$ are definable in both $K^S{}_3$ and Ł_3.

Proof: Negation in $B^I{}_3$ is identical to negation in the other two systems. We can define $B^I{}_3$'s conjunction using the other two system's conjunction, disjunction, and negation (these connectives are defined identically in those two systems) as follows:

$$\mathbf{P} \wedge_{BI} \mathbf{Q} =_{\text{def}} (\mathbf{P} \wedge_{K/L} \mathbf{Q}) \vee_{K/L} ((\mathbf{P} \wedge_{K/L} \neg_{K/L} \mathbf{P}) \vee_{K/L} (\mathbf{Q} \wedge_{K/L} \neg_{K/L} \mathbf{Q}))$$

It is left as an exercise to verify this equivalence. We can then define the other $B^I{}_3$ connectives in terms of these two, using any of the standard classical equivalences. Alternatively, we can give direct definitions for disjunction and the conditional analogous to the preceding definition for conjunction:

$$\mathbf{P} \vee_{BI} \mathbf{Q} =_{\text{def}} (\mathbf{P} \vee_{K/L} \mathbf{Q}) \wedge_{K/L} ((\mathbf{P} \vee_{K/L} \neg_{K/L} \mathbf{P}) \wedge_{K/L} (\mathbf{Q} \vee_{K/L} \neg_{K/L} \mathbf{Q}))$$
$$\mathbf{P} \rightarrow_{BI} \mathbf{Q} =_{\text{def}} (\neg_{K/L}\mathbf{P} \vee_{K/L} \mathbf{Q}) \wedge_{K/L} ((\mathbf{P} \vee_{K/L} \neg_{K/L} \mathbf{P}) \wedge_{K/L} (\mathbf{Q} \vee_{K/L} \neg_{K/L} \mathbf{Q}))$$

On the other hand, not all of Ł_3 is expressible within $K^S{}_3$:

Result 5.29: The Ł_3 conditional is not definable in $K^S{}_3$.

Proof: A formula $\mathbf{P} \rightarrow_L \mathbf{Q}$ has the value **T** when both **P** and **Q** have the value **N**. But every $K^S{}_3$ connective produces a formula with the value **N** when its immediate components (all) have the value **N**, so no combination of $K^S{}_3$ connectives can produce a formula that expresses the Ł_3 conditional.

Nor can any B^{E}_{3} connectives be expressed within K^{S}_{3}:

Result 5.30: The B^{E}_{3} connectives are not definable in K^{S}_{3}.

Proof: No K^{S}_{3} connective produces a formula that has a classical truth-value when its immediate components have the value **N**, so no B^{E}_{3} connective can be defined using K^{S}_{3} connectives.

However, it turns out that every connective of the other three systems is definable in Ł3. We have already shown that this is true of B^{I}_{3}.

Result 5.31: Every K^{S}_{3} connective is definable in Ł3.

Proof: Since K^{S}_{3}'s negation, conjunction, and disjunction are identical to those of Ł3, we need only note that the K^{S}_{3} conditional and biconditional are definable using those connectives.

Result 5.32: Every B^{E}_{3} connective is definable in Ł3.

Proof: It will suffice to show that Bochvar's external assertion is definable in Ł3. The definition $\mathbf{aP} =_{\text{def}} \neg_L(\mathbf{P} \rightarrow_L \neg_L\mathbf{P})$ produces the table for external assertion:

P	aP
T	T
F	F
N	F

All of the other external Bochvar connectives can be defined using external assertion and Bochvar's internal connectives, which we have already shown to be definable in Ł3.

Having shown that Ł3 is powerful enough to define all of the connectives of the other three systems, the question arises, Are all *possible* three-valued truth-functions definable in Ł3? If they are, then Ł3 is a functionally complete system. We showed in Chapter 2 that the classically defined connectives $\neg$ and $\wedge$ form a functionally complete system for classical logic—every possible two-valued truth-function can be defined solely in terms of classical negation and conjunction. Turning to Ł3 we might want to know, for example, whether the connective # with the truth-table

P # Q

P\Q	T	N	F
T	T	N	T
N	T	N	N
F	F	N	F

is definable in Ł3. The answer is *yes*, and we leave it as an exercise to produce a formula that has these truth-conditions—using the algorithm that we are about to present in Result 5.33. More generally, every *regular* three-valued truth-function can be defined in Ł3, where a **regular** truth-function is one that produces classical truth-values when (but not necessarily only when) applied exclusively to classical truth-values:

Result 5.33: All regular three-valued truth-functions are definable in Ł$_3$.

Proof: A regular three-valued n-place truth-function can be described by the truth-table schema

$\mathbf{P}_1$ $\mathbf{P}_2 \ldots \mathbf{P}_n$	
T T ... T	v_1
T T ... N	v_2
...	...
F F ... F	v_{3^n}

where each of $v_1, v_2, \ldots, v_{3^n}$ is one of the values **T**, **N**, **F** and where v_i is **T** or **F** if all of the values to the left of the vertical bar in row i are classical truth-values.

We will first provide, for each row i of the truth-function's table that has the value $v_i = \mathbf{T}$, a formula $\mathbf{Q}_i$ that has the value **T** in that row and **F** in all other rows. We'll be using the external assertion connective, which we have already shown to be definable in Ł$_3$. For each such row of the table, define the formula $\mathbf{Q}_i$ to be $\mathbf{P}_1{}^* \wedge_L \mathbf{P}_2{}^* \wedge_L \ldots \wedge_L \mathbf{P}_n{}^*$ where

$\mathbf{P}_j{}^* = \mathrm{a}\mathbf{P}_j$ if the value of $\mathbf{P}_j$ is **T** in row i,
$\mathrm{a}\neg_L\mathbf{P}_j$ if the value of $\mathbf{P}_j$ is **F** in row i, and
$\neg_L\mathrm{a}\mathbf{P}_j \wedge_L \neg_L\mathrm{a}\neg_L\mathbf{P}_j$ otherwise.

Each of these formulas $\mathbf{P}_j{}^*$ defined for a particular row i will have the value **T** when $\mathbf{P}_j$ has the value it has in row i and will have the value **F** otherwise. So the conjunction $\mathbf{Q}_i$ will have the value **T** in the row i for which it is defined but will be false in each other row since it will have at least one conjunct with the value **F**.

Next we provide, for each row i of the truth-function's table that has the value $v_i = \mathbf{N}$, a formula $\mathbf{Q}_i$ that has the value **N** in that row and **F** in all other rows. Because the truth-function that we are considering is regular, at least one of the $\mathbf{P}_j$ must have the value **N** in such a row. For each such row i, define $\mathbf{Q}_i$ to be $\mathbf{P}_1{}' \wedge_L \mathbf{P}_2{}' \wedge_L \ldots \wedge_L \mathbf{P}_n{}'$ where

$\mathbf{P}_j{}' = \mathrm{a}\mathbf{P}_j$ if the value of $\mathbf{P}_j$ is **T** in row i,
$\mathrm{a}\neg_L\mathbf{P}_j$ if the value of $\mathbf{P}_j$ is **F** in row i, and
$\mathbf{P}_j \wedge_L \neg_L\mathbf{P}_j$ otherwise.

Finally, we form a disjunction of the formulas $\mathbf{Q}_i$ for each row i with $v_i = \mathbf{T}$ or $v_i = \mathbf{N}$. This disjunction expresses the function defined in the truth-table schema: the disjunction will have the value v_i for each row i such that $v_i = \mathbf{T}$ or $v_i = \mathbf{N}$, and **F** for all other rows—the desired result, since all other rows have $v_i = \mathbf{F}$. Except that there is one special case—if the function produces **F** in every row, then that function can be defined in Ł$_3$ using $\mathrm{a}\mathbf{P}_1 \wedge_L \neg_L\mathrm{a}\mathbf{P}_1 \wedge_L \mathbf{P}_2 \wedge_L \ldots \wedge_L \mathbf{P}_n$—this formula always has the value **F**.

In fact we can make a more specific claim than Result 5.33, namely, all *and only* regular three-valued truth-functions are definable in Ł3. For this we establish

Result 5.34: No nonregular truth-function is definable in Ł3.

Proof: All of the Ł3 connectives are regular, so it is impossible to produce a formula that has the value **N** when all of its constituents have values **T** or **F**.

We don't consider it a bad thing that nonregular truth-functions cannot be defined in Ł3, for it is hard to come up with an example where we would want a connective to produce a nonclassical value based on classical values alone for its constituents. However, for the sake of logical theory we note that Ł3 can be made into a functionally complete system with the addition of the nonregular operator

P	%P
T	N
N	N
F	N

This operator was introduced by the Polish logician Jerzy Ł. Słupecki in order to expand Ł3 to a truth-functionally complete system (Słupecki 1936).

5.7 Łukasiewicz's System Expanded

As we noted in Section 5.2, Łukasiewicz's conditional cannot be defined using his other connectives as can be done in classical logic and in the other three-valued systems presented in this chapter. It is customary, in the context of fuzzy logic, to define a second pair of conjunction and disjunction operations for which the interdefinabilities do hold. These two new operations are called *bold* conjunction and disjunction and will be, respectively, symbolized as & and ∇:

$$\mathbf{P} \mathbin{\&} \mathbf{Q} =_{\text{def}} \neg_L(\mathbf{P} \rightarrow_L \neg_L \mathbf{Q})$$
$$\mathbf{P} \nabla \mathbf{Q} =_{\text{def}} \neg_L \mathbf{P} \rightarrow_L \mathbf{Q}$$

The bold connectives have the following truth-tables in Ł3:

Bold Conjunction

P & Q

P\Q	T	N	F
T	T	N	F
N	N	F	F
F	F	F	F

Bold Disjunction

P ∇ Q

P\Q	T	N	F
T	T	T	T
N	T	T	N
F	T	N	F

Note that these differ from the (now called *weak*) conjunction and disjunction operators of Ł3 in the middle position of the truth-table: weak conjunction and weak disjunction both have the value **N** in that position. Rather than define the bold connectives as we did, we could also take them as primitive and define the Ł3 conditional using either of these connectives: $\mathbf{P} \rightarrow_L \mathbf{Q} =_{\text{def}} \neg_L(\mathbf{P} \,\&\, \neg_L\mathbf{Q})$ or $\mathbf{P} \rightarrow_L \mathbf{Q} =_{\text{def}} \neg_L\mathbf{P} \nabla \mathbf{Q}$ (analogous to the classical definitions of the conditional in terms of the other operators).

Using the bold connectives, we have tautologies that are versions of the Law of the Excluded Middle and the Law of Noncontradiction in Ł3: $\mathbf{P} \nabla \neg_L\mathbf{P}$ and $\neg_L(\mathbf{P} \,\&\, \neg_L\mathbf{P})$. Like the weak connectives, the bold connectives meet the minimal requirements that some have proposed for conjunction and disjunction, namely:

1. Conjunction and disjunction are both **associative**: **P** *op* (**Q** *op* **R**) is equivalent to (**P** *op* **Q**) *op* **R**, where *op* is a conjunction or disjunction connective.
2. Conjunction and disjunction are both **commutative**: **P** *op* **Q** is equivalent to **Q** *op* **P**.
3. Conjunction and disjunction are **nondecreasing in both arguments**: if the value of **P** is less than or equal to the value of **R** (using the ranking $\mathbf{T} \geq \mathbf{N} \geq \mathbf{F}$) then the value of **P** *op* **Q** or **Q** *op* **P** is less than or equal to the value of **R** *op* **Q** or **Q** *op* **R**.[8]
4. **P** *conj* **Q** (where *conj* is a conjunction connective) has the value that **Q** has when **P** has the value **T**.
5. **P** *disj* **Q** (where *disj* is a disjunction connective) has the value that **Q** has when **P** has the value **F**.[9]

In the context of infinite-valued logics conjunction and disjunction operations that meet these conditions are called, respectively, ***t-norms*** and ***t-conorms*** (where the *t* is short for *triangular*, terminology that arose because these concepts were first introduced in connection with probabilistic metric spaces in Menger [1942]). It is left as an exercise to prove that t-norm and t-conorm operations that satisfy conditions 1–5 will also satisfy conditions 6 and 7:

6. **P** *conj* **Q** has the value **F** when either **P** or **Q** has the value **F**.
7. **P** *disj* **Q** has the value **T** when either **P** or **Q** has the value **T**.

These seven conditions uniquely define conjunction and disjunction in classical logic (proof is left as an exercise), and so it is natural to view them as the minimal conditions on conjunction and disjunction in other logics.

Now a concluding note. Because we know that all the connectives of the other three-valued systems we have studied, as well as bold conjunction and disjunction, can be defined in Ł3, we can capture all of these using the single system Ł3. If we think that there are really two important interpretations of the English conditional, one being Kleene's and the other being Łukasiewicz's, or perhaps that there are two important types of negation, one being Łukasiewicz's and the other being Bochvar's external negation, then we can capture both within the system Ł3 alone. In the

[8] Note that owing to commutativity, if these operations are nondecreasing in either argument they must be nondecreasing in both. Proof is left as an exercise.

[9] Owing to commutativity, conditions 4 and 5, respectively, hold of **Q** *conj* **P** and **Q** *disj* **P** as well. Proof is similar to the proof of the claim in footnote 8.

following chapters we shall be focusing on Ł3, and in doing so we don't forfeit any virtues that the other systems may have.

5.8 Exercises

SECTION 5.1

1 Using truth-tables, determine whether the following pairs of formulas are equivalent (always have the same truth-value) in K^S_3:

a. $P \wedge Q$ $\neg(\neg P \vee \neg Q)$

b. P $\neg\neg P$

c. $P \rightarrow Q$ $\neg Q \rightarrow \neg P$

2 Show that if $\neg_K$ and $\wedge_K$ are taken as primitive connectives and the other ones are introduced with the definitions

$$\mathbf{P} \vee_K \mathbf{Q} =_{\text{def}} \neg_K (\neg_K \mathbf{P} \wedge_K \neg_K \mathbf{Q})$$
$$\mathbf{P} \rightarrow_K \mathbf{Q} =_{\text{def}} \neg_K (\mathbf{P} \wedge_K \neg_K \mathbf{Q})$$
$$\mathbf{P} \leftrightarrow_K \mathbf{Q} =_{\text{def}} \neg_K (\mathbf{P} \wedge_K \neg_K \mathbf{Q}) \wedge_K \neg_K (\neg_K \mathbf{P} \wedge_K \mathbf{Q})$$

we obtain the correct truth-tables for K^S_3. (You may construct truth-tables for the formulas on the right-hand sides to show this.)

3 Use truth-tables to decide whether the following arguments are valid in K^S_3:

a. P
$P \rightarrow Q$
———
Q

b. $P \wedge \neg Q$
———
$Q \rightarrow P$

c. $P \rightarrow Q$
$\neg Q$
———
$\neg P$

SECTION 5.2

4 Use truth-tables to decide whether each of the following pairs of formulas are equivalent in Ł3:

a. $P \vee Q$ $\neg(\neg P \wedge \neg Q)$

b. P $\neg\neg P$

c. P $P \wedge P$

d. P $P \vee P$

e. $P \rightarrow Q$ $\neg Q \rightarrow \neg P$

f. $P \leftrightarrow Q$ $(P \wedge Q) \vee (\neg P \wedge \neg Q)$

5 Show that if $\neg_L$ and $\rightarrow_L$ are taken as primitive connectives and the other ones are introduced with the definitions

$$\mathbf{P} \vee_L \mathbf{Q} =_{\text{def}} (\mathbf{P} \rightarrow_L \mathbf{Q}) \rightarrow_L \mathbf{Q}$$
$$\mathbf{P} \wedge_L \mathbf{Q} =_{\text{def}} \neg_L (\neg_L \mathbf{P} \vee_L \neg_L \mathbf{Q})$$
$$\mathbf{P} \leftrightarrow_L \mathbf{Q} =_{\text{def}} (\mathbf{P} \rightarrow_L \mathbf{Q}) \wedge_L (\mathbf{Q} \rightarrow_L \mathbf{P})$$

we obtain the correct truth-tables for Ł3.

6 Using truth-tables, show that the following pairs of formulas are equivalent in both $Ł_3$ and $K^S{}_3$, so that the second formula in each pair can be used as a definition of the first in either system.

a. $P \wedge Q$ $\neg(\neg P \vee \neg Q)$

b. $P \leftrightarrow Q$ $(P \rightarrow Q) \wedge (Q \rightarrow P)$

7 Using a truth-table, show that the following formulas are ***not*** equivalent in $Ł_3$, so that the second ***cannot*** be used as a definition of the first.

$P \rightarrow Q$ $\neg(P \wedge \neg Q)$

8 In Exercise 7 you showed that the following formulas are ***not*** equivalent in $Ł_3$, so that the second ***cannot*** be used as a definition of the first in that system:

$P \rightarrow Q$ $\neg(P \wedge \neg Q)$

Are they equivalent in $\mathbf{K}^S{}_3$?

9 Using truth-tables, determine which of the following formulas are tautologies in $Ł_3$ *(note: they are all classical tautologies)*:

a. $\neg P \rightarrow (P \rightarrow Q)$

b. $(P \rightarrow \neg P) \rightarrow \neg P$

c. $(P \leftrightarrow Q) \vee (P \leftrightarrow \neg Q)$

d. $(P \wedge Q) \rightarrow (P \vee Q)$

10 Use truth-tables to decide whether the following arguments are valid in $Ł_3$:

a. P

$\underline{P \rightarrow Q}$

Q

b. $\underline{P \wedge \neg Q}$

$Q \rightarrow P$

c. $P \rightarrow Q$

$\underline{\neg Q}$

$\neg P$

SECTION 5.3

11 Using truth-tables, determine whether the following pairs of formulas are equivalent (always have the same truth-value) in $B^I{}_3$, Bochvar's system of "internal" connectives:

a. $P \wedge Q$ $\neg(\neg P \vee \neg Q)$

b. P $\neg\neg P$

c. $P \rightarrow Q$ $\neg Q \rightarrow \neg P$

12 Do the same, using $B^E{}_3$, Bochvar's system of "external" connectives, in place of the internal connectives in 5a–c.

13 Use truth-tables to decide whether the following arguments are valid in $B^I{}_3$:

a. P

$\underline{P \rightarrow Q}$

Q

b. $\underline{P \wedge \neg Q}$

$Q \rightarrow P$

c. $\dfrac{\begin{array}{l}P \rightarrow Q\\ \neg Q\end{array}}{\neg P}$

14 Let us define a third version of Bochvar's connectives, called the "mixternal" connectives, defined in terms of Bochvar's internal connectives and Bochvar's external assertion operator a:

Connective	*Mixternal form*
Negation	$—\mathbf{P} =_{\text{def}} \mathrm{a}\neg_{BI}\mathbf{P}$
Conjunction	$\mathbf{P} \bullet \mathbf{Q} =_{\text{def}} \mathrm{a}(\mathbf{P} \wedge_{BI} \mathbf{Q})$
Disjunction	$\mathbf{P} + \mathbf{Q} =_{\text{def}} \mathrm{a}(\mathbf{P} \vee_{BI} \mathbf{Q})$
Conditional	$\mathbf{P} \Rightarrow \mathbf{Q} =_{\text{def}} \mathrm{a}(\mathbf{P} \rightarrow_{BI} \mathbf{Q})$
Biconditional	$\mathbf{P} \Leftrightarrow \mathbf{Q} =_{\text{def}} \mathrm{a}(\mathbf{P} \leftrightarrow_{BI} \mathbf{Q})$

Produce truth-tables for this new set of connectives.

15 In which of the four systems K^S_3, $Ł_3$, B^I_3, and B^E_3 are instances of the Law of NonContradiction, $\neg(\mathbf{P} \wedge \neg\mathbf{P})$, tautologies?

SECTION 5.4

16 Prove the following:

a. In each of B^I_3, B^E_3, and K^S_3 the set of quasi-contradictions coincides with the set of classical contradictions.

b. Every $Ł_3$ quasi-contradiction is a classical contradiction.

c. Some classical contradictions are not $Ł_3$ quasi-contradictions.

d. Result 5.20.

17 a. Prove that the argument

$$\frac{P \wedge Q}{P}$$

is quasi-valid in both K^S_3 and $Ł_3$, but not in B^I_3.

b. Prove that the argument

$$\frac{P}{P \vee Q}$$

is quasi-valid in all three of K^S_3, $Ł_3$, and B^I_3.

c. For each of the three systems, give an example of a classically valid argument that is degree-valid in that system.

18 Prove that every degree-entailment that holds in B^I_3, B^E_3, K^S_3, or $Ł_3$ is a classical entailment.

19 Prove that degree-entailment is equivalent to entailment proper plus quasi-entailment.

SECTION 5.5

20 a. Prove that the DeMorgan's Law equivalences hold in K^S_3, $Ł_3$, B^I_3, and B^E_3.

b. Prove that the Distribution equivalences hold in K^S_3, $Ł_3$, B^I_3, and B^E_3.

c. Prove that the Double Negation equivalence holds in K^S_3, $Ł_3$, and B^I_3 but fails in B^E_3.

d. Prove that the Implication equivalences hold in K^{S}_{3}, B^{I}_{3}, and B^{E}_{3} but fail in Ł$_3$.

SECTION 5.6

21 Using truth-tables, verify that the explicit definitions of the B^{I}_{3} conjunction, disjunction, and conditional in Result 5.28 are correct.

22 Using the algorithm in Result 5.33, produce a formula that can be used to define the connective # in Ł$_3$:

P # Q

P \ Q	T	N	F
T	T	N	T
N	T	N	N
F	F	N	F

SECTION 5.7

23 Using truth-tables, show that both **P ∇ ¬P** and ¬(**P** & ¬**P**) are tautologies in Ł$_3$.

24 Prove the claim made in footnote 8, namely, If a (binary) operation is commutative then it must be nondecreasing in both arguments if it is nondecreasing on one of its arguments.

25 Prove that conditions 6 and 7 for t-norm and t-conorm operations follow from one or more of conditions 1–5.

26 Prove the claim that the conditions defining t-norms and t-conorms uniquely define conjunction and disjunction in classical logic, that is, that the conditions produce the classical truth-tables for these operations.

6 Derivation Systems for Three-Valued Propositional Logic

6.1 An Axiomatic System for Tautologies and Validity in Three-Valued Logic

We have introduced three semantic concepts of validity in three-valued logic, validity proper, quasi-validity, and degree validity, along with the corresponding three varieties of tautologousness and contradictoriness. Consequently, we may have different expectations for derivation systems. We'll begin with a system that establishes validity (and tautologousness) proper for $Ł_3$. Derivation systems have also been designed specifically for Kleene's and Bochvar's three-valued logics,[1] but we restrict our attention to systems for $Ł_3$ in this chapter. As we showed in Chapter 5, Kleene's and Bochvar's connectives can all be defined using Łukasiewicz's connectives, so we can represent inferences for those systems within $Ł_3$ axiomatic systems. Moreover, we are anticipating fuzzy logic in which the bulk of formal work is based upon Łukasiewicz's infinite-valued generalization of his three-valued system.

Taking $\neg$ and $\rightarrow$ as primitive connectives, Mordchaj Wajsberg proved in 1931 that the following axiomatic derivation system—which we will call L_3A (for *$Ł_3$ axiomatic system*)—is sound and complete for $Ł_3$ (Wajsberg 1931):[2]

$Ł_3$1. $\mathbf{P \rightarrow (Q \rightarrow P)}$
$Ł_3$2. $\mathbf{(P \rightarrow Q) \rightarrow ((Q \rightarrow R) \rightarrow (P \rightarrow R))}$
$Ł_3$3. $\mathbf{(\neg P \rightarrow \neg Q) \rightarrow (Q \rightarrow P)}$
$Ł_3$4. $\mathbf{((P \rightarrow \neg P) \rightarrow P) \rightarrow P}$

The single derivation rule for this system is Modus Ponens:

MP. From **P** and **P** $\rightarrow$ **Q**, infer **Q**.

1 An axiomatic system for B^I_3 developed by V. K. Finn appears in Bolc and Borowik (1992), ch. 3.3. A (natural deduction) derivation system for a first-order version of K^S_3 appears in Kearns (1979). We should note that $Ł_3$ and the Bochvar systems also have adequate natural deduction systems—see, for example, Beall and van Fraassen (2003), Section 11.6; and Baaz, Fermüller, and Zach (1993).

2 Since we present a system only for $Ł_3$, we will omit the subscript L on the connectives. When we discuss connectives of the *other* three-valued systems, those connectives will retain their subscripts.

The connectives ∨, ∧, and ↔ can be defined as in Section 5.2—so all $Ł_3$ formulas can be expressed using the connectives ¬ and → that appear in the axiom schemata and rule. Axiom schemata $Ł_3$1 and $Ł_3$3 are identical to the schemata CL1 and CL3 presented for classical propositional logic in Chapter 2. The axiom schema CL2, **(P → (Q → R)) → ((P → Q) → (P → R))**, is not derivable within $Ł_3$A. This is as it should be since we showed in Section 5.2 that CL2 isn't a tautology in $Ł_3$. (On the other hand, the new schemata $Ł_3$2 and $Ł_3$4 are derivable in CLA, since they are classical tautologies and the classical system is complete.)

The gist of axiom schema $Ł_3$4 may not be immediately apparent. Recalling that a disjunction **P ∨ Q** can be defined as **(P → Q) → Q** in $Ł_3$, axiom schema $Ł_3$4 may be rewritten as **(P → ¬P) ∨ P**. This formula is closely related to the Law of Excluded Middle. But the Law of Excluded Middle fails to be a tautology in $Ł_3$, while the present formula *is* a tautology in $Ł_3$. If **P** has the truth-value **T** then so does the formula as a whole since **P** is the right disjunct, and if **P** has the truth-value **N** or the truth-value **F** then the left conjunct has the value **T**, so the disjunction as a whole does as well.

Any derivation in the classical system CLA that doesn't involve the axiom schema CL2 counts as a derivation in $Ł_3$A, and any axiom that is derivable in CLA without using CL2 counts as a derived axiom in $Ł_3$A. There are also derived axiom schemata of CLA that are derivable in $Ł_3$A but with different derivations than we used in the classical case. An example is CLD1:

CLD1. P → P

The derivation in CLA used CL2, so that derivation can't be used here. We'll produce a legal $Ł_3$A derivation for **P → P** after introducing some more immediate derived axiom schemata and rules. Derivations in $Ł_3$A can be tricky to construct, so we introduce plenty of derived help. We begin with the derived axioms

$Ł_3$D1. ¬P → (P → Q)
$Ł_3$D2. ¬¬P → P
$Ł_3$D3. P → ¬¬P

($Ł_3$D2 and $Ł_3$D3 are, respectively, CLD3 and CLD4). $Ł_3$D1 is justified as follows:[3]

1	**¬P → (¬Q → ¬P)**	$Ł_3$1, with **¬P** / **P**, **¬Q** / **Q**
2	**(¬Q → ¬P) → (P → Q)**	$Ł_3$3, with **Q** / **P**, **P** / **Q**
3	**(¬P → (¬Q → ¬P)) → (((¬Q → ¬P) → (P → Q)) → (¬P → (P → Q)))**	$Ł_3$2, with **¬P** / **P**, **¬Q → ¬P** / **Q**, **P → Q** / **R**
4	**((¬Q → ¬P) → (P → Q)) → (¬P → (P → Q))**	1,3 MP
5	**¬P → (P → Q))**	2,4 MP

3 Our derivations of $Ł_3$D1–$Ł_3$D7, MCD, and $Ł_3$D11 are due to Minari (2003)—these are the most perspicuous (and generally the shortest) derivations we have seen for these derived axioms.

Note that lines 3–5 yield a formula that is derivable from the formulas on lines 1 and 2 by Hypothetical Syllogism. Since we will follow this pattern often, we introduce Hypothetical Syllogism (HS) as a derived rule for Ł$_3$A:

HS. From $\mathbf{P \rightarrow Q}$ and $\mathbf{Q \rightarrow R}$, infer $\mathbf{P \rightarrow R}$.

HS is derivable by virtue of the pattern in the preceding derivation: introduce an appropriate instance of Ł$_3$2 and then use MP twice. Here is a justification for Ł$_3$D2:

1	$\neg\neg\mathbf{P} \rightarrow (\neg\mathbf{P} \rightarrow \neg(\mathbf{P} \rightarrow \neg\mathbf{P}))$	Ł$_3$D1, with $\neg\mathbf{P}$ / $\mathbf{P}$, $\neg(\mathbf{P} \rightarrow \neg\mathbf{P})$ / $\mathbf{Q}$
2	$(\neg\mathbf{P} \rightarrow \neg(\mathbf{P} \rightarrow \neg\mathbf{P})) \rightarrow ((\mathbf{P} \rightarrow \neg\mathbf{P}) \rightarrow \mathbf{P})$	Ł$_3$3, with $\mathbf{P}$ / $\mathbf{P}$, $\mathbf{P} \rightarrow \neg\mathbf{P}$ / $\mathbf{Q}$
3	$\neg\neg\mathbf{P} \rightarrow ((\mathbf{P} \rightarrow \neg\mathbf{P}) \rightarrow \mathbf{P})$	1,2 HS
4	$((\mathbf{P} \rightarrow \neg\mathbf{P}) \rightarrow \mathbf{P}) \rightarrow \mathbf{P}$	Ł$_3$4, with $\mathbf{P}$ / $\mathbf{P}$
5	$\neg\neg\mathbf{P} \rightarrow \mathbf{P}$	3,4 HS

Ł$_3$D3 is justified as follows:

1	$\neg\neg\neg\mathbf{P} \rightarrow \neg\mathbf{P}$	Ł$_3$D2, with $\neg\mathbf{P}$ / $\mathbf{P}$
2	$(\neg\neg\neg\mathbf{P} \rightarrow \neg\mathbf{P}) \rightarrow (\mathbf{P} \rightarrow \neg\neg\mathbf{P})$	Ł$_3$3, with $\neg\neg\mathbf{P}$ / $\mathbf{P}$, $\mathbf{P}$ / $\mathbf{Q}$
3	$\mathbf{P} \rightarrow \neg\neg\mathbf{P}$	1,2 MP

We can now easily prove CLD1, which we will call

Ł$_3$D4. $\mathbf{P \rightarrow P}$

1	$\mathbf{P} \rightarrow \neg\neg\mathbf{P}$	Ł$_3$D3, with $\mathbf{P}$ / $\mathbf{P}$
2	$\neg\neg\mathbf{P} \rightarrow \mathbf{P}$	Ł$_3$D2, with $\mathbf{P}$ / $\mathbf{P}$
3	$\mathbf{P} \rightarrow \mathbf{P}$	1,2 HS

Ł$_3$D5 will be a useful derived axiom schema for the following proof:

Ł$_3$D5. $\mathbf{((P \rightarrow P) \rightarrow Q) \rightarrow Q}$

1	$(\mathbf{P} \rightarrow \mathbf{P}) \rightarrow ((\mathbf{Q} \rightarrow \neg\mathbf{Q}) \rightarrow (\mathbf{P} \rightarrow \mathbf{P}))$	Ł$_3$1, with $\mathbf{P} \rightarrow \mathbf{P}$ / $\mathbf{P}$, $\mathbf{Q} \rightarrow \neg\mathbf{Q}$ / $\mathbf{Q}$
2	$\mathbf{P} \rightarrow \mathbf{P}$	Ł$_3$D4, with $\mathbf{P}$ / $\mathbf{P}$
3	$(\mathbf{Q} \rightarrow \neg\mathbf{Q}) \rightarrow (\mathbf{P} \rightarrow \mathbf{P})$	1,2 MP
4	$((\mathbf{Q} \rightarrow \neg\mathbf{Q}) \rightarrow (\mathbf{P} \rightarrow \mathbf{P})) \rightarrow (((\mathbf{P} \rightarrow \mathbf{P}) \rightarrow \mathbf{Q}) \rightarrow ((\mathbf{Q} \rightarrow \neg\mathbf{Q}) \rightarrow \mathbf{Q}))$	Ł$_3$2, with $\mathbf{Q} \rightarrow \neg\mathbf{Q}$ / $\mathbf{P}$, $\mathbf{P} \rightarrow \mathbf{P}$ / $\mathbf{Q}$, $\mathbf{Q}$ / $\mathbf{R}$
5	$((\mathbf{P} \rightarrow \mathbf{P}) \rightarrow \mathbf{Q}) \rightarrow ((\mathbf{Q} \rightarrow \neg\mathbf{Q}) \rightarrow \mathbf{Q})$	3,4 MP
6	$((\mathbf{Q} \rightarrow \neg\mathbf{Q}) \rightarrow \mathbf{Q}) \rightarrow \mathbf{Q}$	Ł$_3$4, with $\mathbf{Q}$ / $\mathbf{P}$
7	$((\mathbf{P} \rightarrow \mathbf{P}) \rightarrow \mathbf{Q}) \rightarrow \mathbf{Q}$	5,6 HS

The following axiom is $\mathbf{P \rightarrow (P \vee Q)}$ when rewritten with Ł$_3$ disjunction:

Ł$_3$D6. $\mathbf{P \rightarrow ((P \rightarrow Q) \rightarrow Q)}$

(The closely related $\mathbf{P \rightarrow (Q \vee P)}$ is an instance of Ł$_3$1 when rewritten with disjunction.) Here is a proof of Ł$_3$D6, admittedly complicated but justifying a very useful axiom:

1	$\mathbf{P} \rightarrow ((\mathbf{P} \rightarrow \mathbf{P}) \rightarrow \mathbf{P})$	Ł$_3$1, with $\mathbf{P} / \mathbf{P}$, $\mathbf{P} \rightarrow \mathbf{P} / \mathbf{Q}$
2	$((\mathbf{P} \rightarrow \mathbf{P}) \rightarrow \mathbf{P}) \rightarrow ((\mathbf{P} \rightarrow \mathbf{Q}) \rightarrow ((\mathbf{P} \rightarrow \mathbf{P}) \rightarrow \mathbf{Q}))$	Ł$_3$2, with $\mathbf{P} \rightarrow \mathbf{P} / \mathbf{P}$, $\mathbf{P} / \mathbf{Q}$, $\mathbf{Q} / \mathbf{R}$
3	$\mathbf{P} \rightarrow ((\mathbf{P} \rightarrow \mathbf{Q}) \rightarrow ((\mathbf{P} \rightarrow \mathbf{P}) \rightarrow \mathbf{Q}))$	1,2 HS
4	$((\mathbf{P} \rightarrow \mathbf{Q}) \rightarrow ((\mathbf{P} \rightarrow \mathbf{P}) \rightarrow \mathbf{Q})) \rightarrow ((((\mathbf{P} \rightarrow \mathbf{P}) \rightarrow \mathbf{Q}) \rightarrow \mathbf{Q}) \rightarrow$ $((\mathbf{P} \rightarrow \mathbf{Q}) \rightarrow \mathbf{Q}))$	Ł$_3$2, with $\mathbf{P} \rightarrow \mathbf{Q} / \mathbf{P}$, $(\mathbf{P} \rightarrow \mathbf{P}) \rightarrow \mathbf{Q} / \mathbf{Q}$, $\mathbf{Q} / \mathbf{R}$
5	$\mathbf{P} \rightarrow ((((\mathbf{P} \rightarrow \mathbf{P}) \rightarrow \mathbf{Q})) \rightarrow \mathbf{Q}) \rightarrow ((\mathbf{P} \rightarrow \mathbf{Q}) \rightarrow \mathbf{Q}))$	3,4 HS
6	$((\mathbf{P} \rightarrow \mathbf{P}) \rightarrow \mathbf{Q}) \rightarrow \mathbf{Q}$	Ł$_3$D5, with $\mathbf{P} / \mathbf{P}$, $\mathbf{Q} / \mathbf{Q}$
7	$(((\mathbf{P} \rightarrow \mathbf{P}) \rightarrow \mathbf{Q}) \rightarrow \mathbf{Q}) \rightarrow ((\mathbf{P} \rightarrow \mathbf{P}) \rightarrow (((\mathbf{P} \rightarrow \mathbf{P}) \rightarrow \mathbf{Q}) \rightarrow \mathbf{Q}))$	Ł$_3$1, with $((\mathbf{P} \rightarrow \mathbf{P}) \rightarrow \mathbf{Q}) \rightarrow \mathbf{Q} / \mathbf{P}$, $\mathbf{P} \rightarrow \mathbf{P} / \mathbf{Q}$
8	$(\mathbf{P} \rightarrow \mathbf{P}) \rightarrow (((\mathbf{P} \rightarrow \mathbf{P}) \rightarrow \mathbf{Q}) \rightarrow \mathbf{Q})$	6,7 MP
9	$((\mathbf{P} \rightarrow \mathbf{P}) \rightarrow (((\mathbf{P} \rightarrow \mathbf{P}) \rightarrow \mathbf{Q}) \rightarrow \mathbf{Q})) \rightarrow$ $(((((\mathbf{P} \rightarrow \mathbf{P}) \rightarrow \mathbf{Q}) \rightarrow \mathbf{Q}) \rightarrow ((\mathbf{P} \rightarrow \mathbf{Q}) \rightarrow \mathbf{Q})) \rightarrow$ $((\mathbf{P} \rightarrow \mathbf{P}) \rightarrow ((\mathbf{P} \rightarrow \mathbf{Q}) \rightarrow \mathbf{Q})))$	Ł$_3$1, with $\mathbf{P} \rightarrow \mathbf{P} / \mathbf{P}$, $((\mathbf{P} \rightarrow \mathbf{P}) \rightarrow \mathbf{Q}) \rightarrow \mathbf{Q} / \mathbf{Q}$, $(\mathbf{P} \rightarrow \mathbf{Q}) \rightarrow \mathbf{Q} / \mathbf{R}$
10	$((((\mathbf{P} \rightarrow \mathbf{P}) \rightarrow \mathbf{Q}) \rightarrow \mathbf{Q}) \rightarrow ((\mathbf{P} \rightarrow \mathbf{Q}) \rightarrow \mathbf{Q})) \rightarrow$ $((\mathbf{P} \rightarrow \mathbf{P}) \rightarrow ((\mathbf{P} \rightarrow \mathbf{Q}) \rightarrow \mathbf{Q}))$	8,9 MP
11	$\mathbf{P} \rightarrow ((\mathbf{P} \rightarrow \mathbf{P}) \rightarrow ((\mathbf{P} \rightarrow \mathbf{Q}) \rightarrow \mathbf{Q}))$	5,10 HS
12	$((\mathbf{P} \rightarrow \mathbf{P}) \rightarrow ((\mathbf{P} \rightarrow \mathbf{Q}) \rightarrow \mathbf{Q})) \rightarrow ((\mathbf{P} \rightarrow \mathbf{Q}) \rightarrow \mathbf{Q})$	Ł$_3$D5, with $\mathbf{P} / \mathbf{P}$, $(\mathbf{P} \rightarrow \mathbf{Q}) \rightarrow \mathbf{Q} / \mathbf{Q}$
13	$\mathbf{P} \rightarrow ((\mathbf{P} \rightarrow \mathbf{Q}) \rightarrow \mathbf{Q})$	11,12 HS

With Ł$_3$D6 in hand it's straightforward to justify the transposition axiom schema in Ł$_3$A:

Ł$_3$D7. $(\mathbf{P} \rightarrow (\mathbf{Q} \rightarrow \mathbf{R})) \rightarrow (\mathbf{Q} \rightarrow (\mathbf{P} \rightarrow \mathbf{R}))$

1	$(\mathbf{P} \rightarrow (\mathbf{Q} \rightarrow \mathbf{R})) \rightarrow (((\mathbf{Q} \rightarrow \mathbf{R}) \rightarrow \mathbf{R}) \rightarrow (\mathbf{P} \rightarrow \mathbf{R}))$	Ł$_3$2, with $\mathbf{P} / \mathbf{P}$, $\mathbf{Q} \rightarrow \mathbf{R} / \mathbf{Q}$, $\mathbf{R} / \mathbf{R}$
2	$\mathbf{Q} \rightarrow ((\mathbf{Q} \rightarrow \mathbf{R}) \rightarrow \mathbf{R})$	Ł$_3$D6, with $\mathbf{Q} / \mathbf{Q}$, $\mathbf{R} / \mathbf{R}$
3	$(\mathbf{Q} \rightarrow ((\mathbf{Q} \rightarrow \mathbf{R}) \rightarrow \mathbf{R})) \rightarrow ((((\mathbf{Q} \rightarrow \mathbf{R}) \rightarrow \mathbf{R}) \rightarrow$ $(\mathbf{P} \rightarrow \mathbf{R})) \rightarrow (\mathbf{Q} \rightarrow (\mathbf{P} \rightarrow \mathbf{R})))$	Ł$_3$2, with $\mathbf{Q} / \mathbf{P}$, $(\mathbf{Q} \rightarrow \mathbf{R}) \rightarrow \mathbf{R} / \mathbf{Q}$, $\mathbf{P} \rightarrow \mathbf{R} / \mathbf{R}$
4	$(((\mathbf{Q} \rightarrow \mathbf{R}) \rightarrow \mathbf{R}) \rightarrow (\mathbf{P} \rightarrow \mathbf{R})) \rightarrow (\mathbf{Q} \rightarrow (\mathbf{P} \rightarrow \mathbf{R}))$	2,3 MP
5	$(\mathbf{P} \rightarrow (\mathbf{Q} \rightarrow \mathbf{R})) \rightarrow (\mathbf{Q} \rightarrow (\mathbf{P} \rightarrow \mathbf{R}))$	1,4 HS

In addition to deriving theorems corresponding to Ł$_3$-tautologies (all of the derivations justifying derived axiom schemata establish the theoremhood of the final formulas), we can also derive the conclusions of arguments that are valid in Ł$_3$. For example, $\neg B$ is derivable from $B \rightarrow C$ and $\neg C$ as follows:

1	$B \rightarrow C$	Assumption
2	$\neg C$	Assumption
3	$\neg\neg B \rightarrow B$	Ł$_3$D2, with B / $\mathbf{P}$
4	$(\neg\neg B \rightarrow B) \rightarrow ((B \rightarrow C) \rightarrow (\neg\neg B \rightarrow C))$	Ł$_3$2, with $\neg\neg$B / $\mathbf{P}$, B / $\mathbf{Q}$, C / $\mathbf{R}$
5	$(B \rightarrow C) \rightarrow (\neg\neg B \rightarrow C)$	3,4 MP
6	$\neg\neg B \rightarrow C$	1,5 MP
7	$C \rightarrow \neg\neg C$	Ł$_3$D3, with C / $\mathbf{P}$
8	$\neg\neg B \rightarrow \neg\neg C$	6,7 HS
9	$(\neg\neg B \rightarrow \neg\neg C) \rightarrow (\neg C \rightarrow \neg B)$	Ł$_3$3, with $\neg$B / $\mathbf{P}$, $\neg$C / $\mathbf{Q}$
10	$\neg C \rightarrow \neg B$	8,9 MP
11	$\neg B$	2,10 MP

The pattern of proof on lines 8–10 is common, so we introduce the derived rule Contraposition to capture the pattern:

CON (Contraposition). From $\neg\mathbf{P} \rightarrow \neg\mathbf{Q}$ infer $\mathbf{Q} \rightarrow \mathbf{P}$.

Valid inferences using Kleene's and Bochvar's (internal and external) connectives have corresponding derivations in Ł_3A, provided that we use the Ł_3 definitions for rewriting formulas containing those connectives. For example, the argument

$$\frac{\begin{array}{l}\mathbf{P}\\ \mathbf{P} \rightarrow_K \mathbf{Q}\end{array}}{\mathbf{Q}}$$

is valid in $\mathrm{K}^{\mathrm{S}}_3$. The conditional $P \rightarrow_K Q$ is equivalent to $\neg P \vee Q$ in Ł_3, which is expressible as $(\neg P \rightarrow Q) \rightarrow Q$ using only negation and disjunction. The following derivation establishes the validity of the $\mathrm{K}^{\mathrm{S}}_3$ argument:

1	P	Assumption
2	$(\neg P \rightarrow Q) \rightarrow Q$	Assumption
3	$P \rightarrow (\neg Q \rightarrow P)$	$\text{Ł}_3$1, with P / $\mathbf{P}$, $\neg Q$ / $\mathbf{Q}$
4	$\neg Q \rightarrow P$	1,3 MP
5	$P \rightarrow \neg\neg P$	Ł_3D3, with P / $\mathbf{P}$
6	$\neg Q \rightarrow \neg\neg P$	4,5 HS
7	$\neg P \rightarrow Q$	6, CON
8	Q	2,7 MP

On the other hand, we know that there are no tautologies in $\mathrm{K}^{\mathrm{S}}_3$ and so no theorems of our system correspond to $\mathrm{K}^{\mathrm{S}}_3$ formulas.

The inference

$$\frac{\begin{array}{l}\mathrm{P}\\ \mathrm{P} \rightarrow_{BI} \mathrm{Q}\end{array}}{\mathrm{Q}}$$

is valid in $\mathrm{B}^{\mathrm{I}}_3$. We'll first derive the rule LSIMP in Ł_3A:

LSIMP (Left Conjunct Simplification). From $\mathbf{P} \wedge \mathbf{Q}$ infer $\mathbf{P}$
Justification ($\mathbf{P} \wedge \mathbf{Q}$ is rewritten as $\neg((\neg\mathbf{P} \rightarrow \neg\mathbf{Q}) \rightarrow \neg\mathbf{Q})$):

1	$\neg((\neg\mathbf{P} \rightarrow \neg\mathbf{Q}) \rightarrow \neg\mathbf{Q})$	Assumption
2	$\neg\mathbf{P} \rightarrow ((\neg\mathbf{P} \rightarrow \neg\mathbf{Q}) \rightarrow \neg\mathbf{Q})$	Ł_3D6, with $\neg\mathbf{P}$ / $\mathbf{P}$, $\neg\mathbf{Q}$ / $\mathbf{Q}$
3	$((\neg\mathbf{P} \rightarrow \neg\mathbf{Q}) \rightarrow \neg\mathbf{Q}) \rightarrow \neg\neg((\neg\mathbf{P} \rightarrow \neg\mathbf{Q}) \rightarrow \neg\mathbf{Q})$	Ł_3D3, with $(\neg\mathbf{P} \rightarrow \neg\mathbf{Q}) \rightarrow \neg\mathbf{Q}$ / $\mathbf{P}$
4	$\neg\mathbf{P} \rightarrow \neg\neg((\neg\mathbf{P} \rightarrow \neg\mathbf{Q}) \rightarrow \neg\mathbf{Q})$	2,3 HS
5	$\neg((\neg\mathbf{P} \rightarrow \neg\mathbf{Q}) \rightarrow \neg\mathbf{Q}) \rightarrow \mathbf{P}$	4, CON
6	$\mathbf{P}$	1,5 MP

The second premise $P \rightarrow_{BI} Q$ of the $B^I{}_3$ argument is equivalent to the Ł$_3$ formula $(\neg P \vee Q) \wedge ((P \vee \neg P) \wedge (Q \vee \neg Q))$. The validity of the $B^I{}_3$ argument is established by the following derivation:

1	P	Assumption
2	$(\neg P \vee Q) \wedge ((P \vee \neg P) \wedge (Q \vee \neg Q))$	Assumption
3	$\neg P \vee Q$	2, LSIMP
4	. . . {*the rest of the proof is identical to that for the* $K^S{}_3$ *example earlier, substituting* $(\neg P \rightarrow Q) \rightarrow Q$ *for* $\neg P \vee Q$}	

As with $K^S{}_3$, we know that there are no tautologies in $B^I{}_3$ and so no formulas of $B^I{}_3$ are theorems of Ł$_3$A.

Finally, the inference

$$\begin{array}{l} P \\ \underline{P \rightarrow_{BE} Q} \\ Q \end{array}$$

of Bochvar's external system is also valid in Ł$_3$A. The second premise is expressible in Ł$_3$ with the formula $\neg(P \rightarrow \neg P) \rightarrow \neg(Q \rightarrow \neg Q)$. Here's the derivation:

1	P	Assumption
2	$\neg(P \rightarrow \neg P) \rightarrow \neg(Q \rightarrow \neg Q)$	Assumption
3	$(Q \rightarrow \neg Q) \rightarrow (P \rightarrow \neg P)$	2, CON
4	$((Q \rightarrow \neg Q) \rightarrow (P \rightarrow \neg P)) \rightarrow (P \rightarrow ((Q \rightarrow \neg Q) \rightarrow \neg P)))$	Ł$_3$D7, with $Q \rightarrow \neg Q$ / **P**, P / **Q**, $\neg$P / **R**
5	$P \rightarrow ((Q \rightarrow \neg Q) \rightarrow \neg P)$	3,4 MP
6	$(Q \rightarrow \neg Q) \rightarrow \neg P$	1,5 MP
7	$\neg\neg(Q \rightarrow \neg Q) \rightarrow (Q \rightarrow \neg Q)$	Ł$_3$D2, with $Q \rightarrow \neg Q$ / **P**
8	$\neg\neg(Q \rightarrow \neg Q) \rightarrow \neg P$	6,7 HS
9	$P \rightarrow \neg(Q \rightarrow \neg Q)$	8, CON
10	$\neg(Q \rightarrow \neg Q)$	1,9 MP
11	$\neg Q \rightarrow (Q \rightarrow \neg Q)$	Ł$_3$1, with $\neg$Q / **P**, Q / **Q**
12	$(Q \rightarrow \neg Q) \rightarrow \neg\neg (Q \rightarrow \neg Q)$	Ł$_3$D3, with $Q \rightarrow \neg Q$ / **P**
13	$\neg Q \rightarrow \neg\neg(Q \rightarrow \neg Q)$	11,12 HS
14	$\neg(Q \rightarrow \neg Q) \rightarrow Q$	13, CON
15	Q	10,14 MP

The formula $P \rightarrow_{BE} P$ is a tautology, so we would expect the formula $\neg(P \rightarrow \neg P) \rightarrow \neg(P \rightarrow \neg P)$, which expresses $P \rightarrow_{BE} P$ in Ł$_3$, to be a theorem of Ł$_3$A. It is easy to show that it is, since it is an instance of Ł$_3$D4.

It will not have escaped the reader that (as we commented earlier) derivations in the axiomatic system Ł$_3$A can be difficult to design and construct—and that the addition of derived axioms and rules makes derivations much more manageable. We will continue using axiomatic systems when we turn to fuzzy logic, so we'll take the time here to introduce some further derived axioms and rules—most of which will carry over to fuzzy logic. (The reader may choose to skim or skip the proofs at this point and move on to the last three paragraphs of this section (pp. 113–114), returning to study these proofs when the derived axioms and rules appear in subsequent derivations.)

First we derive the very general and useful rule

SUB (Substitution). From $\mathbf{P} \rightarrow \mathbf{Q}$, $\mathbf{Q} \rightarrow \mathbf{P}$ and a formula **R** that contains **P** as a subformula, infer any formula **R*** that is the result of replacing one or more occurrences of **P** in **R** with **Q**.

For example, we'll use SUB to derive $((\neg\neg Q \rightarrow \neg Q) \rightarrow \neg\neg\, Q) \rightarrow Q$ as follows:

1	$((Q \rightarrow \neg Q) \rightarrow Q) \rightarrow Q$	Ł$_3$4, with Q / **P**
2	$Q \rightarrow \neg\neg\, Q$	Ł$_3$D3, with Q / **P**
3	$\neg\neg Q \rightarrow Q$	Ł$_3$D2, with Q / **P**
4	$((\neg\neg Q \rightarrow \neg Q) \rightarrow \neg\neg\, Q) \rightarrow Q$	1,2,3 SUB

On line 4 we replaced two occurrences of Q in the formula on line 1 with $\neg\neg Q$.

To justify SUB we'll show that if we can derive reciprocal formulas $\mathbf{P} \rightarrow \mathbf{Q}$ and $\mathbf{Q} \rightarrow \mathbf{P}$, then given any formula **R** that contains **P** we can derive both $\mathbf{R} \rightarrow \mathbf{R}^+$ and $\mathbf{R}^+ \rightarrow \mathbf{R}$, where $\mathbf{R}^+$ is identical to **R** except that *one* occurrence of **P** has been replaced with **Q**. It will follow from this that if we can derive **R** we can also derive $\mathbf{R}^+$ by Modus Ponens (and vice versa). Moreover, we can then replace *more* than one occurrence of **P** in **R** with **Q** to obtain *any* **R*** by replacing one occurrence at a time—so SUB will be fully justified.

We will show how we can derive both $\mathbf{R} \rightarrow \mathbf{R}^+$ and $\mathbf{R}^+ \rightarrow \mathbf{R}$ by showing how to derive larger and larger conditionals reflecting the way that **R** has been built up from **P**, and hence the way that $\mathbf{R}^+$ must be built up from **Q**. When we say *reflecting the way that **R** has been built up from **P*** here's what we mean. If **R** is $(\neg(A \rightarrow \mathbf{P}) \rightarrow (A \rightarrow B)) \rightarrow C$, then **R** has been built up from **P** by combining **P** with other formulas and connectives as follows (in accordance with the definition of well-formed formulas from Chapter 2):

$\mathbf{P}$
$A \rightarrow \mathbf{P}$
$\neg(A \rightarrow \mathbf{P})$
$\neg(A \rightarrow \mathbf{P}) \rightarrow (A \rightarrow B)$
$(\neg(A \rightarrow \mathbf{P}) \rightarrow (A \rightarrow B)) \rightarrow C$

and so $\mathbf{R}^+$ will be built up from **Q** as follows:

$\mathbf{Q}$
$A \rightarrow \mathbf{Q}$
$\neg(A \rightarrow \mathbf{Q})$
$\neg(A \rightarrow \mathbf{Q}) \rightarrow (A \rightarrow B)$
$(\neg(A \rightarrow \mathbf{Q}) \rightarrow (A \rightarrow B)) \rightarrow C$

We will show how to derive the reciprocal conditionals that pair off the formulas in each row of the two lists, that is, the reciprocal conditionals

$\mathbf{P} \to \mathbf{Q}, \mathbf{Q} \to \mathbf{P}$

$(A \to \mathbf{P}) \to (A \to \mathbf{Q}), (A \to \mathbf{Q}) \to (A \to \mathbf{P})$

$\neg(A \to \mathbf{P}) \to \neg(A \to \mathbf{Q}), \neg(A \to \mathbf{Q}) \to \neg(A \to \mathbf{P})$

$(\neg(A \to \mathbf{P}) \to (A \to B)) \to (\neg(A \to \mathbf{Q}) \to (A \to B)),$
$\quad(\neg(A \to \mathbf{Q}) \to (A \to B)) \to (\neg(A \to \mathbf{P}) \to (A \to B))$

$((\neg(A \to \mathbf{P}) \to (A \to B)) \to C) \to ((\neg(A \to \mathbf{Q}) \to (A \to B)) \to C),$
$\quad((\neg(A \to \mathbf{P}) \to (A \to B)) \to C) \to ((\neg(A \to \mathbf{Q}) \to (A \to B)) \to C)$

where the last pair are the target conditionals $\mathbf{R} \to \mathbf{R}^+$ and $\mathbf{R}^+ \to \mathbf{R}$ for our example.

The derivability of $\mathbf{P} \to \mathbf{Q}$ and $\mathbf{Q} \to \mathbf{P}$ is given in the statement of the rule SUB. Note that for each pair of conditionals $\mathbf{S}_1 \to \mathbf{S}_2$ and $\mathbf{S}_2 \to \mathbf{S}_1$ in the list of paired conditionals, the following pair, $\mathbf{T}_1 \to \mathbf{T}_2$ and $\mathbf{T}_2 \to \mathbf{T}_1$, have $\mathbf{S}_1$ as an immediate component[4] of $\mathbf{T}_1$, and $\mathbf{T}_2$ results from replacing *one* occurrence of $\mathbf{S}_1$ in $\mathbf{T}_1$ with $\mathbf{S}_2$.[5] Given this general pattern for building up the target formulas, we need only show that given any formulas $\mathbf{S}_1 \to \mathbf{S}_2$ and $\mathbf{S}_2 \to \mathbf{S}_1$ there is a way to derive $\mathbf{T}_1 \to \mathbf{T}_2$ and $\mathbf{T}_2 \to \mathbf{T}_1$ where $\mathbf{S}_1$ is an immediate component of $\mathbf{T}_1$ and $\mathbf{T}_2$ is the result of replacing one occurrence of $\mathbf{S}_1$ in $\mathbf{T}_1$ with $\mathbf{S}_2$. There are three cases, reflecting the structure of $\mathbf{T}_1$ (and therefore of $\mathbf{T}_2$ as well):

Case 1: $\mathbf{T}_1$ is $\mathbf{S}_1 \to \mathbf{U}$ for some formula $\mathbf{U}$, and $\mathbf{T}_2$ is $\mathbf{S}_2 \to \mathbf{U}$. Given $\mathbf{S}_1 \to \mathbf{S}_2$ we can derive $\mathbf{T}_2 \to \mathbf{T}_1$, which is $(\mathbf{S}_2 \to \mathbf{U}) \to (\mathbf{S}_1 \to \mathbf{U})$, as follows:

n	$\mathbf{S}_1 \to \mathbf{S}_2$	*given*
$n+1$	$(\mathbf{S}_1 \to \mathbf{S}_2) \to ((\mathbf{S}_2 \to \mathbf{U}) \to (\mathbf{S}_1 \to \mathbf{U}))$	Ł$_3$2, with $\mathbf{S}_1$ / $\mathbf{P}$, $\mathbf{S}_2$ / $\mathbf{Q}$, $\mathbf{U}$ / $\mathbf{R}$
$n+2$	$(\mathbf{S}_2 \to \mathbf{U}) \to (\mathbf{S}_1 \to \mathbf{U})$	n,$n+1$ MP

$\mathbf{T}_1 \to \mathbf{T}_2$, which is $(\mathbf{S}_1 \to \mathbf{U}) \to (\mathbf{S}_2 \to \mathbf{U})$, is similarly derived from $\mathbf{S}_2 \to \mathbf{S}_1$.

Case 2: $\mathbf{T}_1$ is $\mathbf{U} \to \mathbf{S}_1$ for some formula $\mathbf{U}$, and $\mathbf{T}_2$ is $\mathbf{U} \to \mathbf{S}_2$. Given $\mathbf{S}_1 \to \mathbf{S}_2$ we can derive $\mathbf{T}_1 \to \mathbf{T}_2$, which is $(\mathbf{U} \to \mathbf{S}_1) \to (\mathbf{U} \to \mathbf{S}_2)$, as follows:

n	$\mathbf{S}_1 \to \mathbf{S}_2$	*given*
$n+1$	$(\mathbf{U} \to \mathbf{S}_1) \to ((\mathbf{S}_1 \to \mathbf{S}_2) \to (\mathbf{U} \to \mathbf{S}_2))$	Ł$_3$2, with $\mathbf{U}$ / $\mathbf{P}$, $\mathbf{S}_1$ / $\mathbf{Q}$, $\mathbf{S}_2$ / $\mathbf{R}$
$n+2$	$((\mathbf{U} \to \mathbf{S}_1) \to ((\mathbf{S}_1 \to \mathbf{S}_2) \to (\mathbf{U} \to \mathbf{S}_2))) \to$ $((\mathbf{S}_1 \to \mathbf{S}_2) \to ((\mathbf{U} \to \mathbf{S}_1) \to (\mathbf{U} \to \mathbf{S}_2))$	Ł$_3$D7, with $\mathbf{U} \to \mathbf{S}_1$ / $\mathbf{P}$, $\mathbf{S}_1 \to \mathbf{S}_2$ / $\mathbf{Q}$, $\mathbf{U} \to \mathbf{S}_2$ / $\mathbf{R}$
$n+3$	$(\mathbf{S}_1 \to \mathbf{S}_2) \to ((\mathbf{U} \to \mathbf{S}_1) \to (\mathbf{U} \to \mathbf{S}_2))$	$n+1$,$n+2$ MP
$n+4$	$(\mathbf{U} \to \mathbf{S}_1) \to (\mathbf{U} \to \mathbf{S}_2)$	n,$n+3$ MP

$\mathbf{T}_2 \to \mathbf{T}_1$, which is $(\mathbf{U} \to \mathbf{S}_2) \to (\mathbf{U} \to \mathbf{S}_1)$, is similarly derived from $\mathbf{S}_2 \to \mathbf{S}_1$.

[4] That is, $\mathbf{T}_1$ is formed from $\mathbf{S}_1$ either by prefixing $\mathbf{S}_1$ with a negation operator or by combining $\mathbf{S}_1$ with another formula with one of the binary connectives.

[5] For example, where $\mathbf{S}_1$ and $\mathbf{S}_2$ are the original $\mathbf{P}$ and $\mathbf{Q}$ of the example, $\mathbf{T}_1 \to \mathbf{T}_2$ and $\mathbf{T}_2 \to \mathbf{T}_1$ are $(A \to \mathbf{P}) \to (A \to \mathbf{Q})$, $(A \to \mathbf{Q}) \to (A \to \mathbf{P})$. So $\mathbf{T}_1$ is $(A \to \mathbf{P})$ and $\mathbf{T}_2$ is $(A \to \mathbf{Q})$. Here $\mathbf{P}$ is an immediate component of $(A \to \mathbf{P})$, and $(A \to \mathbf{Q})$ results in replacing one occurrence of $\mathbf{P}$ in $(A \to \mathbf{P})$ with $\mathbf{Q}$.

Case 3: $\mathbf{T}_1$ is $\neg\mathbf{S}_1$ and $\mathbf{T}_2$ is $\neg\mathbf{S}_2$. Given $\mathbf{S}_1 \rightarrow \mathbf{S}_2$ we can derive $\mathbf{T}_2 \rightarrow \mathbf{T}_1$, which is $\neg\mathbf{S}_2 \rightarrow \neg\mathbf{S}_1$, as follows:

n	$\mathbf{S}_1 \rightarrow \mathbf{S}_2$	*given*
n+1	$\neg\neg\,\mathbf{S}_1 \rightarrow \mathbf{S}_1$	Ł$_3$D2, with $\mathbf{S}_1$ / **P**
n+2	$\neg\neg\,\mathbf{S}_1 \rightarrow \mathbf{S}_2$	n,n+1 HS
n+3	$\mathbf{S}_2 \rightarrow \neg\neg\,\mathbf{S}_2$	Ł$_3$D3, with $\mathbf{S}_2$ / **P**
n+4	$\neg\neg\,\mathbf{S}_1 \rightarrow \neg\neg\,\mathbf{S}_2$	n+2,n+3 HS
n+5	$\neg\,\mathbf{S}_2 \rightarrow \neg\,\mathbf{S}_1$	n+4, CON

$\mathbf{T}_1 \rightarrow \mathbf{T}_2$, which is $\neg\mathbf{S}_1 \rightarrow \neg\mathbf{S}_2$, is similarly derived from $\mathbf{S}_2 \rightarrow \mathbf{S}_1$.

And that completes the justification for SUB, since we can use these sequences to fill in any proof using that derived rule by replacing one occurrence of **P** with **Q** in successively larger formulas, repeating the process (if necessary) for replacing additional occurrences of **P** with **Q**.

In Case 3 we used a pattern on lines n+1 through n+5 that immediately justifies

MT (Modus Tollens). From ¬**P** and **Q** → **P** derive ¬**Q**

Here's the derivation:

m	¬**P**	*given*
n	**Q** → **P**	*given*
n+1	¬¬**Q** → **Q**	Ł$_3$D2, with **Q** / **P**
n+2	¬¬**Q** → **P**	n,n+1 HS
n+3	**P** → ¬¬**P**	Ł$_3$D3, with **P** / **P**
n+4	¬¬**Q** → ¬¬**P**	n+2,n+3 HS
n+5	¬**P** → ¬**Q**	n+4, CON
n+6	¬**Q**	m,n+5 MP

As particularly useful special cases of SUB, we also introduce

DN (Double Negation). From any formula **R** that contains **P** as a constituent, infer any formula **R*** that is the result of replacing one or more occurrences of **P** in **R** with ¬¬**P**, and vice versa.

TRAN (Transposition). From any formula **R** that contains **P** → (**Q** → **S**) as a subformula, infer any formula **R*** that is the result of replacing one or more occurrences of **P** → (**Q** → **S**) in **R** with **Q** → (**P** → **S**).

GCON (Generalized Contraposition). From any formula **R** that contains **P** → **Q** as a subformula, infer any formula **R*** that is the result of replacing one or more occurrences of **P** → **Q** in **R** with ¬**Q** → ¬**P**, and vice versa.

Justifications: DN follows from SUB and Ł$_3$D2 and Ł$_3$D3, and TRAN follows from SUB and Ł$_3$D7. GCON follows from SUB, Ł$_3$3, and the fact that every formula of the form (**P** → **Q**) → (¬**Q** → ¬**P**) is a theorem of Ł$_3$A:

1	$(\neg\neg P \rightarrow \neg\neg Q) \rightarrow (\neg Q \rightarrow \neg P)$	Ł3 3, with $\neg P$ / P, $\neg Q$ / Q
2	$(P \rightarrow Q) \rightarrow (\neg Q \rightarrow \neg P)$	1, DN (twice)

Using DN, for example, we can derive an axiom that we'll need later in this section:

Ł$_3$D8. $\neg(P \rightarrow Q) \rightarrow P$

Justification:

1	$\neg P \rightarrow (P \rightarrow Q)$	Ł$_3$D1, with P / P, Q / Q
2	$\neg P \rightarrow \neg\neg(P \rightarrow Q)$	1, DN
3	$\neg(P \rightarrow Q) \rightarrow P$	2, CON

Next we introduce two rules that generalize previous ones. We use the notation $(P_1 \rightarrow (P_2 \rightarrow \cdots \rightarrow (P_{n-1} \rightarrow P_n) \ldots)$ to denote a conditional in which each consequent, except possibly P_n, is itself a conditional. As an example, the formula $(B \rightarrow ((C \rightarrow \neg D) \rightarrow (G \rightarrow (R \rightarrow H))))$ is an instance of $(P_1 \rightarrow (P_2 \rightarrow \cdots \rightarrow (P_{n-1} \rightarrow P_n) \ldots)$ in which B is the antecedent P_1, $(C \rightarrow \neg D)$ is the antecedent P_2, G is the antecedent P_3, R is the antecedent P_4, and H is the consequent P_5. The formula also counts an instance in two other ways because P_n can itself be a conditional—we must take B to be the antecedent P_1 and $(C \rightarrow \neg D)$ to be the antecedent P_2, as before, but then we can either take $(G \rightarrow (R \rightarrow H))$ to be the consequent P_3 or take G to be the antecedent P_3 and $(R \rightarrow H)$ to be the consequent P_4. Here are the rules:

GHS (Generalized Hypothetical Syllogism). From $(P_1 \rightarrow (P_2 \rightarrow \cdots \rightarrow (P_{n-1} \rightarrow P_n) \ldots)$ and $P_n \rightarrow Q$, infer $(P_1 \rightarrow (P_2 \rightarrow \cdots \rightarrow (P_{n-1} \rightarrow Q) \ldots)$

and

GMP (Generalized Modus Ponens). From $(P_1 \rightarrow (P_2 \rightarrow \cdots \rightarrow (P_{n-1} \rightarrow P_n) \ldots)$ and one of the antecedents P_i, $1 \leq i \leq n-1$, infer the conditional that results from deleting P_i, the conditional arrow following P_i, and associated parentheses.

Justification of GHS: We start with $n = 3$, showing how to derive $P_1 \rightarrow (P_2 \rightarrow Q)$ from $P_1 \rightarrow (P_2 \rightarrow P_3)$ and $P_3 \rightarrow Q$:

1	$P_1 \rightarrow (P_2 \rightarrow P_3)$	*given*
2	$P_3 \rightarrow Q$	*given*
3	$(P_2 \rightarrow P_3) \rightarrow ((P_3 \rightarrow Q) \rightarrow (P_2 \rightarrow Q))$	Ł$_3$2, with P_2 / P, P_3 / Q, Q / R
4	$(P_3 \rightarrow Q) \rightarrow ((P_2 \rightarrow P_3) \rightarrow (P_2 \rightarrow Q))$	3, TRAN
5	$(P_2 \rightarrow P_3) \rightarrow (P_2 \rightarrow Q)$	2,4 MP
6	$P_1 \rightarrow (P_2 \rightarrow Q)$	1,5 HS

Given that we have a way to construct the derivation for $n = 3$ we'll show how to construct the derivation when $n = 4$. The derivation begins as follows:

1	$\mathbf{P}_1 \rightarrow ((\mathbf{P}_2 \rightarrow (\mathbf{P}_3 \rightarrow \mathbf{P}_4))$	*given*
2	$\mathbf{P}_4 \rightarrow \mathbf{Q}$	*given*
3	$(\mathbf{P}_3 \rightarrow \mathbf{P}_4) \rightarrow ((\mathbf{P}_4 \rightarrow \mathbf{Q}) \rightarrow (\mathbf{P}_3 \rightarrow \mathbf{Q}))$	Ł$_3$2, with $\mathbf{P}_3$ / $\mathbf{P}$, $\mathbf{P}_4$ / $\mathbf{Q}$, $\mathbf{Q}$ / $\mathbf{R}$
4	$(\mathbf{P}_4 \rightarrow \mathbf{Q}) \rightarrow ((\mathbf{P}_3 \rightarrow \mathbf{P}_4) \rightarrow (\mathbf{P}_3 \rightarrow \mathbf{Q}))$	3, TRAN
5	$(\mathbf{P}_3 \rightarrow \mathbf{P}_4) \rightarrow (\mathbf{P}_3 \rightarrow \mathbf{Q})$	2,4 MP

The formulas on lines 1 and 5 are, respectively, instances of the first two formulas of the preceding derivation, with $(\mathbf{P}_3 \rightarrow \mathbf{P}_4)$ in place of $\mathbf{P}_3$ and $(\mathbf{P}_3 \rightarrow \mathbf{Q})$ in place of $\mathbf{Q}$, so the steps 3–6 of the previous derivation apply here to yield

9 | $\mathbf{P}_1 \rightarrow ((\mathbf{P}_2 \rightarrow (\mathbf{P}_3 \rightarrow \mathbf{Q}))$

in four more steps.

Each subsequent case follows from its previous case in the same way. For arbitrary $n > 3$, the derivation begins as

1	$\mathbf{P}_1 \rightarrow ((\mathbf{P}_2 \rightarrow (\mathbf{P}_3 \rightarrow \cdots \rightarrow (\mathbf{P}_{n-1} \rightarrow \mathbf{P}_n) \ldots)$	*given*
2	$\mathbf{P}_n \rightarrow \mathbf{Q}$	*given*
3	$(\mathbf{P}_{n-1} \rightarrow \mathbf{P}_n) \rightarrow ((\mathbf{P}_n \rightarrow \mathbf{Q}) \rightarrow (\mathbf{P}_{n-1} \rightarrow \mathbf{Q}))$	Ł$_3$2, with $\mathbf{P}_{n-1}$ / $\mathbf{P}$, $\mathbf{P}_n$ / $\mathbf{Q}$, $\mathbf{Q}$ / $\mathbf{R}$
4	$((\mathbf{P}_n \rightarrow \mathbf{Q}) \rightarrow ((\mathbf{P}_{n-1} \rightarrow \mathbf{P}_n) \rightarrow (\mathbf{P}_{n-1} \rightarrow \mathbf{Q}))$	3, TRAN
5	$(\mathbf{P}_{n-1} \rightarrow \mathbf{P}_n) \rightarrow (\mathbf{P}_{n-1} \rightarrow \mathbf{Q})$	2,4 MP

The formulas on lines 1 and 5 are, respectively, instances of the first two formulas $\mathbf{P}_1 \rightarrow ((\mathbf{P}_2 \rightarrow (\mathbf{P}_3 \rightarrow \cdots \rightarrow (\mathbf{P}_{n-2} \rightarrow \mathbf{P}_{n-1}) \ldots)$ and $\mathbf{P}_{n-1} \rightarrow \mathbf{Q}$ of the previous case, with $\mathbf{P}_{n-1} \rightarrow \mathbf{P}_n$ and $\mathbf{P}_{n-1} \rightarrow \mathbf{Q}$, respectively, replacing, $\mathbf{P}_{n-1}$ and $\mathbf{Q}$, so the steps of the previous case will eventually yield

k | $\mathbf{P}_1 \rightarrow ((\mathbf{P}_2 \rightarrow (\mathbf{P}_3 \rightarrow \cdots \rightarrow (\mathbf{P}_{n-1} \rightarrow \mathbf{Q}) \ldots)$

Justification of GMP: By repeated applications of TRAN the antecedents $\mathbf{P}_1, \ldots, \mathbf{P}_{n-1}$ can be permuted in any order. In particular, the antecedent $\mathbf{P}_i$ can be moved to the beginning of the formula, leaving the order of the other antecedents unchanged. At that point a single application of MP will produce the target formula with $\mathbf{P}_i$ removed.

We'll use these two rules in subsequent examples.

In classical logic the inference

$$\frac{\begin{array}{l}P \rightarrow Q \\ \neg P \rightarrow Q\end{array}}{Q}$$

is valid and the corresponding rule, traditionally called *Constructive Dilemma*, is derivable in any adequate axiomatic system for classical propositional logic such as CLA. But the inference is not valid in Ł3. If **P** and **Q** each have the value **N**, then the premises of the inference are true but the conclusion is not. However, the inference

$$\frac{\begin{array}{l}P \rightarrow Q \\ (P \rightarrow \neg P) \rightarrow Q\end{array}}{Q}$$

is valid in Ł3 (left as an exercise), and the corresponding rule is derivable in Ł3A:

MCD (Modified Constructive Dilemma). From **P → Q** and **(P → ¬P) → Q**, infer **Q**.

Justification:

1	**P → Q**	*given*
2	**(P → ¬P) → Q**	*given*
3	**(P → Q) → ((Q → ¬P) → (P → ¬P))**	Ł32, with **P** / **P**, **Q** / **Q**, **¬P** / **R**
4	**(Q → ¬P) → (P → ¬P)**	1,3 MP
5	**(Q → ¬P) → Q**	2,4 HS
6	**(Q → ¬Q) → ((¬Q → ¬ P) → (Q → ¬P))**	Ł32, with **Q** / **P**, **¬Q** / **Q**, **¬P** / **R**
7	**(¬Q → ¬ P) → ((Q → ¬Q) → (Q → ¬P))**	6, TRAN
8	**¬Q → ¬ P**	1, GCON
9	**(Q → ¬Q) → (Q → ¬P)**	7,8 MP
10	**(Q → ¬Q) → Q**	5,9 HS
11	**((Q → ¬Q) → Q) → Q**	Ł34, with **Q** / **P**
12	**Q**	10,11 MP

On the other hand, the rule

DS (Disjunctive Syllogism). From **P ∨ Q**, **P → R** and **Q → R** infer **R**

of classical logic *is* also derivable in Ł3A.

Justification of DS ($\mathbf{P} \vee \mathbf{Q}$ has been rewritten as $(\mathbf{P} \to \mathbf{Q}) \to \mathbf{Q}$):

1	$(\mathbf{P} \to \mathbf{Q}) \to \mathbf{Q}$	*given*
2	$\mathbf{P} \to \mathbf{R}$	*given*
3	$\mathbf{Q} \to \mathbf{R}$	*given*
4	$\neg(\mathbf{P} \to \mathbf{Q}) \to \mathbf{P}$	$Ł_3$D8, with $\mathbf{P} / \mathbf{P}, \mathbf{Q} / \mathbf{Q}$
5	$\neg(\mathbf{P} \to \mathbf{Q}) \to \mathbf{R}$	2,4 HS
6	$(\neg\neg(\mathbf{P} \to \mathbf{Q}) \to \neg\neg\mathbf{Q}) \to (\neg\mathbf{Q} \to \neg(\mathbf{P} \to \mathbf{Q}))$	$Ł_3$3, with $\neg(\mathbf{P} \to \mathbf{Q}) / \mathbf{P}, \neg\mathbf{Q} / \mathbf{Q}$
7	$((\mathbf{P} \to \mathbf{Q}) \to \mathbf{Q}) \to (\neg\mathbf{Q} \to \neg(\mathbf{P} \to \mathbf{Q}))$	6, DN (twice)
8	$((\mathbf{P} \to \mathbf{Q}) \to \mathbf{Q}) \to (\neg\mathbf{Q} \to \mathbf{P})$	4,7 GHS
9	$(\neg\mathbf{Q} \to \mathbf{P}) \to ((\mathbf{P} \to \neg\mathbf{P}) \to (\neg\mathbf{Q} \to \neg\mathbf{P}))$	$Ł_3$2, with $\neg\mathbf{Q} / \mathbf{P}, \mathbf{P} / \mathbf{Q}, \neg\mathbf{P} / \mathbf{R}$
10	$(\neg\mathbf{Q} \to \neg\mathbf{P}) \to (\mathbf{P} \to \mathbf{Q})$	$Ł_3$3, with $\mathbf{Q} / \mathbf{P}, \mathbf{P} / \mathbf{Q}$
11	$(\neg\mathbf{Q} \to \mathbf{P}) \to ((\mathbf{P} \to \neg\mathbf{P}) \to (\mathbf{P} \to \mathbf{Q})$	9,10 GHS
12	$(\mathbf{P} \to \mathbf{Q}) \to (((\mathbf{P} \to \mathbf{Q}) \to \neg(\mathbf{P} \to \mathbf{Q})) \to \neg(\mathbf{P} \to \mathbf{Q}))$	$Ł_3$D6, with $\mathbf{P} \to \mathbf{Q} / \mathbf{P}, \neg(\mathbf{P} \to \mathbf{Q}) / \mathbf{Q}$
13	$(\neg\mathbf{Q} \to \mathbf{P}) \to ((\mathbf{P} \to \neg\mathbf{P}) \to (((\mathbf{P} \to \mathbf{Q}) \to \neg(\mathbf{P} \to \mathbf{Q})) \to \neg(\mathbf{P} \to \mathbf{Q})))$	11,12 GHS
14	$(\neg\mathbf{Q} \to \mathbf{P}) \to ((\mathbf{P} \to \neg\mathbf{P}) \to (((\mathbf{P} \to \mathbf{Q}) \to \neg(\mathbf{P} \to \mathbf{Q})) \to \mathbf{R}))$	5,13 GHS
15	$((\mathbf{P} \to \mathbf{Q}) \to \mathbf{Q}) \to ((\mathbf{P} \to \neg\mathbf{P}) \to (((\mathbf{P} \to \mathbf{Q}) \to \neg(\mathbf{P} \to \mathbf{Q})) \to \mathbf{R}))$	8,14 HS
16	$((\mathbf{P} \to \mathbf{Q}) \to \mathbf{Q}) \to (((\mathbf{P} \to \mathbf{Q}) \to \neg(\mathbf{P} \to \mathbf{Q})) \to ((\mathbf{P} \to \neg\mathbf{P}) \to \mathbf{R}))$	15, TRAN
17	$((\mathbf{P} \to \mathbf{Q}) \to \neg(\mathbf{P} \to \mathbf{Q})) \to (((\mathbf{P} \to \mathbf{Q}) \to \mathbf{Q}) \to ((\mathbf{P} \to \neg\mathbf{P}) \to \mathbf{R}))$	16, TRAN
18	$((\mathbf{P} \to \mathbf{Q}) \to \neg(\mathbf{P} \to \mathbf{Q})) \to ((\mathbf{P} \to \neg\mathbf{P}) \to (((\mathbf{P} \to \mathbf{Q}) \to \mathbf{Q}) \to \mathbf{R}))$	17, TRAN
19	$(\mathbf{P} \to \mathbf{Q}) \to (((\mathbf{P} \to \mathbf{Q}) \to \mathbf{Q}) \to \mathbf{Q})$	$Ł_3$D6, with $\mathbf{P} \to \mathbf{Q} / \mathbf{P}, \mathbf{Q} / \mathbf{Q}$
20	$(\mathbf{P} \to \mathbf{Q}) \to (((\mathbf{P} \to \mathbf{Q}) \to \mathbf{Q}) \to \mathbf{R})$	3,19 GHS
21	$(((\mathbf{P} \to \mathbf{Q}) \to \mathbf{Q}) \to \mathbf{R}) \to ((\mathbf{P} \to \neg\mathbf{P}) \to (((\mathbf{P} \to \mathbf{Q}) \to \mathbf{Q}) \to \mathbf{R}))$	$Ł_3$1, with $((\mathbf{P} \to \mathbf{Q}) \to \mathbf{Q}) \to \mathbf{R} / \mathbf{P}$, $\mathbf{P} \to \neg\mathbf{P} / \mathbf{Q}$
22	$(\mathbf{P} \to \mathbf{Q}) \to ((\mathbf{P} \to \neg\mathbf{P}) \to (((\mathbf{P} \to \mathbf{Q}) \to \mathbf{Q}) \to \mathbf{R}))$	20,21 HS
23	$(\mathbf{P} \to \neg\mathbf{P}) \to (((\mathbf{P} \to \mathbf{Q}) \to \mathbf{Q}) \to \mathbf{R})$	18,22 MCD
24	$\mathbf{R} \to (((\mathbf{P} \to \mathbf{Q}) \to \mathbf{Q}) \to \mathbf{R})$	$Ł_3$1, with $\mathbf{R} / \mathbf{P}$ and $(\mathbf{P} \to \mathbf{Q}) \to \mathbf{Q} / \mathbf{R}$
25	$\mathbf{P} \to (((\mathbf{P} \to \mathbf{Q}) \to \mathbf{Q}) \to \mathbf{R})$	2,24 HS
26	$((\mathbf{P} \to \mathbf{Q}) \to \mathbf{Q}) \to \mathbf{R}$	23,25 MCD
27	$\mathbf{R}$	1,26 MP

Restricting our attention to lines 2–26 of the previous derivation, we have also justified the rule

DC (Disjunctive Consequence). From $\mathbf{P} \to \mathbf{R}$ and $\mathbf{Q} \to \mathbf{R}$ infer $(\mathbf{P} \vee \mathbf{Q}) \to \mathbf{R}$

and as a consequence we have

$Ł_3$D9. $(\mathbf{P} \vee \mathbf{Q}) \to (\mathbf{Q} \vee \mathbf{P})$

Justification (rewriting the formula as $((\mathbf{P} \to \mathbf{Q}) \to \mathbf{Q}) \to ((\mathbf{Q} \to \mathbf{P}) \to \mathbf{P})$):

1	$\mathbf{P} \to ((\mathbf{Q} \to \mathbf{P}) \to \mathbf{P})$	$Ł_3$1, with $\mathbf{P} / \mathbf{P}, \mathbf{Q} \to \mathbf{P} / \mathbf{Q}$
2	$\mathbf{Q} \to ((\mathbf{Q} \to \mathbf{P}) \to \mathbf{P})$	$Ł_3$D6, with $\mathbf{Q} / \mathbf{P}, \mathbf{P} / \mathbf{Q}$
3	$((\mathbf{P} \to \mathbf{Q}) \to \mathbf{Q}) \to ((\mathbf{Q} \to \mathbf{P}) \to \mathbf{P})$	1,2 DC

One more very useful derived axiom is

$Ł_3$D10. $(\mathbf{P} \to \mathbf{Q}) \vee (\mathbf{Q} \to \mathbf{P})$

Justification: We'll derive the formula $((\mathbf{P} \rightarrow \mathbf{Q}) \rightarrow (\mathbf{Q} \rightarrow \mathbf{P})) \rightarrow (\mathbf{Q} \rightarrow \mathbf{P}))$ in which the wedge has been rewritten in terms of the conditional. (This is again a fairly complicated derivation, but the resulting axiom is quite useful.)[6]

1	$((P \rightarrow Q) \rightarrow (Q \rightarrow P)) \rightarrow ((Q \rightarrow P) \rightarrow P) \rightarrow ((P \rightarrow Q) \rightarrow P))$	Ł$_3$2, with $P \rightarrow Q$ / P, $Q \rightarrow P$ / Q, P / R
2	$((P \rightarrow Q) \rightarrow Q) \rightarrow ((Q \rightarrow P) \rightarrow P)$	Ł$_3$D9, with P / P, Q / Q
3	$((Q \rightarrow P) \rightarrow P) \rightarrow ((P \rightarrow Q) \rightarrow Q)$	Ł$_3$D9, with Q / P, P / Q
4	$((P \rightarrow Q) \rightarrow (Q \rightarrow P)) \rightarrow ((P \rightarrow Q) \rightarrow Q) \rightarrow ((P \rightarrow Q) \rightarrow P))$	1,2,3 SUB
5	$(((P \rightarrow Q) \rightarrow (Q \rightarrow P)) \rightarrow ((P \rightarrow Q) \rightarrow Q) \rightarrow ((P \rightarrow Q) \rightarrow P))) \rightarrow$ $((((P \rightarrow Q) \rightarrow Q) \rightarrow ((P \rightarrow Q) \rightarrow P)) \rightarrow$ $((Q \rightarrow (P \rightarrow Q)) \rightarrow (Q \rightarrow P)) \rightarrow$ $(((P \rightarrow Q) \rightarrow (Q \rightarrow P)) \rightarrow ((Q \rightarrow (P \rightarrow Q)) \rightarrow (Q \rightarrow P))))$	Ł$_3$2, with $(P \rightarrow Q) \rightarrow (Q \rightarrow P)$ / P, $((P \rightarrow Q) \rightarrow Q) \rightarrow ((P \rightarrow Q) \rightarrow P)$ / Q, $(Q \rightarrow (P \rightarrow Q)) \rightarrow (Q \rightarrow P)$ / R
6	$(((P \rightarrow Q) \rightarrow Q) \rightarrow ((P \rightarrow Q) \rightarrow P)) \rightarrow$ $((Q \rightarrow (P \rightarrow Q)) \rightarrow (Q \rightarrow P)) \rightarrow$ $(((P \rightarrow Q) \rightarrow (Q \rightarrow P)) \rightarrow ((Q \rightarrow (P \rightarrow Q)) \rightarrow (Q \rightarrow P)))$	4,5 MP
7	$((\neg Q \rightarrow \neg(P \rightarrow Q)) \rightarrow (\neg P \rightarrow \neg(P \rightarrow Q))) \rightarrow$ $((\neg Q \rightarrow \neg(P \rightarrow Q)) \rightarrow (\neg P \rightarrow \neg(P \rightarrow Q)))$	Ł$_3$D4, with $(\neg Q \rightarrow \neg(P \rightarrow Q)) \rightarrow (\neg P \rightarrow \neg(P \rightarrow Q))$ / P
8	$((\neg Q \rightarrow \neg(P \rightarrow Q)) \rightarrow (\neg P \rightarrow \neg(P \rightarrow Q))) \rightarrow$ $(\neg P \rightarrow ((\neg Q \rightarrow \neg(P \rightarrow Q)) \rightarrow \neg(P \rightarrow Q)))$	7, TRAN
9	$((\neg Q \rightarrow \neg(P \rightarrow Q)) \rightarrow \neg(P \rightarrow Q)) \rightarrow ((\neg(P \rightarrow Q) \rightarrow \neg Q) \rightarrow \neg Q)$	Ł$_3$D9, with $\neg Q$ / P, $\neg(P \rightarrow Q)$ / Q
10	$((\neg Q \rightarrow \neg(P \rightarrow Q)) \rightarrow (\neg P \rightarrow \neg(P \rightarrow Q))) \rightarrow$ $(\neg P \rightarrow ((\neg(P \rightarrow Q) \rightarrow \neg Q) \rightarrow \neg Q))$	8,9 GHS
11	$((\neg Q \rightarrow \neg(P \rightarrow Q)) \rightarrow (\neg P \rightarrow \neg(P \rightarrow Q))) \rightarrow$ $((\neg(P \rightarrow Q) \rightarrow \neg Q) \rightarrow (\neg P \rightarrow \neg Q))$	10, TRAN
12	$((P \rightarrow Q) \rightarrow Q) \rightarrow ((P \rightarrow Q) \rightarrow P) \rightarrow$ $((Q \rightarrow (P \rightarrow Q)) \rightarrow (Q \rightarrow P))$	11, GCON (four times)
13	$((P \rightarrow Q) \rightarrow (Q \rightarrow P)) \rightarrow ((Q \rightarrow (P \rightarrow Q)) \rightarrow (Q \rightarrow P))$	4,12 HS
14	$(Q \rightarrow (P \rightarrow Q)) \rightarrow (((P \rightarrow Q) \rightarrow (Q \rightarrow P)) \rightarrow (Q \rightarrow P))$	13, TRAN
15	$Q \rightarrow (P \rightarrow Q)$	Ł$_3$1, with Q / P, P / Q
16	$((P \rightarrow Q) \rightarrow (Q \rightarrow P)) \rightarrow (Q \rightarrow P)$	14,15 MP

In closing this section, we note two important general results about Ł$_3$A. The first is that the set of theorems of Ł$_3$A is decidable. The reason is similar to that for the decidability of the axiomatic system for classical propositional logic, CLA: Ł$_3$A is a sound and complete system for Ł$_3$, and the set of tautologies in Ł$_3$ is decidable on the basis of the construction of truth-tables.[7]

The second result is negative: the Deduction Theorem for classical logic—the metatheorem that states that $\mathbf{P} \rightarrow \mathbf{Q}$ is a theorem if and only if **Q** is derivable from **P**—does not hold in Ł$_3$A. It is true that if $\mathbf{P} \rightarrow \mathbf{Q}$ is a theorem then **Q** is derivable from **P** (by MP), but the converse does not hold. Given the soundness and completeness

[6] I owe the outline of this derivation to Meredith (1928, p. 54), where the axiom is derived for Łukasiewicz's *infinite*-valued logic, a logic that will serve as our first fuzzy logic.

[7] The tautologies of other derivation systems for any of our four three-valued propositional logics are similarly decidable.

of Ł_3A, the converse is equivalent to: if **Q** is true whenever **P** is then $\mathbf{P} \rightarrow \mathbf{Q}$ is a tautology. This is not the case; for example, whenever $P \wedge ((P \rightarrow Q) \wedge (P \rightarrow (Q \rightarrow R)))$ is true in Ł_3 so is R, but $(P \wedge ((P \rightarrow Q) \wedge (P \rightarrow (Q \rightarrow R)))) \rightarrow R$ isn't an Ł_3-tautology. However, the following **Modified Deduction Theorem** does hold:

> *Result 6.1:* **Q** is derivable from **P** in Ł_3A if and only if $\mathbf{P} \rightarrow (\mathbf{P} \rightarrow \mathbf{Q})$ is a theorem of Ł_3A.[8]

Given completeness we know that R is derivable from $P \wedge ((P \rightarrow Q) \wedge (P \rightarrow (Q \rightarrow R)))$ in Ł_3A, so it follows from the Modified Deduction Theorem that $(P \wedge ((P \rightarrow Q) \wedge (P \rightarrow (Q \rightarrow R)))) \rightarrow ((P \wedge ((P \rightarrow Q) \wedge (P \rightarrow (Q \rightarrow R)))) \rightarrow R)$ is a theorem!

We leave proof of the Modified Deduction Theorem as an exercise, but will prove here a final derived axiom that will be useful when the reader turns to that exercise:

Ł_3D11. $\mathbf{(P \rightarrow (P \rightarrow (P \rightarrow Q))) \rightarrow (P \rightarrow (P \rightarrow Q))}$

Justification:

1	$\neg P \rightarrow (P \rightarrow Q)$	Ł_3D1, with P / P, Q / Q
2	$(P \rightarrow \neg P) \rightarrow ((\neg P \rightarrow (P \rightarrow Q)) \rightarrow (P \rightarrow (P \rightarrow Q)))$	$\text{Ł}_3$2, with P / P, $\neg$P / Q, P $\rightarrow$ Q / R
3	$(P \rightarrow \neg P) \rightarrow (P \rightarrow (P \rightarrow Q))$	1,2 GMP
4	$((P \rightarrow \neg P) \rightarrow (P \rightarrow (P \rightarrow Q))) \rightarrow (((P \rightarrow (P \rightarrow Q)) \rightarrow P) \rightarrow ((P \rightarrow \neg P) \rightarrow P))$	$\text{Ł}_3$2, with P $\rightarrow$ $\neg$P / P, P $\rightarrow$ (P $\rightarrow$ Q) / Q, P / R
5	$((P \rightarrow (P \rightarrow Q)) \rightarrow P) \rightarrow ((P \rightarrow \neg P) \rightarrow P)$	3,4 MP
6	$((P \rightarrow \neg P) \rightarrow P) \rightarrow P$	$\text{Ł}_3$4, with P / P
7	$((P \rightarrow (P \rightarrow Q)) \rightarrow P) \rightarrow P$	5,6 HS
8	$(((P \rightarrow (P \rightarrow Q)) \rightarrow P) \rightarrow P) \rightarrow ((P \rightarrow (P \rightarrow (P \rightarrow Q))) \rightarrow (P \rightarrow (P \rightarrow Q)))$	Ł_3D9, with P $\rightarrow$ (P $\rightarrow$ Q) / P, P / Q
9	$(P \rightarrow (P \rightarrow (P \rightarrow Q))) \rightarrow (P \rightarrow (P \rightarrow Q))$	7,8 MP

6.2 A Pavelka-Style Derivation System for Ł_3

In this section we anticipate derivation systems for fuzzy logic by developing a "Pavelka-style" axiomatic derivation system for Ł_3.[9] In Pavelka-style systems we introduce constant names for truth-values and we annotate formulas in derivations with truth-values. With this added expressive power we'll be able to use derivations not only to establish (given soundness and completeness) tautologousness and validity but also quasi-tautologousness, quasi-validity, and degree-validity for three-valued logics.

8 This is called the *Stutterer's Deduction Theorem* in Goldberg, LeBlanc, and Weaver (1974).

9 This system is a three-valued restriction of a special type of fuzzy derivation system first developed by Jan Pavelka (1979).

We augment the language Ł3 with three special atomic formulas **t**, **f**, and **n** with the stipulation that on every truth-value assignment these formulas, respectively, have the truth-values **T**, **F**, and **N**. We assume the ordering of truth-values: **F** < **N** < **T**—that is, **F** is the least true value and **N** is truer than **F** but not as true as **T**, which is the most true. Given a truth-value constant **v**, the formula **v** → **P** will then mean: ***P** has at least the value **v***. To see why, examine the truth-tables

t P	t → P
T T	T
T N	N
T F	F

n P	n → P
N T	T
N N	T
N F	N

f P	f → P
F T	T
F N	T
F F	T

The formula in the first table has the value **T** only when **P** has the value **T**, the formula in the second table has the value **T** when the value of **P** is **T** or **N**, and the formula in the third table always has the value **T**. These entries justify the reading of **v** → **P** as ***P** has at least the value **v***. On the other hand, the formula **P** → **v** means: ***P** has at most the value **v***, or: ***P** is no truer than **v***. This is confirmed by the following truth-tables:

t P	P → t
T T	T
T N	T
T F	T

n P	P → n
N T	N
N N	T
N F	T

f P	P → f
F T	F
F N	N
F F	T

Combining these, we can see that **v** ↔ **P** will mean: ***P** has exactly the truth-value **v***.

The pair [**P**, *v*], where **P** is any formula and *v* is one of the three truth-values **T**, **F**, or **N**, is called a *graded formula*. The value *v* in the graded formula indicates that the formula **P** has *at least* the value *v*. We repeat: [**P**, *v*] means that the formula **P** has *at least—but not necessarily exactly*—the value *v*. Thus, as Hájek (1998b, p. 80) notes, [**P**, *v*] is shorthand for the ungraded formula **v** → **P**, where **v** is the constant name for the truth-value *v*. Every derivation in a Pavelka-style system consists of graded formulas.

We add axioms and a new rule involving the truth-value constants to the system Ł3A to produce a new axiomatic system that we call Ł3PA (for *Ł3 Pavelka-style axiomatic system*). Every axiom will be graded with the value **T**, and each rule will specify the grades involved in its application. Ł3PA takes Łukasiewicz's negation and conditional as the primitive connectives and consists of the following axioms, where Ł3P1-Ł3P4 are graded versions of the axioms of Ł3A:

Ł3P1. [**P** → (**Q** → **P**), **T**]
Ł3P2. [(**P** → **Q**) → ((**Q** → **R**) → (**P** → **R**)), **T**]
Ł3P3. [(¬**P** → ¬**Q**) → (**Q** → **P**), **T**]
Ł3P4. [((**P** → ¬**P**) → **P**) → **P**, **T**]
Ł3P5.1.1. [(**t** → **t**) → **t**, **T**]
Ł3P5.1.2. [**t** → (**t** → **t**), **T**]
Ł3P5.2.1. [(**t** → **n**) → **n**, **T**]

Ł$_3$P5.2.2. [**n → (t → n), T**]
Ł$_3$P5.3.1. [**(t → f) → f, T**]
Ł$_3$P5.3.2. [**f → (t → f), T**]
Ł$_3$P5.4.1. [**(n → t) → t, T**]
Ł$_3$P5.4.2. [**t → (n → t), T**]
Ł$_3$P5.5.1. [**(n → n) → t, T**]
Ł$_3$P5.5.2. [**t → (n → n), T**]
Ł$_3$P5.6.1. [**(n → f) → n, T**]
Ł$_3$P5.6.2. [**n → (n → f), T**]
Ł$_3$P5.7.1. [**(f → t) → t, T**]
Ł$_3$P5.7.2. [**t → (f → t), T**]
Ł$_3$P5.8.1. [**(f → n) → t, T**]
Ł$_3$P5.8.2. [**t → (f → n), T**]
Ł$_3$P5.9.1. [**(f → f) → t, T**]
Ł$_3$P5.9.2. [**t → (f → f), T**]
Ł$_3$P6.1.1. [**¬t → f, T**]
Ł$_3$P6.1.2. [**f → ¬t, T**]
Ł$_3$P6.2.1. [**¬n → n, T**]
Ł$_3$P6.2.2. [**n → ¬n, T**]
Ł$_3$P6.3.1. [**¬f → t, T**]
Ł$_3$P6.3.2. [**t → ¬f, T**]
Ł$_3$P7.1. [**t, T**]
Ł$_3$P7.2. [**n, N**]
Ł$_3$P7.3. [**f, F**]

The axioms Ł$_3$P5 and Ł$_3$P6 give truth-conditions for the connectives. Ł$_3$P5.1.1 and Ł$_3$P5.2.2, for example, say that if the antecedent and the consequent of a conditional both have the value **T**, then the conditional has at most and at least the value **T**; that is, it is true; while Ł$_3$P5.2.1 and Ł$_3$P5.2.2 together say that a conditional whose antecedent has the value **T** and whose consequent has the value **N** itself has the value **N**. Ł$_3$P6.1.1 and Ł$_3$P6.1.2 say that the negation of a true sentence is false, while Ł$_3$P6.2.1 and Ł$_3$P6.2.2 say that the negation of a sentence that has the value **N** itself has the value **N**. The .2 versions of the Ł$_3$P5 axioms are in fact not needed, since they can all be derived—for example, Ł$_3$P5.1.2 is derivable by virtue of being an instance of Ł$_3$P1. Nor are all of the Ł$_3$P6 axioms needed—for example, Ł$_3$P6.3.1 and Ł$_3$P6.3.2 are derivable given Ł$_3$P6.1.1 and Ł$_3$P6.1.2, and vice versa. These derivations are all left as exercises.

The Ł$_3$P7 axioms reflect the truth-values of the three atomic formulas **t**, **n**, and **f** by placing lower bounds on those values. The axiom Ł$_3$P7.1 is derivable from the other axioms of Ł$_3$PA—again, proof is left as an exercise.

Ł$_3$PA contains two graded derivation rules:

MP. From [**P**, v_1] and [**P → Q**, v_2], infer [**Q**, v_3], where v_3 is defined in terms of v_1 and v_2 as specified in the following table:

v_1	v_2	v_3
T	T	T
T	N	N
T	F	F
N	T	N
N	N	F
N	F	F
F	T	F
F	N	F
F	F	F

TCI (Truth-value Constant Introduction). From [**P**, v] infer [**v** → **P**, **T**] where **v** is the constant name for the value v.

In MP the idea is to associate with the derived formula **Q** the *least* truth-value that it could have, on the basis of the least values assigned to the conditional and its antecedent. Thus, the first row reflects the fact that **Q** must have (at least) the value **T** if both **P** and **P** → **Q** have (at least) the value **T**. The fourth row states that if **P** has at least the value **N** (so it has one of the values **T**, **N**) and **P** → **Q** has (at least) the value **T**, then **Q** has at least the value **N**. This is correct, for consider the possibilities represented by this row: either **P** has the value **N** and **P** → **Q** has the value **T**, in which case **Q** must have the value **N** or the value **T**, or **P** has the value **T** and **P** → **Q** has the value **T**, in which case **Q** has the value **T**. Thus, we know that in either case **Q** must have at least the value **N**. (It is left as an exercise to justify the values of **Q** represented in the other rows.) The rule TCI allows us to move from a graded formula **P** to a formula that states what the graded value is.

Note that all of the axioms that we derived for Ł3A in Section 6.1 can be derived here as well—since the axioms of Ł3A are included in Ł3PA—and their graded values will all be **T** because Ł3PA's axioms are graded with **T** and when MP is applied to formulas graded with **T** it produces another formula with grade **T**. Thus every formula in the justification for a derived axiom schema will be graded with the value **T**. Here we will prefix the derived axiom numbers with *L3P* rather than simply *L3*, to emphasize that we are now working within the Pavelka-style system. In addition, and for the same reasons, all of the rules that we derived for Ł3A can be used in Ł3PA to derive formulas graded with **T** from formulas that are themselves graded with **T**. (Later we will derive fully graded versions of those rules.)

All of the formulas in our first derivations illustrating Ł3PA will be graded with **T**; our emphasis here will be on using the axioms specifying truth-conditions for complex formulas. First we'll derive a formula that says that if P has the value **T**, then $\neg P$ has the value **F**. Note that saying that P has the value **T** is equivalent to saying that it has at least the value **T**, which is symbolized as $\boldsymbol{t} \rightarrow P$, and saying that $\neg P$ has the value **F** is equivalent to saying that $\neg P$ has at most the value **F**, which is symbolized as $\neg P \rightarrow f$. Here's the derivation:

1	$[(\mathbf{t} \rightarrow \mathrm{P}) \rightarrow (\mathbf{t} \rightarrow \mathrm{P}), \mathbf{T}]$	Ł$_3$PD4, with $\mathbf{t} \rightarrow \mathrm{P} / \mathbf{P}$
2	$[(\mathbf{t} \rightarrow \mathrm{P}) \rightarrow (\neg\mathrm{P} \rightarrow \neg\mathbf{t}), \mathbf{T}]$	1, GCON
3	$[\neg\mathbf{t} \rightarrow \mathbf{f}, \mathbf{T}]$	Ł$_3$P6.1.1
4	$[(\mathbf{t} \rightarrow \mathrm{P}) \rightarrow (\neg\mathrm{P} \rightarrow \mathbf{f}), \mathbf{T}]$	2,3 GHS

Next we'll derive a formula that says that if P has the value **F**, then $P \rightarrow Q$ has the value **T**:

1	$[(\mathrm{P} \rightarrow \mathbf{f}) \rightarrow (\mathrm{P} \rightarrow \mathbf{f}), \mathbf{T}]$	Ł$_3$PD4, with $\mathrm{P} \rightarrow \mathbf{f} / \mathbf{P}$
2	$[\mathbf{t} \rightarrow \neg\mathbf{f}, \mathbf{T}]$	Ł$_3$P6.3.2
3	$[(\mathbf{t} \rightarrow \neg\mathbf{f}) \rightarrow ((\neg\mathbf{f} \rightarrow \neg\mathrm{P}) \rightarrow (\mathbf{t} \rightarrow \neg\mathrm{P})), \mathbf{T}]$	Ł$_3$P3, with $\mathbf{t} / \mathbf{P}, \neg\mathbf{f} / \mathbf{Q}, \neg\mathrm{P} / \mathbf{R}$
4	$[(\neg\mathbf{f} \rightarrow \neg\mathrm{P}) \rightarrow (\mathbf{t} \rightarrow \neg\mathrm{P}), \mathbf{T}]$	2,3 MP
5	$[(\mathrm{P} \rightarrow \mathbf{f}) \rightarrow (\mathbf{t} \rightarrow \neg\mathrm{P}), \mathbf{T}]$	4, GCON
6	$[\neg\mathrm{P} \rightarrow (\mathrm{P} \rightarrow \mathrm{Q}), \mathbf{T}]$	Ł$_3$PD1, with $\mathrm{P} / \mathbf{P}, \mathrm{Q} / \mathbf{Q}$
7	$[(\mathrm{P} \rightarrow \mathbf{f}) \rightarrow (\mathbf{t} \rightarrow (\mathrm{P} \rightarrow \mathrm{Q})), \mathbf{T}]$	5,6 GHS

Here's a derivation of a formula that says that if P has at least the value **T**, then it has at least the value **N**:

1	$[\mathbf{t} \rightarrow \mathbf{t}, \mathbf{T}]$	Ł$_3$PD4, with $\mathbf{t} / \mathbf{P}$
2	$[(\mathbf{t} \rightarrow \mathbf{t}) \rightarrow \mathbf{t}, \mathbf{T}]$	Ł$_3$P5.1.1
3	$[\mathbf{t}, \mathbf{T}]$	1,2 MP
4	$[\mathbf{t} \rightarrow (\mathbf{n} \rightarrow \mathbf{t}), \mathbf{T}]$	Ł$_3$P5.4.2
5	$[\mathbf{n} \rightarrow \mathbf{t}, \mathbf{T}]$	3,4 MP
6	$[(\mathbf{n} \rightarrow \mathbf{t}) \rightarrow ((\mathbf{t} \rightarrow \mathrm{P}) \rightarrow (\mathbf{n} \rightarrow \mathrm{P})), \mathbf{T}]$	Ł$_3$P2, with $\mathbf{n} / \mathbf{P}, \mathbf{t} / \mathbf{Q}, \mathrm{P} / \mathbf{R}$
7	$[(\mathbf{t} \rightarrow \mathrm{P}) \rightarrow (\mathbf{n} \rightarrow \mathrm{P}), \mathbf{T}]$	5,6 MP

And here's a derivation of a formula that *says* that $P \rightarrow P$ is a tautology in Ł$_3$; that is, it is at least true:

1	$[\mathrm{P} \rightarrow \mathrm{P}, \mathbf{T}]$	Ł$_3$PD4, with $\mathrm{P} / \mathbf{P}$
2	$[(\mathrm{P} \rightarrow \mathrm{P}) \rightarrow (\mathbf{t} \rightarrow (\mathrm{P} \rightarrow \mathrm{P})), \mathbf{T}]$	Ł$_3$P1, with $\mathrm{P} \rightarrow \mathrm{P} / \mathbf{P}, \mathbf{t} / \mathbf{Q}$
3	$[\mathbf{t} \rightarrow (\mathrm{P} \rightarrow \mathrm{P}), \mathbf{T}]$	1, 2 MP

Now we'll do some derivations containing formulas graded with values other than **T**. We can add graded assumptions to derivations, for example, to derive $[\neg B, \mathbf{N}]$ from $[B \rightarrow C, \mathbf{T}]$ and $[\neg C, \mathbf{N}]$:

1	$[\mathrm{B} \rightarrow \mathrm{C}, \mathbf{T}]$	Assumption
2	$[\neg\mathrm{C}, \mathbf{N}]$	Assumption
3	$[\neg\mathrm{C} \rightarrow \neg\mathrm{B}, \mathbf{T}]$	1, GCON
4	$[\neg\mathrm{B}, \mathbf{N}]$	2,3 MP

That is, if $\neg C$ has at least the value **N**, and $B \rightarrow C$ is true, then $\neg B$ has at least the value **N**. Note that we have applied the derived rule GCON only to a formula graded with **T**. TCI will then allow us to derive $\mathbf{n} \rightarrow \neg B$ graded with **T**:

5	$[\mathbf{n} \rightarrow \neg\mathrm{B}, \mathbf{T}]$	4, TCI

The three axiom schemata Ł$_3$G7.1–Ł$_3$G7.3 guarantee the converse of TCI, that is, we can go from a graded formula [**v** → **P**, **T**], where **v** is the name of the truth-value v, to the graded formula [P, v] as in

1	[**n** → P, **T**]	*given*
2	[**n**, **N**]	Ł$_3$P7.2
3	[P, **N**]	1,2 MP

Obviously there are analogous inferences for **t** → **P** and **f** → **P**. Here is a derivation of [P, **N**] from [P, **T**]:

1	[P, **T**]	Assumption
2	[**t**, **T**]	Ł$_3$P7.1
3	[**t** → (**n** → **t**), **T**]	Ł$_3$P5.4.2
4	[**n** → **t**, **T**]	2,3 MP
5	[**t** → P, **T**]	1, TCI
6	[**n** → P, **T**]	4,5 HS
7	[**n**, **N**]	Ł$_3$P7.2
8	[P, **N**]	6,7 MP

The quasi-tautologousness of $A \vee \neg A$ is expressed by the graded formula [*(A ∨ ¬A)*, **N**] or [*((A → ¬A) → ¬A)*, **N**], which we can derive as follows:

1	[¬A → ((A → ¬A) → ¬A), **T**]	Ł$_3$P1, with ¬A / **P**, A → ¬A / **Q**
2	[(**n** → ¬A) → ((¬A → ((A → ¬A) → ¬A)) → (**n** → ((A → ¬A) → ¬A))), **T**]	Ł$_3$P2, with **n** / **P**, ¬A / **Q**, (A → ¬A) → ¬A / **R**
3	[(**n** → ¬A) → (**n** → ((A → ¬A) → ¬A)), **T**]	1,2 GMP
4	[A → ((A → ¬A) → ¬A), **T**]	Ł$_3$PD6, with A / **P**, ¬A / **Q**
5	[(**n** → A) → ((A → ((A → ¬A) → ¬A)) → (**n** → ((A → ¬A) → ¬A))), **T**]	Ł$_3$P2, with **n** / **P**, A / **Q**, (A → ¬A) → ¬A / **R**
6	[(**n** → A) → (**n** → ((A → ¬A) → ¬A)), **T**]	4,5 GMP
7	[(**n** → A) ∨ (A → **n**), **T**]	Ł$_3$PD10, with **n** / **P**, A / **Q**
8	[(**n** → A) ∨ (¬**n** → ¬A), **T**]	7, GCON
9	[¬**n** → **n**, **T**]	Ł$_3$P6.2.1
10	[**n** → ¬**n**, **T**]	Ł$_3$P6.2.2
11	[(**n** → A) ∨ (**n** → ¬A), **T**]	8,9,10 SUB
12	[**n** → ((A → ¬A) → ¬A), **T**]	3,6,11 DS
13	[**n**, **N**]	Ł$_3$P7.2
14	[((A → ¬A) → ¬A), **N**]	12,13 MP

Whenever we derive a graded formula without making any assumptions, we may regard the graded formula as a derived axiom. Thus we can have derived axioms with values other than **T**; for example, we have just justified

Ł$_3$PD12: [**P** ∨ ¬**P**, **N**]

This in turn allows derivations like

1	[(P ∨ ¬P) → Q, **T**]	Assumption
2	[P ∨ ¬P, **N**]	Ł$_3$PD12, with P / **P**
3	[Q, **N**]	1,2 MP

Whereas in classical logic the truth of *(P ∨ ¬P) → Q* would guarantee the truth of *Q*, in Ł$_3$ the best we can say is that the truth of *(P ∨ ¬P) → Q* guarantees that *Q* has *at least* the value **N**, and this derivation reflects that.

We must modify the definitions of theoremhood and entailment for system Ł$_3$PA. First, we will say that a formula **P** is a ***theorem to degree v*** in Ł$_3$PA if there is a proof (a derivation without assumptions) of [**P**, *v*] and there is *no* proof of [**P**, *w*] with *w greater* than *v*. We need to add the second condition because it is possible to have two proofs that give **P** different values. For example, we can derive [*(P → ¬¬Q) → (¬Q → ¬P)*, **T**] as follows:

1	[(P → ¬¬Q) → (P → ¬¬Q), **T**]	Ł$_3$PD4, with P → ¬¬Q / **P**
2	[(P → ¬¬Q) → (P → Q), **T**]	1, DN
3	[[(P → ¬¬Q) → (¬Q→ ¬P), **T**]	2, GCON

This derivation is sufficient to establish that the formula *(P → ¬¬Q) → (¬Q → ¬P)* is a theorem to degree **T**, since there is no greater truth-value than **T**. But note the following derivation:

1	[**t**, **T**]	Ł$_3$P7.1
2	[**t** → (¬((P → ¬¬Q) → (¬Q → ¬P)) → **t**), **T**]	Ł$_3$P1, with **t** / **P**, ¬((P → ¬¬Q) → (¬Q → ¬P)) / **Q**
3	[¬((P → ¬¬Q) → (¬Q → ¬P)) → **t**, **T**]	1,2 MP
4	[¬**t** → ¬¬((P → ¬¬Q) → (¬Q → ¬P)), **T**]	3, GCON
5	[¬**t** → ((P → ¬¬Q) → (¬Q → ¬P)), **T**]	4, DN
6	[**f** → ¬**t**, **T**]	Ł$_3$P6.1.2
7	[**f** → ((P → ¬¬Q) → (¬Q → ¬P)), **T**]	5,6 HS
8	[**f**, **F**]	Ł$_3$P7.3
9	[(P → ¬¬Q) → (¬Q → ¬P), **F**]	7,8 MP

This derivation establishes that the formula *(P → ¬¬Q) → (¬Q → ¬P)* has at least the value **F**—but we certainly don't want to conclude on the basis of this derivation that that's all we can say about *(P → ¬¬Q) → (¬Q → ¬P)*, which is a tautology of Ł$_3$! So we need to consider that there may be—and indeed are, as we showed previously—other derivations that grade the formula *(P → ¬¬Q) → (¬Q → ¬P)* with **T**. On the other hand, although we will not prove this, there is *no* proof that ends with the graded formula [*A ∨ ¬A*, **T**]—this formula is only a theorem to degree **N**. Ł$_3$PA is weakly sound and complete in the following sense: a formula of Ł$_3$ is a tautology if and only if it is a theorem to degree **T** in Ł$_3$PA, and a formula of Ł$_3$ is a quasi-tautology if and only if it is a theorem to either degree **T** or degree **N** in Ł$_3$PA.

Second, we will say that a formula **Q** is ***derivable to degree v*** in Ł3PA from a *graded* set of formulas Γ if there is a derivation of the graded formula [**Q**, v] from graded formulas [$\mathbf{P}_i$, v_i] where each [$\mathbf{P}_i$, v_i] is a member of Γ, and for no value w that is *greater* than v is there a derivation of [**Q**, w] from the graded formulas in Γ. Ł3PA is strongly sound and complete: in addition to weak completeness it is also the case that a set of formulas Γ of Ł3 entails a formula **P** of Ł3 if and only if **P** is derivable to degree **T** from the graded set in which each member of Γ occurs with the grade **T** and that Γ quasi-entails **P** if and only if **P** is derivable either to degree **T** or to degree **N** from any graded set in which each member of Γ occurs with either the grade **T** or the grade **N**. We will explore the connection between the strong completeness of Ł3A and the strong completeness of Ł3PA in an exercise. It follows from (weak) soundness and completeness that the set of theorems to degree **T** (or **N** or **F**) of Ł3PA is decidable, by virtue of truth-table tests for tautologousness (quasi-tautologousness, contradictoriness)—remembering that **t**, **n**, and **f** must always be assigned the values **T**, **N**, and **F**, respectively.

Given soundness and completeness, we can also use Ł3PA to establish quasi-validity and degree-validity in addition to validity proper. For example, the argument

P
———
P ∨ Q

is quasi-valid in Ł3. That means that given the assumption that *P* has at least the value **N**, we should be able to derive the conclusion that *P* ∨ *Q* has at least the value **N**. More generally, an argument

$\mathbf{P}_1$
$\mathbf{P}_2$
. . .
$\mathbf{P}_n$
———
Q

is quasi-valid if and only if [**Q**, **N**] can be derived from [$\mathbf{P}_1$, **N**], [$\mathbf{P}_2$, **N**], . . . , and [$\mathbf{P}_n$, **N**]. Here's a derivation for the preceding argument in which we derive [(*P* → *Q*) → *Q*, **N**] from [*P*, **N**]:

1	[P, **N**]	Assumption
2	[P → ((P → Q) → Q), **T**]	Ł3PD6, with P / **P**, Q / **Q**
3	[(P → Q) → Q, **N**]	1,2 MP

This derivation in conjunction with the following one establishes the degree-validity of the argument from *P* to *P* ∨ *Q*:

1	[P, **T**]	Assumption
2	[P → ((P → Q) → Q), **T**]	Ł3PD6, with P / **P**, Q / **Q**
3	[(P → Q) → Q, **T**]	1,2 MP

Generally, an argument

$$\begin{array}{l}\mathbf{P}_1\\ \mathbf{P}_2\\ \ldots\\ \underline{\mathbf{P}_n}\\ \mathbf{Q}\end{array}$$

is degree-valid if and only if it is valid and also quasi-valid, that is, if and only if $[\mathbf{Q}, \mathbf{T}]$ can be derived from $[\mathbf{P}_1, \mathbf{T}]$, $[\mathbf{P}_2, \mathbf{T}], \ldots,$ and $[\mathbf{P}_n, \mathbf{T}]$ and, in addition, $[\mathbf{Q}, \mathbf{N}]$ can be derived from $[\mathbf{P}_1, \mathbf{N}]$, $[\mathbf{P}_2, \mathbf{N}], \ldots,$ and $[\mathbf{P}_n, \mathbf{N}]$.

Given the definability of the connectives for K^S_3, B^I_3, and B^E_3 in the system Ł$_3$, the expressive power of Ł$_3$PA also allows us to prove quasi-tautologousness, quasi-validity, and degree-validity, along with tautologousness and validity proper, for all four systems. As an example, we will establish that $(A \vee_{BI} \neg_{BI} A)$ is a quasi-tautology. This B^I_3 formula translates into $(A \vee \neg A) \wedge ((A \vee \neg A) \wedge (\neg A \vee \neg\neg A))$ using the connectives of Ł$_3$, so we derive $[((A \vee \neg A) \wedge ((A \vee \neg A) \wedge (\neg A \vee \neg\neg A))), \mathbf{N}]$ to establish the formula's quasi-tautologousness. The derivation begins with lines 1–12 of the previous proof of $[((A \rightarrow \neg A) \rightarrow \neg A), \mathbf{N}]$:

1	...	
	...	
12	$[\mathbf{n} \rightarrow ((A \rightarrow \neg A) \rightarrow \neg A), \mathbf{T}]$	3,6,11 DS
13	$[((\neg(A \vee \neg A) \rightarrow \neg(A \vee \neg A)) \rightarrow \neg(A \vee \neg A)) \rightarrow$ $((\neg(A \vee \neg A) \rightarrow \neg(A \vee \neg A)) \rightarrow \neg(A \vee \neg A)), \mathbf{T}]$	Ł$_3$PD4, with $(\neg(A \vee \neg A) \rightarrow \neg(A \vee \neg A)) \rightarrow \neg(A \vee \neg A)$ / $\mathbf{P}$
14	$[\neg(A \vee \neg A) \rightarrow \neg(A \vee \neg A), \mathbf{T}]$	Ł$_3$PD4, with $\neg(A \vee \neg A)$ / $\mathbf{P}$
15	$[((\neg(A \vee \neg A) \rightarrow \neg(A \vee \neg A)) \rightarrow \neg(A \vee \neg A)) \rightarrow \neg(A \vee \neg A), \mathbf{T}]$	13,14 GMP
16	$[\neg\neg((\neg(A \vee \neg A) \rightarrow \neg(A \vee \neg A)) \rightarrow \neg(A \vee \neg A)) \rightarrow \neg(A \vee \neg A), \mathbf{T}]$	15, DN
17	$[(A \vee \neg A) \rightarrow \neg((\neg(A \vee \neg A) \rightarrow \neg(A \vee \neg A)) \rightarrow \neg(A \vee \neg A)), \mathbf{T}]$	16, GCON
	(i.e., $[(A \vee \neg A) \rightarrow ((A \vee \neg A) \wedge (A \vee \neg A)), T]$—recall that $\mathbf{P} \wedge \mathbf{Q}$ is definable in L_3 as $\neg(\neg\mathbf{P} \vee \neg\mathbf{Q})$ which in turn is definable as $\neg((\neg\mathbf{P} \rightarrow \neg\mathbf{Q}) \rightarrow \neg\mathbf{Q})$)	
18	$[\neg(A \vee \neg A) \rightarrow (\neg(A \vee \neg A) \vee \neg(A \vee \neg A)), \mathbf{T}]$	Ł$_3$PD6, with $\neg(A \vee \neg A)$ / $\mathbf{P}$, $\neg(A \vee \neg A)$ / $\mathbf{Q}$
19	$[\neg(\neg(A \vee \neg A) \vee \neg(A \vee \neg A)) \rightarrow \neg\neg(A \vee \neg A), \mathbf{T}]$	18, GCON
20	$[\neg(\neg(A \vee \neg A) \vee \neg(A \vee \neg A)) \rightarrow (A \vee \neg A), \mathbf{T}]$	19, DN
	(i.e., $((A \vee \neg A) \wedge (A \vee \neg A)) \rightarrow (A \vee \neg A), \mathbf{T}$)	
21	$[\mathbf{n} \rightarrow ((A \vee \neg A) \wedge (A \vee \neg A)), \mathbf{T}]$	12,17,20 SUB
	(we have used the "that is" formulas in making this substitution)	
22	$[\mathbf{n} \rightarrow ((A \vee \neg A) \wedge ((A \vee \neg A) \wedge (A \vee \neg A))), \mathbf{T}]$	21,17,20 SUB
23	$[\mathbf{n} \rightarrow ((A \vee \neg A) \wedge ((A \vee \neg A) \wedge (\neg\neg A \vee \neg A))), \mathbf{T}]$	22, DN
24	$[(\neg\neg A \vee \neg A) \rightarrow (\neg A \vee \neg\neg A), \mathbf{T}]$	Ł$_3$PD9, with $\neg\neg A$ / $\mathbf{P}$, $\neg A$ / $\mathbf{Q}$
25	$[(\neg A \vee \neg\neg A) \rightarrow (\neg\neg A \vee \neg A), \mathbf{T}]$	Ł$_3$PD9, with $\neg A$ / $\mathbf{P}$, $\neg\neg A$ / $\mathbf{Q}$
26	$[\mathbf{n} \rightarrow ((A \vee \neg A) \wedge ((A \vee \neg A) \wedge (\neg A \vee \neg\neg A))), \mathbf{T}]$	23,24,25 SUB
27	$[\mathbf{n}, \mathbf{N}]$	Ł$_3$PD7.2
28	$((A \vee \neg A) \wedge ((A \vee \neg A) \wedge (\neg A \vee \neg\neg A)), \mathbf{N}]$	26,27 MP

So far we have applied rules derived in Ł$_3$A only to formulas that are graded with **T**. This restriction is inconvenient, so we will now introduce graded versions of derived rules. The first is

HS. From [**P** → **Q**, v_1] and [**Q** → **R**, v_2] infer [**P** → **R**, v_3] where v_3 is defined in terms of v_1 and v_2 as specified in the following table:

v_1	v_2	v_3
T	T	T
T	N	N
T	F	F
N	T	N
N	N	F
N	F	F
F	T	F
F	N	F
F	F	F

Justification:

1	[**P** → **Q**, v_1]	*given*
2	[**Q** → **R**, v_2]	*given*
3	[(**P** → **Q**) → ((**Q** → **R**) → (**P** → **R**)), **T**]	Ł$_3$P2
4	[(**Q** → **R**) → (**P** → **R**), v_1]	1,3 MP
5	[**P** → **R**, v_3]	2,4 MP

Note that the formula on line 4 is graded with v_1, since the value of the conditional on line 3 is **T**, and in this case the rule MP assigns the same value as the antecedent to the consequent. The value v_3 on line 5 is computed by finding the row in the table for the MP rule with v_2 as the value of the antecedent **P** (v_2 is the value of the formula on line 2), and v_1 as the value of the conditional **P** → **Q** (v_1 is the value of the conditional on line 4). In this case we end up with the same table as we did for the rule MP!

The graded version of CON is

CON. From [¬**P** → ¬**Q**, v] infer [**Q** → **P**, v]

Justification:

1	[¬**P** → ¬**Q**, v]	*given*
2	[(¬**P** → ¬**Q**) → (**Q** → **P**), **T**]	Ł$_3$P3
3	[**Q** → **P**, v]	1,2 MP

Recall that if a conditional **P** → **Q** is true, the graded MP rule assigns to **Q** the same value as **P**. Thus the graded value v is justified given the graded value **T** for the conditional on line 2. The reader will be asked in the exercises to derive graded versions of other rules. There is, however, one rule that is tricky to derive and so we'll do so here.

The rule is Modified Constructive Dilemma: from $\mathbf{P} \rightarrow \mathbf{Q}$ and $(\mathbf{P} \rightarrow \neg\mathbf{P}) \rightarrow \mathbf{Q}$ derive $\mathbf{Q}$. If we use the derivation of the ungraded version in Section 6.1 as the basis for our justification, we will end up with a rule that is not graded as strongly as it could be:

1	$[\mathbf{P} \rightarrow \mathbf{Q}, v_1]$	*given*
2	$[(\mathbf{P} \rightarrow \neg\mathbf{P}) \rightarrow \mathbf{Q}, v_2]$	*given*
3	$[(\mathbf{P} \rightarrow \mathbf{Q}) \rightarrow ((\mathbf{Q} \rightarrow \neg\mathbf{P}) \rightarrow (\mathbf{P} \rightarrow \neg\mathbf{P})), \mathbf{T}]$	Ł$_3$P2, with $\mathbf{P} / \mathbf{P}, \mathbf{Q} / \mathbf{Q}, \neg\mathbf{P} / \mathbf{R}$
4	$[(\mathbf{Q} \rightarrow \neg\mathbf{P}) \rightarrow (\mathbf{P} \rightarrow \neg\mathbf{P}), v_1]$	1,3 MP
5	$[(\mathbf{Q} \rightarrow \neg\mathbf{P}) \rightarrow \mathbf{Q}, v_3]$	2,4 HS
	where v_3 is a function of v_1 and v_2 as indicated in the table for graded HS	
6	$[(\mathbf{Q} \rightarrow \neg\mathbf{Q}) \rightarrow ((\neg\mathbf{Q} \rightarrow \neg\,\mathbf{P}) \rightarrow (\mathbf{Q} \rightarrow \neg\mathbf{P})), \mathbf{T}]$	Ł$_3$P2, with $\mathbf{Q} / \mathbf{P}, \neg\mathbf{Q} / \mathbf{Q}, \neg\mathbf{P} / \mathbf{R}$
7	$[(\neg\mathbf{Q} \rightarrow \neg\,\mathbf{P}) \rightarrow ((\mathbf{Q} \rightarrow \neg\mathbf{Q}) \rightarrow (\mathbf{Q} \rightarrow \neg\mathbf{P})), \mathbf{T}]$	6, TRAN
8	$[\neg\mathbf{Q} \rightarrow \neg\,\mathbf{P}, v_1]$	1, GCON
9	$[(\mathbf{Q} \rightarrow \neg\mathbf{Q}) \rightarrow (\mathbf{Q} \rightarrow \neg\mathbf{P}), v_1]$	7,8 MP
10	$[(\mathbf{Q} \rightarrow \neg\mathbf{Q}) \rightarrow \mathbf{Q}, v_4]$	5,9 HS
	where v_4 is a function of v_1 and v_3 as indicated in the table for graded HS	
11	$[((\mathbf{Q} \rightarrow \neg\mathbf{Q}) \rightarrow \mathbf{Q}) \rightarrow \mathbf{Q}, \mathbf{T}]$	Ł$_3$P4, with $\mathbf{Q} / \mathbf{P}$
12	$[\mathbf{Q}, v_4]$	10,11 MP

(We have used graded GCON, which is assigned as an exercise; after doing the exercise the reader may confirm that we have assigned the correct value to the formula derived on line 8.) The problem with this derivation is that on the basis of the values v_1 and v_2 we'll get the following values for v_4:

v_1	v_2	v_4
T	T	T
T	N	N
T	F	F
N	T	F
N	N	F
N	F	F
F	T	F
F	N	F
F	F	F

The value when $v_1 = \mathbf{N}$ and $v_2 = \mathbf{T}$ is too low; v_4 should be **N** in this case because (as the reader can verify with a truth-table) that is the least value that **Q** can have when $\mathbf{P} \rightarrow \mathbf{Q}$ has the value **N** and $(\mathbf{P} \rightarrow \neg\mathbf{P}) \rightarrow \mathbf{Q}$ has the value **T**. Generally, the least value that $\mathbf{Q}$ can have is the minimum of the least values that $\mathbf{P} \to \mathbf{Q}$ and $(\mathbf{P} \to \neg\mathbf{P}) \to \mathbf{Q}$ have, and we state this in our derived rule:

MCD (Modified Constructive Dilemma). From $[\mathbf{P} \to \mathbf{Q}, v_1]$ and $[(\mathbf{P} \to \neg\mathbf{P}) \to \mathbf{Q}, v_2]$, infer $[\mathbf{Q}, \min(v_1, v_2)]$.

Correct justification:
We begin the derivation as

1	$[\mathbf{P} \to \mathbf{Q}, v_1]$	*given*
2	$[(\mathbf{P} \to \neg\mathbf{P}) \to \mathbf{Q}, v_2]$	*given*
3	$[\mathbf{v}_1 \to (\mathbf{P} \to \mathbf{Q}), \mathbf{T}]$	1, TCI
4	$[\mathbf{v}_2 \to ((\mathbf{P} \to \neg\mathbf{P}) \to \mathbf{Q}), \mathbf{T}]$	3, TCI

where $\mathbf{v}_1$ stands for the atomic formula that denotes the value v_1 and $\mathbf{v}_2$ stands for the atomic formula that denotes the value v_2. Now, note that either $v_1 \leq v_2$ or $v_2 < v_1$. In the former case, we will continue the derivation with

5	$[\mathbf{t} \to (\mathbf{v}_1 \to \mathbf{v}_2), \mathbf{T}]$	Ł$_3$P5.*x.2*
	{This is axiom L$_3$G5.x.2 for some value of x, because we are assuming that $v_1 \leq v_2$}	
6	$[\mathbf{t}, \mathbf{T}]$	Ł$_3$P7.1
7	$[\mathbf{v}_1 \to \mathbf{v}_2, \mathbf{T}]$	5,6 MP
8	$[\mathbf{v}_1 \to ((\mathbf{P} \to \neg\mathbf{P}) \to \mathbf{Q}), \mathbf{T}]$	4,7 HS

Because $v_1 \leq v_2$ in this case, $v_1 = \min(v_1, v_2)$. We can therefore rewrite the derivation so far as

1	$[\mathbf{P} \to \mathbf{Q}, v_1]$	*given*
2	$[(\mathbf{P} \to \neg\mathbf{P}) \to \mathbf{Q}, v_2]$	*given*
3	$[\mathbf{min}\,(\mathbf{v}_1, \mathbf{v}_2) \to (\mathbf{P} \to \mathbf{Q}), \mathbf{T}]$	1, TCI
4	$[\mathbf{v}_2 \to ((\mathbf{P} \to \neg\mathbf{P}) \to \mathbf{Q}), \mathbf{T}]$	2, TCI
5	$[\mathbf{t} \to (\mathbf{min}\,(\mathbf{v}_1, \mathbf{v}_2) \to \mathbf{v}_2), \mathbf{T}]$	Ł$_3$P5.*x.2*
6	$[\mathbf{t}, \mathbf{T}]$	Ł$_3$P7.1
7	$[\mathbf{min}\,(\mathbf{v}_1, \mathbf{v}_2) \to \mathbf{v}_2), \mathbf{T}]$	5,6 MP
8	$[\mathbf{min}\,(\mathbf{v}_1, \mathbf{v}_2) \to ((\mathbf{P} \to \neg\mathbf{P}) \to \mathbf{Q}), \mathbf{T}]$	4,7 HS

(where $\mathbf{min}\,(\mathbf{v}_1, \mathbf{v}_2)$ stands for the atomic formula denoting the value $\min(v_1, v_2)$. In the case where $v_2 < v_1$ we begin the derivation with

1	$[\mathbf{P} \to \mathbf{Q}, v_1]$	*given*
2	$[(\mathbf{P} \to \neg\mathbf{P}) \to \mathbf{Q}, v_2]$	*given*
3	$[\mathbf{v}_1 \to (\mathbf{P} \to \mathbf{Q}), \mathbf{T}]$	1, TCI
4	$[\mathbf{min}\,(\mathbf{v}_1, \mathbf{v}_2) \to ((\mathbf{P} \to \neg\mathbf{P}) \to \mathbf{Q}), \mathbf{T}]$	2, TCI
5	$[\mathbf{t} \to (\mathbf{min}\,(\mathbf{v}_1, \mathbf{v}_2) \to \mathbf{v}_1), \mathbf{T}]$	Ł$_3$P5.*x.2*
6	$[\mathbf{t}, \mathbf{T}]$	Ł$_3$P7.1
7	$[\mathbf{min}\,(\mathbf{v}_1, \mathbf{v}_2) \to \mathbf{v}_1), \mathbf{T}]$	5,6 MP
8	$[\mathbf{min}\,(\mathbf{v}_1, \mathbf{v}_2) \to (\mathbf{P} \to \mathbf{Q}), \mathbf{T}]$	3,7 HS

In either case the derivation continues as follows (with $\{x, y\}$ meaning line x if we began the first way, and line y if we began the second way):

9	$[(\mathbf{P} \to (\min(\mathbf{v}_1, \mathbf{v}_2) \to \mathbf{Q})), \mathbf{T}]$	$\{3,8\}$ TRAN
10	$[(\mathbf{P} \to \neg\mathbf{P}) \to (\min(\mathbf{v}_1, \mathbf{v}_2) \to \mathbf{Q}), \mathbf{T}]$	$\{8,4\}$ TRAN
11	$[(\mathbf{P} \to (\min(\mathbf{v}_1, \mathbf{v}_2) \to \mathbf{Q})) \to (((\min(\mathbf{v}_1, \mathbf{v}_2) \to \mathbf{Q}) \to \neg\mathbf{P}) \to (\mathbf{P} \to \neg\mathbf{P})), \mathbf{T}]$	Ł$_3$P2, with $\mathbf{P}$ / P, $\min(\mathbf{v}_1, \mathbf{v}_2) \to \mathbf{Q}$ / Q, $\neg\mathbf{P}$ / R
12	$[((\min(\mathbf{v}_1, \mathbf{v}_2) \to \mathbf{Q}) \to \neg\mathbf{P}) \to (\mathbf{P} \to \neg\mathbf{P}), \mathbf{T}]$	9,11 MP
13	$[((\min(\mathbf{v}_1, \mathbf{v}_2) \to \mathbf{Q}) \to \neg\mathbf{P}) \to (\min(\mathbf{v}_1, \mathbf{v}_2) \to \mathbf{Q}), \mathbf{T}]$	10,12 HS
14	$[((\min(\mathbf{v}_1, \mathbf{v}_2) \to \mathbf{Q}) \to \neg(\min(\mathbf{v}_1, \mathbf{v}_2) \to \mathbf{Q})) \to ((\neg(\min(\mathbf{v}_1, \mathbf{v}_2) \to \mathbf{Q}) \to \neg\mathbf{P}) \to ((\min(\mathbf{v}_1, \mathbf{v}_2) \to \mathbf{Q}) \to \neg\mathbf{P})), \mathbf{T}]$	Ł$_3$2, with $\min(\mathbf{v}_1, \mathbf{v}_2) \to \mathbf{Q}$ / P, $\neg(\min(\mathbf{v}_1, \mathbf{v}_2) \to \mathbf{Q})$ / Q, $\neg\mathbf{P}$ / R
15	$[(\neg(\min(\mathbf{v}_1, \mathbf{v}_2) \to \mathbf{Q}) \to \neg\mathbf{P}) \to (((\min(\mathbf{v}_1, \mathbf{v}_2) \to \mathbf{Q}) \to \neg(\min(\mathbf{v}_1, \mathbf{v}_2) \to \mathbf{Q})) \to ((\min(\mathbf{v}_1, \mathbf{v}_2) \to \mathbf{Q}) \to \neg\mathbf{P})), \mathbf{T}]$	14, TRAN
16	$[\neg(\min(\mathbf{v}_1, \mathbf{v}_2) \to \mathbf{Q}) \to \neg\mathbf{P}, \mathbf{T}]$	9, GCON
17	$[((\min(\mathbf{v}_1, \mathbf{v}_2) \to \mathbf{Q}) \to \neg(\min(\mathbf{v}_1, \mathbf{v}_2) \to \mathbf{Q})) \to ((\min(\mathbf{v}_1, \mathbf{v}_2) \to \mathbf{Q}) \to \neg\mathbf{P}), \mathbf{T}]$	15,16 MP
18	$[((\min(\mathbf{v}_1, \mathbf{v}_2) \to \mathbf{Q}) \to \neg(\min(\mathbf{v}_1, \mathbf{v}_2) \to \mathbf{Q})) \to (\min(\mathbf{v}_1, \mathbf{v}_2) \to \mathbf{Q}), \mathbf{T}]$	13,17 HS
19	$[(((\min(\mathbf{v}_1, \mathbf{v}_2) \to \mathbf{Q}) \to \neg(\min(\mathbf{v}_1, \mathbf{v}_2) \to \mathbf{Q})) \to (\min(\mathbf{v}_1, \mathbf{v}_2) \to \mathbf{Q})) \to (\min(\mathbf{v}_1, \mathbf{v}_2) \to \mathbf{Q}), \mathbf{T}]$	Ł$_3$P4, with $\min(\mathbf{v}_1, \mathbf{v}_2) \to \mathbf{Q}$ / P
20	$[(\min(\mathbf{v}_1, \mathbf{v}_2) \to \mathbf{Q}), \mathbf{T}]$	18,19 MP
21	$[\min(\mathbf{v}_1, \mathbf{v}_2), \min(\upsilon_1, \upsilon_2)]$	Ł$_3$P7.x
22	$[\mathbf{Q}, \min(\upsilon_1, \upsilon_2)]$	20,21 MP

Thus, for example, we can derive $[\mathrm{P} \vee \mathrm{Q}, \mathbf{N}]$ from $[(\mathrm{P} \to \neg\mathrm{P}) \to (\mathrm{P} \vee \mathrm{Q}), \mathbf{N}]$ as follows

1	$[(\mathrm{P} \to \neg\mathrm{P}) \to (\mathrm{P} \vee \mathrm{Q}), \mathbf{N}]$	Assumption
2	$[\mathrm{P} \to (\mathrm{P} \vee \mathrm{Q}), \mathbf{T}]$	Ł$_3$PD6, with P / **P**, Q / **Q**
3	$[\mathrm{P} \vee \mathrm{Q}, \mathbf{N}]$	1,2 MCD

because $\mathbf{N} = \min(\mathbf{T}, \mathbf{N})$.

6.3 Exercises

SECTION 6.1

1 Prove that the rule RSIMP is derivable in Ł$_3$A:

RSIMP (Right Conjunct Simplification). From $\mathbf{P} \wedge \mathbf{Q}$ infer $\mathbf{Q}$.

2 Show that the inference

$$\frac{\begin{array}{l}\mathrm{P} \to \mathrm{Q} \\ (\mathrm{P} \to \neg\mathrm{P}) \to \mathrm{Q}\end{array}}{\mathrm{Q},}$$

which is captured by the derived rule MCD in Ł$_3$A, is valid in Ł$_3$.

3 Produce derivations to show that the following formulas are theorems of Ł$_3$A:

a. $(P \vee Q) \rightarrow (\neg P \rightarrow Q)$

b. $(P \wedge Q) \rightarrow (Q \wedge P)$

c. $\neg(P \rightarrow Q) \rightarrow \neg Q$

d. $\neg(P \rightarrow Q) \rightarrow (Q \rightarrow P)$

e. $\neg(P \rightarrow Q) \rightarrow (P \rightarrow \neg Q)$

f. $P \rightarrow (\neg Q \rightarrow \neg(P \rightarrow Q))$

4 Prove the Modified Deduction Theorem for Ł$_3$A in two parts:

a. Show that if $\mathbf{P} \rightarrow (\mathbf{P} \rightarrow \mathbf{Q})$ is a theorem then $\mathbf{Q}$ is derivable from $\mathbf{P}$ (this part is pretty easy).

b. Show that if $\mathbf{Q}$ is derivable from $\mathbf{P}$ then $\mathbf{P} \rightarrow (\mathbf{P} \rightarrow \mathbf{Q})$ is a theorem. You can show this by proving that if we have a derivation

1	$\mathbf{P}$	Assumption
2	$\mathbf{R}_1$	
3	$\mathbf{R}_2$	
...	...	
n	$\mathbf{R}_n$	
$n+1$	$\mathbf{Q}$	

in Ł$_3$A, where $\mathbf{R}_1, \ldots, \mathbf{R}_n$ are the intermediate formulas in the derivation, then every one of the following formulas is a theorem of Ł$_3$A:

$\mathbf{P} \rightarrow (\mathbf{P} \rightarrow \mathbf{P})$

$\mathbf{P} \rightarrow (\mathbf{P} \rightarrow \mathbf{R}_1)$

$\mathbf{P} \rightarrow (\mathbf{P} \rightarrow \mathbf{R}_2)$

...

$\mathbf{P} \rightarrow (\mathbf{P} \rightarrow \mathbf{R}_n)$

$\mathbf{P} \rightarrow (\mathbf{P} \rightarrow \mathbf{Q})$

Specifically, show that

i. $\mathbf{P} \rightarrow (\mathbf{P} \rightarrow \mathbf{P})$ is a theorem.

ii. If any of the $\mathbf{R}_i$—or $\mathbf{Q}$—are axioms of Ł$_3$A, then $\mathbf{P} \rightarrow (\mathbf{P} \rightarrow \mathbf{R}_i)$ (or $\mathbf{P} \rightarrow (\mathbf{P} \rightarrow \mathbf{Q})$) is a theorem.

iii. If any of the $\mathbf{R}_i$—or $\mathbf{Q}$—follows from previous formulas $\mathbf{R}_j$ and $\mathbf{R}_k$ by the rule MP, then $\mathbf{P} \rightarrow (\mathbf{P} \rightarrow \mathbf{R}_i)$ (or $\mathbf{P} \rightarrow (\mathbf{P} \rightarrow \mathbf{Q})$) follows from $\mathbf{P} \rightarrow (\mathbf{P} \rightarrow \mathbf{R}_j)$ and $\mathbf{P} \rightarrow (\mathbf{P} \rightarrow \mathbf{R}_k)$ (although not, of course, by a single application of MP—you will find Ł$_3$D11 useful in this part).

SECTION 6.2

5 a. Show that the rules Ł$_3$P5.2.2, Ł$_3$5.3.2, Ł$_3$5.4.2, and Ł$_3$5.7.2 are derivable from other rules in the system Ł$_3$PA, *excluding all* of the .2 versions of the Ł$_3$P5 rules.

b. Show that the rule Ł$_3$P5.5.2 is derivable from other axioms in the system Ł$_3$PA, *excluding all* of the .2 versions of the Ł$_3$P5 axioms.

c. Show that the rule Ł$_3$P5.6.2 is derivable from other axioms in the system Ł$_3$PA, *excluding all* of the .2 versions of the Ł$_3$P5 axioms.
d. Show that the rule Ł$_3$P5.8.2 is derivable from other axioms in the system Ł$_3$PA, *excluding all* of the .2 versions of the Ł$_3$P5 axioms.
e. Show that the rule Ł$_3$P5.9.2 is derivable from other axioms in the system Ł$_3$PA, *excluding all* of the .2 versions of the Ł$_3$P5 axioms.

6 Show that Ł$_3$P6.3.1 and Ł$_3$P6.3.2 are derivable in Ł$_3$PA given Ł$_3$P6.1.1 and Ł$_3$P6.1.2, and vice versa.

7 Show that Ł$_3$P7.1 is derivable from the other axioms of Ł$_3$PA.

8 Justify the truth-values for v_3 in each row of the truth-value table for the MP rule in Ł$_3$PA.

9 Derive the following theorems in Ł$_3$PA:

a. [(**n** → P) → (¬P → **n**), **T**]
b. [(P → **f**) → (**t** → (Q → (P → R))), **T**]
c. [(**n** → P) → (**n** → (Q → P)), **T**]
d. [(**t** → P) → ((**n** → Q) → (**n** → (P → Q))), **T**]
e. [(**t** → (P → Q))→ (P → Q), **T**]
f. [(**t** → (P → Q)) → ((**t** → P) → (**t** → Q)), **T**]

10 Derive graded versions of the following rules for Ł$_3$PA. In each case, make sure that you have assigned the strongest possible graded value to the inferred formula.

a. ŁSIMP
b. RSIMP
c. SUB

Note: The graded rule should look like this:
SUB (Substitution). From [**P** → **Q**, **T**], [**Q** → **P**, **T**], and a graded formula [**R**, v] such that **R** contains **P** as a subformula, infer any graded formula [**R***, x] in which **R*** is the result of replacing one or more occurrences of **P** in **R** with **Q** and x is . . .*(fill in the blank)*.

d. MT
e. DN
f. TRAN
g. GCON
h. GHS
i. GMP
j. DS
k. DC

11 Prove that if we can derive a graded formula [**Q**, v] from some (possible empty) set of graded assumptions in system Ł$_3$PA, then for any value v' less than v, we can also derive [**Q**, v'] from those same graded assumptions.

12 a. Derive [(¬P → **n**), **T**] from [P, **N**] in Ł$_3$PA.
b. Derive [Q → P, **N**] from [P, **N**] in Ł$_3$PA.
c. Derive [**t** → (P ∨ ¬P), **N**] in Ł$_3$PA.

13 Given the (weak or strong) completeness of Ł3A for Ł3, we can prove that if a formula **P** (without truth-value constants) is a quasi-tautology of Ł3 then [**P**, **N**] is a theorem of Ł3PA as follows:

If **P** is a quasi-tautology of Ł3 then the formula ¬**P** → **P** is always true (given that P always has the value **T** or **N**) and is therefore a tautology. Because Ł3A is complete, ¬**P** → **P** is a theorem of Ł3A and so [¬**P** → **P**, **T**] is a theorem of Ł3PA. [(¬**P** → **P**) → **P**, **N**] is also a theorem:

1	[(¬P → ¬¬P)→ ¬¬P, N]	Ł3PD12, with ¬P / P
2	[(¬P → P) → P, N]	1, DN

and so we can derive [**P**, **N**] from these two theorems by Modus Ponens.

Given the strong completeness of Ł3A for Ł3, prove that if a set of formulas Γ quasi-entails **P** (where none of the formulas contain truth-value constants) then [**P**, **N**] is derivable in Ł3PA from the graded set of formulas Γ_G in which each member of Γ has the grade **N**.

7 Three-Valued First-Order Logics: Semantics

7.1 A First-Order Generalization of Ł3

We now introduce full three-valued first-order logical systems—systems in which we can evaluate the Sorites paradox and Black's Problem of the Fringe. In classical first-order logic, predicates are interpreted simply by defining their extensions. The **extension** of a predicate consists of those objects (or tuples of objects) of which the predicate is true. It's implicit in classical semantics that all objects (or tuples of objects) that do not fall within the extension of a predicate are in its *counterextension*. The **counterextension** of a predicate consists of those objects/tuples of objects of which the predicate is false. In three-valued semantics we draw a finer distinction. We will now associate with each predicate *three* sets of objects (tuples of objects): those of which the predicate is true, those of which the predicate is false, and those of which the predicate is neither true nor false. Let us call this third set the ***fringe***.

An **interpretation I** for three-valued first-order logic consists of

1. A nonempty set D (the *domain*)
2. An assignment of three sets ext(**P**), cxt(**P**), and fge(**P**) (for *extension, counterextension, and fringe*) to each predicate **P** of arity *n*, meeting the following requirements:
   - the three sets, one or two of which may be empty, consist of *n*-tuples of members of D:
     $\mathrm{ext}(\mathbf{P}) \subseteq \mathrm{D}^n$, $\mathrm{cxt}(\mathbf{P}) \subseteq \mathrm{D}^n$, and $\mathrm{fge}\ (\mathbf{P}) \subseteq \mathrm{D}^n$
   - the three sets are *mutually exclusive*:
     $\mathrm{ext}(\mathbf{P}) \cap \mathrm{cxt}(\mathbf{P}) = \varnothing$
     $\mathrm{ext}(\mathbf{P}) \cap \mathrm{fge}(\mathbf{P}) = \varnothing$
     $\mathrm{cxt}(\mathbf{P}) \cap \mathrm{fge}(\mathbf{P}) = \varnothing$
   - the three sets are *mutually exhaustive* of D^n:
     $\mathrm{ext}(\mathbf{P}) \cup \mathrm{cxt}(\mathbf{P}) \cup \mathrm{fge}(\mathbf{P}) = \mathrm{D}^n$
3. An assignment of a member of D to each individual constant **a**:
   $\mathrm{I}(\mathbf{a}) \in \mathrm{D}$

We also need *variable assignments*: as in classical first-order logic, a **variable assignment v** assigns a member of D to each individual variable **x**: $\mathrm{v}(\mathbf{x}) \in \mathrm{D}$, and an

x-variant v′ of a variable assignment v is an assignment such that v′(**y**) = v(**y**) for every variable **y** other than **x**.

The truth-conditions for formulas on three-valued interpretations are defined in terms of *satisfaction* and *dissatisfaction* by variable assignments. Here are the truth-conditions for the first-order generalization of Ł3, which we will call *Ł3*∀:[1]

1. An atomic formula $\mathbf{Pt}_1 \ldots \mathbf{t}_n$ is
   - **satisfied** by a variable assignment v on an interpretation I if $<I^*(\mathbf{t}_1), \ldots, I^*(\mathbf{t}_n)> \in$ ext(**P**), where $I^*(\mathbf{t}_i)$ is $I(\mathbf{t}_i)$ if $\mathbf{t}_i$ is a constant and is $v(\mathbf{t}_i)$ if $\mathbf{t}_i$ is a variable,
   - **dissatisfied** if $<I^*(\mathbf{t}_1), \ldots, I^*(\mathbf{t}_n)> \in$ cxt(**P**), and
   - **neither satisfied nor dissatisfied** if $<I^*(\mathbf{t}_1), \ldots, I^*(\mathbf{t}_n)> \in$ fge(**P**).

When a formula is neither satisfied nor dissatisfied by a variable assignment, we will say that it is ***undetermined*** by that assignment. Clauses 2–6 reflect the Łukasiewicz truth-tables for propositional logic:

2. A formula ¬**P** is
   - **satisfied** by a variable assignment v on an interpretation I if **P** is dissatisfied by v on I,
   - **dissatisfied** if **P** is satisfied by v on I, and
   - **undetermined** otherwise.
3. A formula **P** ∧ **Q** is
   - **satisfied** by a variable assignment v on an interpretation I if both **P** and **Q** are satisfied by v on I,
   - **dissatisfied** if either **P** or **Q** is dissatisfied by v on I, and
   - **undetermined** otherwise.
4. A formula **P** ∨ **Q** is
   - **satisfied** by a variable assignment v on an interpretation I if either **P** or **Q** is satisfied by v on I,
   - **dissatisfied** if both **P** and **Q** are dissatisfied by v on I, and
   - **undetermined** otherwise.
5. A formula **P** → **Q** is
   - **satisfied** by a variable assignment v on an interpretation I if **P** is dissatisfied by v on I, or **Q** is satisfied by v on I, or **P** and **Q** are both undetermined by v on I,
   - **dissatisfied** if **P** is satisfied by v on I and **Q** is dissatisfied by v on I, and
   - **undetermined** otherwise.
6. A formula **P** ↔ **Q** is
   - **satisfied** by a variable assignment v on an interpretation I if either **P** and **Q** are both satisfied by v on I, or **P** and **Q** are both dissatisfied by v on I, or **P** and **Q** are both undetermined by v on I,

[1] As in Chapter 6, we will omit the subscript *L* on the Łukasiewicz connectives but include them on the connectives for the other first-order systems.

- **dissatisfied** if either **P** is satisfied by v on I and **Q** is dissatisfied by v on I or **P** is dissatisfied by v on I and **Q** is satisfied by v on I, and
- **undetermined** otherwise.

The satisfaction conditions for quantified formulas are based on the idea that a universally quantified formula is like an extended conjunction and an existentially quantified formula is like an extended disjunction. That is, if *everything* is P then *this* is P *and that* is P *and...*, while if *something* is P then either *this* is P *or that* is P *or*.... Since a conjunction is true in Łukasiewicz's three-valued system if both conjuncts are true, false if at least one conjunct is false, and neither true nor false otherwise, we will want a universal quantification to be true if what it says is true of everything, false if what it says is false of at least one thing, and neither true nor false otherwise. Thus:

7. A formula $(\forall \mathbf{x})\mathbf{P}$ is
   - **satisfied** by a variable assignment v on I if **P** is satisfied by *every* **x**-variant of v on I,
   - **dissatisfied** if **P** is dissatisfied by *at least one* **x**-variant of v on I, and
   - **undetermined** otherwise.

A disjunction is true in Łukasiewicz's three-valued system if at least one disjunct is true, false if both disjuncts are false, and neither true nor false otherwise, so we will want an existentially quantified formula to be true if what it says is true of at least one thing, false if what it says is false of everything, and neither true nor false otherwise:

8. A formula $(\exists \mathbf{x})\mathbf{P}$ is
   - **satisfied** by v on I if **P** is satisfied by *at least one* **x**-variant of v on I,
   - **dissatisfied** if **P** is dissatisfied by *every* **x**-variant of v on I, and
   - **undetermined** otherwise.

Finally, a formula has the value **T** on I if it is satisfied by every variable assignment on I, the value **F** if it is dissatisfied by every variable assignment on I, and the value **N** if it is undetermined by every variable assignment on I. Under our definitions, closed formulas (but not necessarily open ones) will receive one of the three values **T**, **F**, or **N** on any interpretation.

Here's an example of an interpretation for $Ł_3\forall$:

D: set of heights between 4′ 7″ and 6′ 7″ by $^1/_8$″ increments, inclusive
$\text{ext}(T) = \{\langle h\rangle: h \in D \text{ and } h \geq 5' \, 11''\}$
$\text{fge}(T) = \{\langle h\rangle: h \in D \text{ and } 5' \, 3'' < h < 5' \, 11''\}$
$\text{cxt}(T) = \{\langle h\rangle: h \in D \text{ and } h \leq 5' \, 3''\}$
$\text{ext}(V) = \varnothing$
$\text{fge}(V) = \{\langle h\rangle: h \in D \text{ and } h \leq 4' \, 9''\}$
$\text{cxt}(V) = \{\langle h\rangle: h \in D \text{ and } h > 4' \, 9''\}$

ext(N) = {$<h_1, h_2>$: $h_1 \in D$, $h_2 \in D$, and $h_1 - h_2 \geq 3''$}
fge(N) = {$<h_1, h_2>$: $h_1 \in D$, $h_2 \in D$, and $1'' < h_1 - h_2 < 3''$}
cxt(N) = {$<h_1, h_2>$: $h_1 \in D$, $h_2 \in D$, and $h_1 - h_2 \leq 1''$}
I(a) = 6′
I(b) = 5′ 6″
I(c) = 4′ 11″
I(d) = 4′ 10″
I(e) = 4′ 9″

We'll call this interpretation *VH,* for ***vague heights***. Intuitively, *T* means *is tall*; we have somewhat arbitrarily—but not unreasonably—chosen the cutoff points for being tall and being not tall. *V* means *very tiny* and *N* means *is noticeably taller than*—again, the cutoffs are somewhat arbitrary but seem okay. Note that we've decided that none of the heights in the domain are very tiny. The formula *Ta* has the value **T** on interpretation VH because I(*a*) is a member of ext(*T*); *Tc, Td,* and *Te* have the value **F** since I(*c*), I(*d*), and I(*e*) are members of cxt(*T*), and *Tb* has the value **N** because I(*b*) is a member of fge(*T*). *Va, Vb, Vc,* and *Vd* all have the value **F** while *Ve* has the value **N.** *Naa* has the value **F**, while *Nab, Nac, Nad,* and *Nae* all have the value **T.** *Ncd* and *Nde* have the value **F**, and *Nce* has the value **N.**

The truth-values of compound formulas without quantifiers are as expected for a Łukasiewicz system; for example, $Ta \vee Tb$ has the value **T** on VH, $Tb \vee Tc$ has the value **N**, and $Tb \wedge Tc$ has the value **F**. $Ta \rightarrow Ta$, $Tb \rightarrow Tb$, $Tc \rightarrow Tc$, $Td \rightarrow Td$, and $Te \rightarrow Te$ all have the value **T**. $Ta \rightarrow Tb$ and $Tb \rightarrow Tc$ both have the value **N**, while $Tc \rightarrow Td$ and $Td \rightarrow Te$ both have the value **T**, as do $Tb \rightarrow Ta$, $Tc \rightarrow Tb$, $Td \rightarrow Tc$, and $Te \rightarrow Td$.

The universally quantified formula $(\forall x)Tx$ has the value **F** on interpretation VH because at least one *x*-variant of each variable assignment will dissatisfy *Tx*, namely, any *x*-variant that assigns a member of cxt(*T*) to *x*. $(\forall x)Vx$ has the value **F** for a similar reason. The existentially quantified formula $(\exists x)Tx$ has the value **T** because at least one *x*-variant of each variable assignment does satisfy *Tx*: any *x*-variant that assigns a member of ext(*T*) to *x* will do so. But the formula $(\exists x)Vx$ has the value **N** because no *x*-variant of any variable assignment satisfies *Vx* (so $(\exists x)Vx$ isn't true), but not all *x*-variants dissatisfy *Vx*: those that assign a member of fge(*V*) to *x* neither satisfy nor dissatisfy *Vx* (so $(\exists x)Vx$ isn't false either). $(\exists x)\neg Vx$ has the value **T** because the counterextension of *V* is nonempty; *x*-variants that assign members of cxt(*V*) to *x* will satisfy $\neg Vx$.

$(\forall x)(\exists y)Nyx$ has the value **F** on interpretation VH, precisely because it's false that for every height in the domain you can find one that's noticeably taller. Consider any variable assignment v: the *x*-variant v′ of v that assigns 6′ 7″ to *x* dissatisfies the formula $(\exists y)Nyx$ (as do all *x*-variants that assign any height greater than 6′ 4″ to *x*). v′ dissatisfies this formula because every *y*-variant v″ of v′ dissatisfies *Nyx*: for every height in the domain that we can assign to *y*, $<v''(y), v''(x)>$ ($= <v''(y), 6' 7''>$) is a member of cxt(*N*) because v″(*y*) – 6′ 7″ is less than 1′.

The formula $(\forall x)(\neg Vx \rightarrow (\exists y)Nxy)$, which we may read as *every height that isn't very tiny is noticeably taller than some height*, has the value **N** on VH. We'll first show that for any variable assignment v,

i. the x-variants of v that assign members of cxt(V) to x either satisfy or neither satisfy nor dissatisfy the formula $\neg Vx \rightarrow (\exists y)Nxy$, and
ii. the x-variants of v that assign members of fge(V) to x cannot dissatisfy $\neg Vx \rightarrow (\exists y)Nxy$.

For (i), consider the x-variants v′ that assign members of cxt(V) to x. These x-variants will all satisfy $\neg Vx$. Some of them, those that assign heights $\geq$ 4′ 10″, also satisfy $(\exists y)Nxy$—because each height $\geq$ 4′ 10″ is noticeably taller than at least one height in the domain. But among these heights those that lie strictly between 4′ 9″ and 4′ 10″ neither satisfy nor dissatisfy $(\exists y)Nxy$—for each of these heights v′(x) there is a y-variant v″ of v′ such that $<$v′(x), v″(y)$>$ $\in$ fge(N), but there is no y-variant v″ such that $<$v′(x), v″(y)$>$ $\in$ ext(N). So some x-variants that assign members of cxt(V) to x satisfy the conditional $\neg Vx \rightarrow (\exists y)Nxy$ while others neither satisfy nor dissatisfy the conditional. For (ii), consider the x-variants that assign members of fge(V) to x: none of these can dissatisfy $\neg Vx \rightarrow (\exists y)Nxy$ because none of them satisfy $\neg Vx$.

Finally, there are no x-variants that assign members of ext(V) to x, because ext(V) is empty, so between (i) and (ii) we have considered all of the x-variants of v, with the result that some x-variants of v satisfy the conditional $\neg Vx \rightarrow (\exists y)Nxy$, some x-variants leave the conditional undetermined, and no x-variants dissatisfy it. Thus the universally quantified formula $(\forall x)(\neg Vx \rightarrow (\exists y)Nxy)$ has the value **N**.

The formula $(\forall x)(Tx \rightarrow Tx)$ has the value **T** on interpretation VH, as well as on all other interpretations for Ł$_3\forall$. Every variable assignment satisfies the universal quantification because every x-variant will satisfy $Tx \rightarrow Tx$—the identical antecedent and consequent are either both satisfied or both dissatisfied or both undetermined on any variable assignment. On the other hand, the *tall* version of the Law of Excluded Middle, $(\forall x)(Tx \vee \neg Tx)$, has the value **N** on interpretation VH. This is because the formula $Tx \vee \neg Tx$ is undetermined by variable assignments that assign a value between 5′ 3$^1/_8$″ and 5′ 10$^7/_8$″ (inclusive) to x—so $(\forall x)(Tx \vee \neg Tx)$ doesn't have the value **T**, but $Tx \vee \neg T$ is not dissatisfied by any—so $(\forall x)(Tx \vee \neg Tx)$ doesn't have the value **F**. No variable assignment can *dis*satisfy $Tx \vee \neg Tx$ because no variable assignment can dissatisfy both disjuncts.

The negation $\neg(\forall x)(Tx \vee \neg Tx)$ of the Law of Excluded Middle is equivalent to $(\exists x)(\neg Tx \wedge \neg\neg Tx)$ in Ł$_3\forall$. This is the formula at issue in Max Black's Problem of the Fringe: we would like to affirm that a predicate is vague by stating that there is at least one object in its fringe, and it would seem that we can express this with the formula $(\exists x)(\neg Tx \wedge \neg\neg Tx)$: there is at least one object that is neither tall nor not tall. But by the Principle of Double Negation, which holds in Ł$_3\forall$ as it does in classical logic, the formula $(\exists x)(\neg Tx \wedge \neg\neg Tx)$ is equivalent to $(\exists x)(\neg Tx \wedge Tx)$—and that formula seems to be a contradiction. So in claiming that the predicate *tall* is vague we would seem to be committed to the truth of a contradiction—that's the problem. In Chapter 3 we noted that the fringe formula is indeed a contradiction of classical first-order logic.

Neither $(\exists x)(\neg Tx \wedge \neg\neg Tx)$ nor $(\exists x)(\neg Tx \wedge Tx)$ is a contradiction (i.e., a formula that always has the value **F**) in $Ł_3\forall$; both formulas have the value **N** on interpretation VH. We'll just consider the formula $(\exists x)(\neg Tx \wedge Tx)$. This formula is undetermined on every variable assignment v: no *x*-variant v′ of v can satisfy $\neg Tx \wedge Tx$, so $(\exists x)(\neg Tx \wedge Tx)$ isn't satisfied by v; but $(\exists x)(\neg Tx \wedge Tx)$ is not *dis*satisfied by any v, either, because not every *x*-variant v′ of v dissatisfies $\neg Tx \wedge \neg\neg Tx$ (any assignment that assigns a member of fge(*T*) to *x*, for example, neither satisfies nor dissatisfies $\neg Tx \wedge Tx$). The formula $(\exists x)(\neg Tx \wedge Tx)$ therefore has the value **N** on interpretation VH.

But the fact that $(\exists x)(\neg Tx \wedge \neg\neg Tx)$ isn't a contradiction in $Ł_3\forall$ doesn't address Black's problem, because this formula was supposed to affirm the existence of borderline cases and consequently to assert that the predicate *T* is vague. The formula has the value **N** on VH, so it can't assert the vagueness of the predicate *T* there, and more importantly, there is *no* interpretation on which the formula is *true*. (Extending the terminology of Chapter 5, the formula is quasi-contradictory: it can only have one of the values **T**, **N**). So it is impossible for this formula truly to assert the vagueness of the predicate *T*.

Does this mean that we have no way within $Ł_3\forall$ to assert the vagueness of predicates? Not at all. Recall that Bochvar's external connectives are all definable in $Ł_3$, and so they are also definable in $Ł_3\forall$. In particular, external negation is definable here, where $\neg_{BE}\mathbf{P}$ is satisfied by a variable assignment if **P** is either dissatisfied or undetermined and $\neg_{BE}\mathbf{P}$ is dissatisfied otherwise. We can use the formula $(\exists x)(\neg_{BE} Tx \wedge \neg_{BE}\neg Tx)$, in which the first two negations are Bochvar's external negation and the third is Lukasiewicz's negation, to assert that the predicate *tall* is vague. If we regard Łukasiewicz's negation as expressing the English prefix *un*, and take *short* to mean the same as *untall*, then this formula asserts: *At least one height is neither tall nor short.* Since *Tx* and $\neg Tx$ are both undetermined on any variable assignment that assigns a value in fge(*T*) to *x*, both $\neg_{BE} Tx$ and $\neg_{BE}\neg Tx$ will be satisfied by these assignments and the formula $(\exists x)(\neg_{BE} Tx \wedge \neg_{BE}\neg Tx)$ is therefore true on interpretation VH.

The Sorites paradox also has a solution in $Ł_3\forall$. The argument

Ts_1
Es_2s_1
Es_3s_2
Es_4s_3
...
$Es_{193}s_{192}$
$(\forall x)(\forall y)((Tx \wedge Eyx) \rightarrow Ty)$
———
Ts_{193}

is valid in $Ł_3\forall$. The proof is exactly like the proof showing that the argument is valid in classical first-order logic; the key point is that *if* the premise $(\forall x)(\forall y)((Tx \wedge Eyx) \rightarrow Ty)$

has the value **T** in Ł$_3$∀ then *Ty* must be satisfied by every variable assignment that satisfies $Tx \wedge Ey$. But the validity of the Sorites argument is not enough to generate the *paradox*: the paradox depends on the argument's premises' actually being true, for it is only when the premises are in fact true that the conclusion must be true as well. So let's determine the truth-values of the premises and conclusion of the argument on the following interpretation, which we will call *VH**:

D: set of heights between 4′ 7″ and 6′ 7″ by $^1/_8$″ increments, inclusive
ext(T) $= \{<h>: h \in D \text{ and } h \geq 5' \, 11''\}$
cxt(T) $= \{<h>: h \in D \text{ and } h \leq 5' \, 3''\}$
fge(T) $= \{<h>: h \in D \text{ and } 5' \, 3'' < h < 5' \, 11''\}$
ext(E) $= \{<h_1,h_2>: h_1 \text{ is } ^1/_8'' \text{ less than } h_2\}$
cxt(E) $= \{<h_1,h_2>: h_1 \text{ is not } ^1/_8'' \text{ less than } h_2\}$
fge (E) $= \varnothing$
I(s_1) = 6′ 7″
I(s_2) = 6′ 6$^7/_8$″
. . .
I(s_{193}) = 4′ 7″

The formula Ts_1 has the value **T** on this interpretation, since I(s_1) is in ext(*T*). All of the premises $Es_{i+1}s_i$ in this argument have the value **T**, because each pair <I(s_{i+1}), I(s_i)> is a member of ext(*E*). The conclusion of the argument, Ts_{193}, has the value **F** since I(s_{193}) is in cxt(*T*). But the premise $(\forall x)\,(\forall y)\,((Tx \wedge Eyx) \rightarrow Ty)$ is not true on VH*; it has the value **N** because it is undetermined by every variable assignment. Consider any variable assignment v. We will show that

i. some *x*-variants v′ of v satisfy $(\forall y)\,((Tx \wedge Eyx) \rightarrow Ty)$, and
ii. some *x*-variants v′ of v neither satisfy nor dissatisfy $(\forall y)\,((Tx \wedge Eyx) \rightarrow Ty)$, but
iii. no *x*-variants v′ of v dissatisfy $(\forall y)\,((Tx \wedge Eyx) \rightarrow Ty)$.

For (i), consider the *x*-variant v′ such that v′(*x*) = 6′ 7″ (some other *x*-variants will work as well). v′ satisfies $(\forall y)\,((Tx \wedge Eyx) \rightarrow Ty)$ because every *y*-variant v″ of v′ satisfies $(Tx \wedge Eyx) \rightarrow Ty$. Why? First consider the variant with v″(*y*) = 6′ 6$^7/_8$″ (and, of course, v″(*x*) is still 6′ 7″). v″ satisfies *Tx*, *Ey*, and *Ty*—and so it satisfies the conditional $(Tx \wedge Eyx) \rightarrow Ty$. Now consider any other *y*-variant v″ of v′—these are variants that assign values other than 6′ 6$^7/_8$″ to *y*. Such a *y*-variant must dissatisfy *Eyx* since no height in the domain other than 6′ 6$^7/_8$″ is $^1/_8$″ less than 6′ 7″, and so it will also dissatisfy $Tx \wedge Eyx$ and therefore satisfy $(Tx \wedge Eyx) \rightarrow Ty$.

To show (ii), consider the *x*-variant v′ such that v′(*x*) = 5′ 11″. The *y*-variant v″ of v′ that assigns 5′ 10$^7/_8$″ to *y* satisfies both *Tx* and *Eyx* but fails to satisfy or dissatisfy *Ty* and so it fails to satisfy or dissatisfy $(Tx \wedge Eyx) \rightarrow Ty$. Thus v′ doesn't satisfy $(\forall y)\,((Tx \wedge Eyx) \rightarrow Ty)$. All other *y*-variants of v′ satisfy $(Tx \wedge Eyx) \rightarrow Ty$ because they dissatisfy *Eyx* – so v′ doesn't dissatisfy $(\forall y)\,((Tx \wedge Eyx) \rightarrow Ty)$ either: v′ neither satisfies nor dissatisfies $(\forall y)\,((Tx \wedge Eyx) \rightarrow Ty)$.

Finally, to show (iii), we note that for no value of x can we dissatisfy $(\forall y)((Tx \wedge Eyx) \rightarrow Ty)$. To dissatisfy the formula there would have to be a value of y such that Tx and Eyx are satisfied but Ty is dissatisfied. However, there is no pair of heights h_1 and h_2 such that h_1 is in the extension of T, h_2 is $1/8''$ less than h_1, and h_2 is in the counterextension of T.

On interpretation VH*, therefore, the Principle of Charity premise is not true. Despite the Sorites' semantic validity in $Ł_3\forall$ we are not forced to accept its conclusion on this interpretation. Moreover, in denying the truth of the Principle of Charity we have not run afoul of an ancillary problem that we noted in Chapter 1. There we said that when we deny the Principle of Charity we must accept its negation, which says that there are two heights, $1/8''$ apart, such that it is true that one is tall and false that the other is—a claim that is certainly not true! We may now add that this problem arises only within the framework of classical logic, where denying the truth of a claim commits us to its falsity and hence to the truth of its negation. But when the Principle of Charity has the value **N** in $Ł_3\forall$, the principle's negation will also have the value **N**. So this solution to the Sorites paradox in $Ł_3\forall$ doesn't endorse the ludicrous claim that $1/8''$ can take us from a height that is tall to a height of which it is *false* that it is tall.[2]

Is VH* a reasonable interpretation for the Sorites argument? Well, yes. Even if we haven't got the cutoff points for the extension and counterextension of *tall* exactly right, the conditions that diffuse the paradox here will undoubtedly be met by any reasonable interpretation of *tall*. The paradox is dissolved as long as the premise $(\forall x)(\forall y)((Tx \wedge Eyx) \rightarrow Ty)$ can turn out to be neither true nor false, and this will be the case as long as there is at least one pair of heights x and y such that $(Tx \wedge Eyx) \rightarrow Ty$ is neither satisfied nor dissatisfied but no pair of heights x and y for which the formula is *dis*satisfied. As long as we have a nonempty fringe along with a nonempty extension (or a nonempty counterextension) for the predicate T there will be a pair of heights x and y such that $(Tx \wedge Eyx) \rightarrow Ty$ is neither satisfied nor dissatisfied—we naturally assume that the interpretation of E remains the same—and that seems right for vague predicates. And as long as there are no two consecutive heights such that one is in the extension of the predicate T while the other is in its *counter*extension, a situation also guaranteed by a nonempty fringe, the formula $(Tx \wedge Eyx) \rightarrow Ty$ can't be *dis*satisfied by any values of x and y.

7.2 Quantifiers Based on the Other Three-Valued Systems

Because conjunction and disjunction are defined in K^S_3 in exactly the same way as in $Ł_3$, the quantifier clauses for the two systems are the same. The only difference

[2] The astute reader may still worry, however, that here $1/8''$ can take us from a height that is tall to a height of which it is neither true nor false that it is tall. We will return to this very legitimate concern in Chapter 10.

between $K^S{}_3\forall$ and $Ł_3\forall$ occurs in the satisfaction clauses for conditionals and biconditionals, which for $K^S{}_3\forall$ will stipulate the satisfaction conditions capturing the semantics of the $K^S{}_3$ conditional and biconditional, for example,

5. A formula $\mathbf{P} \rightarrow_K \mathbf{Q}$ is
   - satisfied by v on I if either **P** is dissatisfied by v on I or **Q** is satisfied by v on I,
   - dissatisfied if **P** is satisfied by v on I and **Q** is dissatisfied by v on I, and
   - undetermined otherwise.

Black's formula $(\exists x)(\neg Tx \wedge \neg\neg Tx)$ will have the value **N** on interpretation VH in $K^S{}_3\forall$ as it does in $Ł_3\forall$, and the same holds for the Law of Excluded Middle because the connectives in these formulas are identical to Lukasiewicz's. And although the Kleene conditional differs from Lukasiewicz's, the Sorites argument remains valid in $K^S{}_3\forall$ while its Principle of Charity premise has the value **N** on interpretation VH* (the reader will be asked in the exercises to explain why).

New pairs of quantifiers occur in the first-order systems $B^I{}_3\forall$ and $B^E{}_3\forall$. Concerning the former, recall that in $B^I{}_3\forall$ the value **N** is contagious. So if we model the quantifiers on the truth-conditions for conjunctions and disjunctions in $B^I{}_3$, quantified formulas should have the value **N** whenever there is at least one variable assignment that fails to satisfy or dissatisfy the formula following the quantifier. Thus the satisfaction clauses for quantifiers in $B^I{}_3\forall$ are

7. A formula $(\forall_{BI}\mathbf{x})\mathbf{P}$ is
   - satisfied by a variable assignment v on I if **P** is satisfied by *every* **x**-variant of v on I,
   - dissatisfied if **P** is dissatisfied by *at least one* **x**-variant of v on I and there is no **x**-variant of v that fails to satisfy or dissatisfy **P** on I, and
   - undetermined otherwise.
8. A formula $(\exists_{BI}\mathbf{x})\mathbf{P}$ is
   - satisfied by v on I if **P** is satisfied by *at least one* **x**-variant of v on I and there is no **x**-variant of v that fails to satisfy or dissatisfy **P** on I,
   - dissatisfied if **P** is dissatisfied by *every* **x**-variant of v on I, and
   - undetermined otherwise.

The satisfaction clauses for formulas formed with the binary propositional connectives will also differ from those for $Ł_3\forall$, to reflect the fact that the value **N** is contagious for all connectives in Bochvar's internal system. For example, the clause for disjunction is:

4. A formula $\mathbf{P} \vee_{BI} \mathbf{Q}$ is
   - satisfied if both **P** and **Q** are satisfied, or one is satisfied and the other is dissatisfied,
   - dissatisfied if both **P** and **Q** are dissatisfied, and
   - undetermined otherwise.

The universally quantified formula $(\forall_{BI}x)Tx$ has the value **N** on interpretation VH. This is because at least one value that can be assigned to *x* will cause *Tx* to be neither satisfied nor dissatisfied, and that is sufficient for concluding that the universal quantification has neither the value **T** nor the value **F** in $B^I{}_3\forall$. This contrasts with $L_3\forall$ (and $K^S{}_3\forall$), where $(\forall x)Tx$ has the value **F** because there is at least one value that can be assigned to *x* that *dis*satisfies *Tx*. For a similar reason the existentially quantified formula $(\exists_{BI}x)Tx$ has the value **N** on interpretation VH in $B^I{}_3\forall$, whereas in the other two systems the existential quantification $(\exists x)Tx$ has the value **T**. The Bochvarian internal Law of Excluded Middle $(\forall_{BI}x)(Tx \vee_{BI} \neg_{BI}Tx)$ also has the value **N**. In fact, *any* (closed) quantified formula that contains *Tx* will have the value **N** in $B^I{}_3\forall$ on interpretation VH since at least one value of *x* fails to satisfy or dissatisfy *Tx* and this failure is contagious. So the Principle of Charity premise of the Sorites argument, along with Black's fringe formula, also have the value **N** in $B^I{}_3\forall$ on interpretations VH* and VH (while the Sorites argument remains valid).

In Bochvar's external system a conjunction is true if both conjuncts are true and is false otherwise, and a disjunction is true if at least one conjunct is true and is false otherwise. So a universally quantified formula should be true if every variable assignment satisfies the formula following the quantifier and should be false otherwise, while an existentially quantified formula should be true if there is at least one variable assignment that satisfies the formula following the quantifier and should be false otherwise. The satisfaction clause for the universal quantifier in $B^E{}_3\forall$ is thus

7. A formula $(\forall_{BE}\mathbf{x})\mathbf{P}$ is
   - **satisfied** by a variable assignment v on I if **P** is satisfied by every **x**-variant of v on I,
   - **dissatisfied** otherwise.

The satisfaction clause for the $B^E{}_3\forall$ existential quantifier is left as an exercise. Clauses for formulas formed with the propositional connectives in $B^E{}_3\forall$ will reflect their truth-conditions in $B^E{}_3$. For example, the clause for negated formulas is

2. A formula $\neg_{BE}\mathbf{P}$ is
   - satisfied by a variable assignment v on I if **P** is not satisfied by v on I, and
   - dissatisfied otherwise.

The formula $(\forall_{BE}x)Tx$ has the value **F** on interpretation VH since any variable assignment that assigns to *x* a height less than 5′ 11″ will fail to satisfy *Tx*. The Law of Excluded Middle formula $(\forall_{BE}x)(Tx \vee_{BE} \neg_{BE}Tx)$ has the value **T** on VH, since every variable assignment will satisfy $Tx \vee_{BE} \neg_{BE}Tx$. Indeed, the Law of Excluded Middle is true on *every* interpretation in $B^E{}_3\forall$, since $\neg_{BE}Tx$ must be satisfied by any variable assignment that fails to satisfy *Tx*. Black's fringe formula $(\exists_{BE}x)(\neg_{BE}Tx \wedge_{BE} \neg_{BE}\neg_{BE}Tx)$ has the value **F** in $B^E{}_3\forall$ on interpretation VH and indeed on every interpretation because the external negation clause guarantees that it is always the case that one of the two conjuncts is false. The Sorites argument remains valid in $B^E{}_3\forall$. The Principle of Charity premise is false on interpretation VH*, so its denial is true, and that—as we have noted—seems to be paradoxical, for the denial asserts that $^1/_8''$ can make a difference.

The four Bochvar quantifiers are definable in $Ł_3\forall$. It is left as an exercise to confirm that $(\forall_{BI}\mathbf{x})\mathbf{P}$ is equivalent to Łukasiewicz's $(\forall\mathbf{x})\mathbf{P} \vee (\exists\mathbf{x})(\mathbf{P} \wedge \neg\mathbf{P})$, and since Bochvar's internal negation is identical to $Ł_3$ negation we can define Bochvar's internal existential quantifier in $Ł_3\forall$ using internal negation and the (defined) universal quantifier. Bochvar's external quantifiers can be defined using the external assertion operator (which is definable in $Ł_3\forall$) and Łukasiwicz's quantifiers, for example, $(\forall_{BE}\mathbf{x})\mathbf{P} =_{\text{def}} (\forall\mathbf{x})\mathbf{aP}$. The reader will be asked in the exercises to verify the correctness of these definitions.

7.3 Tautologies, Validity, and "Quasi-"Semantic Concepts

Before introducing first-order semantic concepts we extend the concept of normality to apply to first-order systems. Let us call a three-valued interpretation ***classical*** if it assigns an empty fringe to every predicate. We will say that *a three-valued first-order system is* ***normal*** if its propositional subsystem is normal (that is, the propositional connectives form a normal propositional logic) and, in addition, the truth-conditions for quantified formulas on classical interpretations are the same as in classical logic. On such interpretations, every atomic formula is either satisfied or dissatisfied, and only those parts of the semantic clauses for compound formulas that involve satisfied or dissatisfied components are applicable. $Ł_3\forall$ is normal, so that the following lemma applies:

> *First-order Normality Lemma:* In a normal three-valued first-order system, a classical interpretation behaves exactly as it does in classical first-order logic—every formula that is true on that interpretation in the three-valued system is also true on that interpretation in classical logic, and every formula that is false on that interpretation in the three-valued system is also false on that interpretation in classical logic.[3]
>
> *Proof:* The lemma follows from the fact that the connectives and quantifiers in a normal system behave exactly as they do in classical logic whenever a classical interpretation is used.

It follows that

> *Result 7.1:* Every formula that is a tautology in $Ł_3\forall$ is also a tautology in classical first-order logic, and every formula that is a contradiction in $Ł_3\forall$ is also a contradiction in classical first-order logic.

[3] Recall that the subscripts *K*, and so on, are not part of the connectives or quantifiers in formulas. They are merely a notational convention indicating which system we are working in. So it makes sense to talk of the behavior of the same formulas in the three-valued systems and classical logic.

Result 7.2: Not every formula that is a tautology in classical first-order logic is also a tautology in Ł$_3$∀, and not every formula that is a contradiction in classical first-order logic is also a contradiction in Ł$_3$∀.[4]

Result 7.1 holds because Ł$_3$∀ is a normal system. An example of a classical tautology that is also a tautology in Ł$_3$∀ is $(\forall x)(Tx \rightarrow Tx)$, as discussed earlier. An example of a classical tautology that is not a tautology in Ł$_3$∀ is the Law of Excluded Middle $(\forall x)(Tx \vee \neg Tx)$, which we have shown has the value **N** on some Ł$_3$∀ interpretations.

We have seen that the Sorites paradox is semantically valid in Ł$_3$∀. Because Ł$_3$∀ is normal, *any* argument that is semantically valid in Ł$_3$∀ is also semantically valid in classical first-order logic. More generally:

Result 7.3: Every entailment that holds in Ł$_3$∀ also holds in classical first-order logic.

But the converse does not hold:

Result 7.4: Not every entailment that holds in classical first-order logic also holds in Ł$_3$∀.

These results just generalize the results for Ł$_3$. In fact, all of the semantic results for the various propositional systems in Chapter 5 generalize for their corresponding first-order systems.

We define a **quasi-tautology** in a three-valued first-order system to be a closed formula that never has the value **F** on an interpretation, and a **quasi-contradiction** to be a closed formula that never has the value **T**. Because Ł$_3$∀ is normal, every quasi-tautology in Ł$_3$∀ is also a tautology in first-order classical logic. On the other hand, some classical tautologies are not quasi-tautologies in Ł$_3$∀; the formula $(\forall x)(\neg(Px \rightarrow \neg Px) \vee \neg(\neg Px \rightarrow Px))$ is an example (proof that this is not a quasi-tautology is left as an exercise). Similar results hold for quasi-contradictions:

Result 7.5: Every quasi-tautology in Ł$_3$∀ is a tautology in classical first-order logic, and every quasi-contradiction in Ł$_3$∀ is a contradiction in classical first-order logic.

Result 7.6: Not every formula that is a tautology in classical first-order logic is a quasi-tautology in Ł$_3$∀, and not every formula that is a contradiction in classical first-order logic is a quasi-contradiction in Ł$_3$∀.

A set Γ of closed first-order formulas **quasi-entails** a closed first-order formula **P** if, whenever each member of Γ has either the value **T** or the value **N**, so does **P**. A first-order argument is **quasi-valid** in a three-valued system if the set consisting of its premises quasi-entails its conclusion. Here the generalized results are

Result 7.7: Every quasi-entailment in Ł$_3$∀ is an entailment in classical first-order logic.

4 As in Chapter 3, we define the concept of a tautology and a contradiction as well as all other semantic concepts to apply only to closed formulas.

Result 7.8: Not every entailment in classical first-order logic is a quasi-entailment in $Ł_3\forall$.

Interpretation VH* suffices to show that the Sorites argument is not quasi-valid in $Ł_3\forall$. On that interpretation, the premises of the argument all have the value **T** except the Principle of Charity premise, which has the value **N**, but the conclusion has the value **F**. So we now have two things that we can say about the Sorites argument in $Ł_3\forall$. First, although it is logically valid, there are reasonable interpretations on which not all of the premises are true—those on which the Principle of Charity premise has the value **N**. Second, the argument is not quasi-valid—the conclusion may be false even if none of the premises are false. Anticipating fuzzy logic, we may say that the argument isn't *degree-of-truth-preserving*; the conclusion can be less true than the least true premise. Similar comments hold for the first-order versions of the other three-valued systems.

Interpretation VH* also shows that the Sorites argument isn't quasi-valid in $K_3\forall$or $B^I{}_3\forall$. In both systems the Principle of Charity premise has the value **N**, the remaining premises have the value **T**, and the conclusion is false. The following interpretation shows that the argument isn't quasi-valid in $B^E{}_3\forall$ either (VH* won't work here, since the $B^E{}_3\forall$ Principle of Charity premise is false on that interpretation):

D: set of heights between 4′ 7″ and 6′ 7″ by $^1/_8$″ increments, inclusive
ext(T) = ∅
cxt(T) = {<h>: h ∈ D and h < 6′ 7″}
fge(T) = {<h>: h ∈ D and h = 6′ 7″}
(i.e., {<6′ 7″>})
ext(E) = {<h_1,h_2>: h_1 is $^1/_8$″ less than h_2}
cxt(E) = {<h_1,h_2>: h_1 is not $^1/_8$″ less than h_2}
fge(E) = ∅
I(s_1) = 6′ 7″
I(s_2) = 6′ $6^7/_8$″
. . .
I(s_{193}) = 4′ 7″

It is obvious that on this interpretation the first premise of the Sorites argument, Ts_1, has the value **N**, and all of the $Es_{i+1}s_i$ premises have the value **T**. The external Bochvarian Principle of Charity premise also has the value **T**—every variable assignment dissatisfies $Tx \wedge_{BE} Eyx$ (since no variable assignment satisfies Tx) and so the conditional $(Tx \wedge_{BE} Eyx) \rightarrow_{BE} Ty$ along with its universal generalization are satisfied by every variable assignment. This establishes that all of the premises have either the value **T** or the value **N**. But the conclusion has the value **F** on this interpretion, and that is sufficient to establish that the argument is not quasi-valid in $B^E{}_3\forall$.

Finally, a set Γ of closed first-order formulas **degree-entails** the closed first-order formula **P** if the rank of **P** is no lower than the rank of any of the members of

Γ, and an argument is **degree-valid** in a three-valued first-order system if the set consisting of the premises degree-entails the conclusion in that system. The results for $L_3\forall$ also carry over from the propositional case

> *Result 7.9:* Every degree-entailment in $Ł_3\forall$ is an entailment in classical first-order logic.

> *Result 7.10:* Not every entailment in classical first-order logic is a degree-entailment in $Ł_3\forall$.

as do the results for the other three-valued first-order systems.

7.4 Exercises

SECTION 7.1

1 Determine the truth-values of the following formulas in $Ł_3\forall$ on an interpretation that makes the following assignments:

D: set of positive integers
$\text{ext}(O) = \{<i>: i \in D \text{ and } i \text{ is odd}\}$
$\text{cxt}(O) = \{<i>: i \in D \text{ and } i \text{ is even}\}$
$\text{fge}(O) = \varnothing$
$\text{ext}(B) = \{<i>: i \in D \text{ and } i \geq 10{,}000\}$
$\text{cxt}(B) = \{<i>: i \in D \text{ and } i \leq 9\}$
$\text{fge}(B) = \{<i>: i \in D \text{ and } 9 < i < 10{,}000\}$
Read Bx as: x is big
$\text{ext}(G) = \{<i, j>: i \in D, j \in D, \text{ and } i \geq j + 10{,}000\}$
$\text{cxt}(G) = \{<i, j>: i \in D, j \in D, \text{ and } i \leq j + 9\}$
$\text{fge}(G) = \{<i, j>: i \in D, j \in D, \text{ and } j + 9 < i < j + 10{,}000\}$
Read Gxy as: x is much greater than y
$I(a) = 1$
$I(b) = 2$
$I(c) = 9$
$I(d) = 100$
$I(e) = 10{,}000$
$I(f) = 10{,}001$

a. $Oa \wedge Ba$
b. $Oe \wedge Be$
c. $Of \wedge Bf$
d. $Oa \rightarrow Ba$
e. $Ba \rightarrow Oa$
f. $Ba \rightarrow Ba$
g. $Bf \rightarrow \neg Of$
h. $(\forall x)Bx$

i. $(\forall x)\neg Gxx$
j. $(\exists x)Gxb$
k. $(\exists x)Gdx$
l. $(\exists x)(\exists y)((Ox \wedge Oy) \wedge Gxy)$
m. $(\exists x)(\exists y)((Bx \wedge By) \wedge Gxy)$
n. $(\forall x)(\forall y)(Gxy \vee Gyx)$
o. $(\exists x)(\forall y)(\neg Bx \wedge Gxy)$
p. $(\forall x)(\forall y)(Gxy \rightarrow \neg Gxy)$
q. $(\exists x)(\exists y)(Gxy \wedge Gyx)$
r. $(\forall x)(\exists y)Gxy$
s. $(\exists y)(\forall x)Gxy$

2 Determine the truth-values of the following formulas in Ł$_3$∀ on interpretation VH:

a. $(\forall x)(Vx \rightarrow \neg Tx)$
b. $(\forall x)(\forall y)(Nxy \vee Nyx)$
c. $(\forall x)(\forall y)(\neg Nxy \vee \neg Nyx)$
d. $(\exists x)(\exists y)(Nxy \wedge Nyx)$
e. $(\forall x)((\exists y)Nxy \rightarrow Tx)$
f. $(\forall x)(\forall y)((Tx \wedge Vy) \rightarrow Nxy)$
g. $(\forall x)(\forall y)((Tx \wedge \neg Ty) \rightarrow Nxy)$
h. $(\forall x)(\forall y)((Vx \wedge Vy) \rightarrow \neg Nxy)$

3 Justify the following two claims, which were made at the end of Section 7.1, regarding reasonable alternative interpretations of *T* on VH*:

a. As long as we have a nonempty fringe along with a nonempty extension for the predicate *T* there will be a pair of heights x and y such that $(Tx \wedge Eyx) \rightarrow Ty$ is neither satisfied nor dissatisfied.
b. As long as there are no two consecutive heights such that one is in the extension of the predicate *T* while the other is in its *counter*extension, the formula $(Tx \wedge Eyx) \rightarrow Ty$ can't be dissatisfied by any values of *x* and *y*.

SECTION 7.2

4 Explain why the Principle of Charity premise of the Sorites paradox using Kleene's rather than Łukasiewicz's definition of the conditional will have the value **N** on the interpretation VH* of Section 7.1.

5 Produce a correct clause 8 for the definition of satisfaction for existentially quantified formulas in $B^E{}_3\forall$.

6 Prove that $(\forall_{BI}\mathbf{x})\mathbf{P}$ is equivalent to Łukasiewicz's $(\forall \mathbf{x})\mathbf{P} \vee (\exists \mathbf{x})(\mathbf{P} \wedge \neg \mathbf{P})$.

7 Prove that the definition $(\forall_{BE}\mathbf{x})\mathbf{P} =_{\text{def}} (\forall \mathbf{x})a\mathbf{P}$, where the latter universal quantifier is Łukasiewicz's, gives the correct satisfaction conditions for Bochvar's external universal quantifier. Be sure to include the satisfaction clauses for the external assertion operator.

8 Show how to define Bochvar's external existential quantifier in Ł$_3$∀, and prove that your definition gives the correct truth-conditions.

SECTION 7.3

9 Prove that the formula $(\forall x)(\neg\ (Px \rightarrow \neg Px) \vee \neg\ (\neg Px \rightarrow Px))$ is not a quasi-tautology in $\text{Ł}_3\forall$.

10 For each of the following formulas, decide whether it is a tautology and/or a quasi-tautology in $\text{Ł}_3\forall$, and show that your answer is correct.

a. $(\forall x)Px \rightarrow (\exists x)Px$

b. $(\forall x)Px \vee (\exists x)\neg Px$

c. $Pa \rightarrow (\exists x)Px$

d. $(\exists x)Px \rightarrow Pa$

e. $(\exists x)Px \vee (\exists x)\neg Px$

11 For each of the following arguments, decide whether it is valid and/or quasi-valid in $\text{Ł}_3\forall$, and show that your answer is correct.

a. $\underline{(\forall x)(Bx \rightarrow Cx)}$
$\neg(\exists x)(Bx \wedge \neg Cx)$

b. $(\exists x)Bx \vee (\exists x)Cx$
$Ga \rightarrow (\forall x)\neg Bx$
$\underline{Ga \rightarrow (\forall x)\neg Cx}$
$\neg Ga$

c. $\underline{(\forall x)(\forall y)(Tx \rightarrow Ty)}$
$(\exists x)Tx \wedge (\exists x)\neg Tx$

d. $\underline{(\forall x)(\exists y)Lxy}$
$(\exists y)(\forall x)Lxy$

8 Derivation Systems for Three-Valued First-Order Logic

8.1 An Axiomatic System for Tautologies and Validity in Three-Valued First-Order Logic

We can extend the axiomatic systems presented in Chapter 6 for Łukasiewicz's three-valued propositional logics to sound and complete systems for three-valued first-order logic by adding axioms and rules for the quantifiers. In this section we'll develop $Ł_3\forall A$, an extension of the Wajsberg axiomatic system $Ł_3A$ in which derivability coincides with validity proper. We include $Ł_3A$'s axioms and rule:

Axiom schemata:

$Ł_3\forall 1$. $\mathbf{P \rightarrow (Q \rightarrow P)}$

$Ł_3\forall 2$. $\mathbf{(P \rightarrow Q) \rightarrow ((Q \rightarrow R) \rightarrow (P \rightarrow R))}$

$Ł_3\forall 3$. $\mathbf{(\neg P \rightarrow \neg Q) \rightarrow (Q \rightarrow P)}$

$Ł_3\forall 4$. $\mathbf{((P \rightarrow \neg P) \rightarrow P) \rightarrow P}$

Derivation rule:

MP. From **P** and **P → Q**, infer **Q**.

and we add the following axioms and rule for the universal quantifier (the same ones we used for classical quantificational logic) to obtain the full first-order system $Ł_3\forall A$:

$Ł_3\forall 5$. $\mathbf{(\forall x)(P \rightarrow Q) \rightarrow (P \rightarrow (\forall x)Q)}$

where **P** is a formula in which **x** does not occur free

$Ł_3\forall 6$. $\mathbf{(\forall x)P \rightarrow P(a/x)}$

where **a** is any individual constant and the expression **P(a/x)** means: *the result of substituting the constant **a** for the variable **x** wherever **x** occurs free in **P***

UG. From **P(a/x)**, *infer* **(∀x)P**

where **x** is any individual variable, provided that no assumption contains the constant **a** and that **P** itself does not contain the constant **a**.

This system is sound and complete system for $Ł_3\forall$.[1]

[1] Axioms $Ł_3\forall 5$ and $Ł_3\forall 5$ and the rule UG are sufficient to derive the quantificational axioms of a system presented in LeBlanc (1977), and so, because that system and $Ł_3\forall A$ are otherwise identical, the completeness proof in that paper establishes the completeness of $Ł_3\forall A$ as well. As was the

As for CL∀A, we stipulate that only *closed* formulas occur in derivations. Here's a derivation of the formula *(∀x)Px → (∃x)Px* rewritten as *(∀x)Px → ¬(∀x)¬Px* (we define the existential quantifier as **(∃x)P** $=_{\text{def}}$ **¬(∀x)¬P**):

1	(∀x)Px → Pa	Ł$_3$∀6, with (∀x)Px / **(∀x)Px**, a / **a**
2	(∀x)¬Px → ¬Pa	Ł$_3$∀6, with (∀x)¬Px / **(∀x)Px**, a / **a**
3	¬¬(∀x)¬Px → ¬Pa	2, DN
4	Pa → ¬(∀x)¬Px	3, GCON
5	(∀x)Px → ¬(∀x)¬Px	1,4 HS

Because the axioms and rules of Ł$_3$A are included in the quantificational system Ł$_3$∀A, all of the derived rules and axioms from Section 6.1 of Chapter 6 carry directly over to Ł$_3$∀A.

A derived axiom schema related to Ł$_3$∀5 is

Ł$_3$∀D12. (∀x)(P → Q) → ((∀x)P → (∀x)Q)

(We have numbered the derived axiom schema as *12* since we already have eleven derived axiom schemata from Ł$_3$A.)

Justification for Ł$_3$∀D12: In this derivation the constant **a** is chosen to be a constant that does not occur in **P** or **Q**; it is always possible to find such a constant because the language contains infinitely many constants and each formula is only finitely long.

1	**(∀x)(P → Q) → (P(a/x) → Q(a/x))**	Ł$_3$∀6, with **(∀x)(P → Q)** / **(∀x)P, a / a**
2	**P(a/x) → ((∀x)(P → Q) → Q(a/x))**	1, TRAN
3	**(∀x)P → P(a/x)**	Ł$_3$∀6, with **(∀x)P** / **(∀x)P, a / a, x / x**
4	**(∀x)P → ((∀x)(P → Q) → Q(a/x))**	2,3 HS
5	**(∀x) ((∀x)P → ((∀x)(P → Q) → Q))**	4, UG
6	**(∀x) ((∀x)P → ((∀x)(P → Q) → Q)) → ((∀x)P → (∀x)((∀x)(P → Q) → Q))**	Ł$_3$∀5, with **(∀x) ((∀x)P → ((∀x)(P → Q) → Q))** / **(∀x)(P → Q)**
7	**(∀x)P → (∀x)((∀x)(P → Q) → Q)**	5,6 MP
8	**(∀x)((∀x)(P → Q) → Q) → ((∀x)(P → Q) → (∀x)Q))**	Ł$_3$∀5, with **(∀x)((∀x)(P → Q) → Q)** / **(∀x)(P → Q)**
9	**(∀x)P → ((∀x)(P → Q) → (∀x)Q)**	7,8 HS
10	**(∀x)(P → Q) → ((∀x)P → (∀x)Q)**	9, TRAN

On line 1 we anticipated the use of TRAN on line 2 and wrote the consequent as **(P(a/x) → Q(a/x))** rather than **(P → Q)(a/x)**; it is the same formula. On line 5 we were allowed to generalize on the constant **a** because **a** does not occur in **P** or in **Q**—and of course there are no assumptions, and hence none involving **a**, in the derivation. The instance of axiom schemata Ł$_3$∀5 on line 6 is allowable because it satisfies the constraint that **x** does not occur *free* in the antecedent, **(∀x)P**, and the instance on line 8 is similarly allowable.

case in classical logic, the set of theorems of Ł$_3$∀A, or equivalently, the set of tautologies of Ł$_3$∀, is undecidable.

We can use Ł$_3\forall$D12 to show that the formula $((\exists x)Px \vee (\exists x)Qx) \rightarrow (\exists x)(Px \vee Qx)$ is a theorem. In the derivation the formula is rewritten without the existential quantifier as $(\neg(\forall x)\neg Px \vee \neg(\forall x)\neg Qx) \rightarrow \neg(\forall x)\neg(Px \vee Qx)$:

1	$Pa \rightarrow (Pa \vee Qa)$	Ł$_3\forall$D6, with Pa / **P**, Qa / **Q**
2	$\neg(Pa \vee Qa) \rightarrow \neg Pa$	1, GCON
3	$(\forall x)(\neg(Px \vee Qx) \rightarrow \neg Px)$	2, UG
4	$(\forall x)(\neg(Px \vee Qx) \rightarrow \neg Px) \rightarrow$ $((\forall x)\neg(Px \vee Qx) \rightarrow (\forall x)\neg Px)$	Ł$_3\forall$D12, with $(\forall x)(\neg(Px \vee Qx) \rightarrow \neg Px)$ / $(\forall \mathbf{x})(\mathbf{P} \rightarrow \mathbf{Q})$
5	$(\forall x)\neg(Px \vee Qx) \rightarrow (\forall x)\neg Px$	3,4 MP
6	$\neg(\forall x)\neg Px \rightarrow \neg(\forall x)\neg(Px \vee Qx)$	5, GCON
7	$Qa \rightarrow (Pa \vee Qa)$	Ł$_3\forall$1, with Qa / **P**, Pa → Qa/ **Q**
	On line 7 we have rewritten the formula Qa → ((Pa → Qa) → Qa) using disjunction	
8	$\neg(Pa \vee Qa) \rightarrow \neg Qa$	7, GCON
9	$(\forall x)(\neg(Px \vee Qx) \rightarrow \neg Qx)$	8, UG
10	$(\forall x)(\neg(Px \vee Qx) \rightarrow \neg Qx) \rightarrow$ $((\forall x)\neg(Px \vee Qx) \rightarrow (\forall x)\neg Qx)$	Ł$_3\forall$D12, with $(\forall x)(\neg(Px \vee Qx) \rightarrow \neg Qx)$ / $(\forall \mathbf{x})(\mathbf{P} \rightarrow \mathbf{Q})$
11	$(\forall x)\neg(Px \vee Qx) \rightarrow (\forall x)\neg Qx$	9,10 MP
12	$\neg(\forall x)\neg Qx \rightarrow \neg(\forall x)\neg(Px \vee Qx)$	11, GCON
13	$(\neg(\forall x)\neg Px \vee \neg(\forall x)\neg Qx) \rightarrow \neg(\forall x)\neg(Px \vee Qx)$	6,12 DC

A derived axiom schema similar to Ł$_3\forall$5 for the existential quantifier is

Ł$_3\forall$D13. $(\exists \mathbf{x})(\mathbf{P} \rightarrow \mathbf{Q}) \rightarrow (\mathbf{P} \rightarrow (\exists \mathbf{x})\mathbf{Q})$
where **P** is a formula in which **x** does not occur free

Justification: We derive the formula $\neg(\forall \mathbf{x})\neg(\mathbf{P} \rightarrow \mathbf{Q}) \rightarrow (\mathbf{P} \rightarrow \neg(\forall \mathbf{x})\neg \mathbf{Q})$ in which the existential quantifier is defined in terms of negation and the universal quantifier. On line 1, **a** is chosen to be a constant that does not occur in **P** or in **Q**.

1	$\mathbf{P} \rightarrow ((\mathbf{P} \rightarrow \mathbf{Q(a/x)}) \rightarrow \mathbf{Q(a/x)})$	Ł$_3\forall$D6, with **P** / **P**, **Q(a/x)** / **Q**
2	$\mathbf{P} \rightarrow (\neg\mathbf{Q(a/x)} \rightarrow \neg(\mathbf{P} \rightarrow \mathbf{Q(a/x)}))$	1, GCON
3	$\neg\mathbf{Q(a/x)} \rightarrow (\mathbf{P} \rightarrow \neg(\mathbf{P} \rightarrow \mathbf{Q(a/x)}))$	2, TRAN
4	$(\forall \mathbf{x})(\neg\mathbf{Q} \rightarrow (\mathbf{P} \rightarrow \neg(\mathbf{P} \rightarrow \mathbf{Q})))$	3, UG
5	$(\forall \mathbf{x})(\neg\mathbf{Q} \rightarrow (\mathbf{P} \rightarrow \neg(\mathbf{P} \rightarrow \mathbf{Q}))) \rightarrow$ $((\forall \mathbf{x})\neg\mathbf{Q} \rightarrow (\forall \mathbf{x})(\mathbf{P} \rightarrow \neg(\mathbf{P} \rightarrow \mathbf{Q})))$	Ł$_3\forall$D12, with $(\forall \mathbf{x})(\neg\mathbf{Q} \rightarrow (\mathbf{P} \rightarrow \neg(\mathbf{P} \rightarrow \mathbf{Q})))$ / $(\forall \mathbf{x})(\mathbf{P} \rightarrow \mathbf{Q})$
6	$(\forall \mathbf{x})\neg\mathbf{Q} \rightarrow (\forall \mathbf{x})(\mathbf{P} \rightarrow \neg(\mathbf{P} \rightarrow \mathbf{Q}))$	4,5 MP
7	$(\forall \mathbf{x})(\mathbf{P} \rightarrow \neg(\mathbf{P} \rightarrow \mathbf{Q})) \rightarrow (\mathbf{P} \rightarrow (\forall \mathbf{x})\neg(\mathbf{P} \rightarrow \mathbf{Q}))$	Ł$_3\forall$5, with $(\forall \mathbf{x})(\mathbf{P} \rightarrow \neg(\mathbf{P} \rightarrow \mathbf{Q}))$ / $(\forall \mathbf{x})(\mathbf{P} \rightarrow \mathbf{Q})$
8	$(\forall \mathbf{x})\neg\mathbf{Q} \rightarrow (\mathbf{P} \rightarrow (\forall \mathbf{x})\neg(\mathbf{P} \rightarrow \mathbf{Q}))$	6,7 HS
9	$\mathbf{P} \rightarrow ((\forall \mathbf{x})\neg\mathbf{Q} \rightarrow (\forall \mathbf{x})\neg(\mathbf{P} \rightarrow \mathbf{Q}))$	8, TRAN
10	$\mathbf{P} \rightarrow (\neg(\forall \mathbf{x})\neg(\mathbf{P} \rightarrow \mathbf{Q}) \rightarrow \neg(\forall \mathbf{x})\neg\mathbf{Q})$	9, GCON
11	$\neg(\forall \mathbf{x})\neg(\mathbf{P} \rightarrow \mathbf{Q}) \rightarrow (\mathbf{P} \rightarrow \neg(\forall \mathbf{x})\neg\mathbf{Q})$	10, TRAN

Note that because by hypothesis **x** does not occur in **P**, the formula on line 3 is $(\neg\mathbf{Q} \rightarrow (\mathbf{P} \rightarrow \neg(\mathbf{P} \rightarrow \mathbf{Q})))(\mathbf{a/x})$, which is required for UG on line 4. The conditions are met for the use of UG on this line since no assumptions have been made, about **a** or anything else, and **a** does not occur in either **P** or **Q**. And the condition for the use of Ł$_3$∀5 on line 7 is met since the condition on Ł$_3$∀D13 stipulates that **x** does not occur free in **P**.

We can also derive a rule that introduces a universal quantifier in the consequent of a conditional:

UGC (Universal Generalization in the Consequent). From **P → Q(a/x)** infer **P → (∀x)Q**
where **x** is any individual variable, provided that no assumption contains the constant **a** and that **P** itself does not contain the constant **a**.

Justification: Let **P → Q(a/x)** be a formula in a derivation that meets the stated conditions: that is, no assumptions in the derivation contain the constant **a**, and the subformula **P** also doesn't contain the constant **a**. Then we may continue the derivation as follows:

n	**P → Q(a/x)**	
n+1	**(∀x)(P → Q)**	*n*, UG
n+2	**(∀x)(P → Q) → (P → (∀x)Q)**	Ł$_3$∀5, with **(∀x)(P → Q)** / **(∀x)(P → Q)**
n+3	**P → (∀x)Q**	*n*+1, *n*+2 MP

We must be sure that we have met the conditions for the rules used in this derivation sequence. First, by hypothesis no assumptions contain the constant **a**, and by hypothesis **P** does not contain the constant **a**, so the conditions for the use of UG to obtain the formula on line *n*+1 have been met. Similarly, because by hypothesis **P** does not contain the constant **a**, the condition for using Ł$_3$∀5 to generate the formula on line *n*+2 has also been met.

We noted in Chapter 7 that the formula *(∀x)(Tx → Tx)* is a tautology in Ł$_3$∀. Given the soundness and completeness of our current axiomatization, we would therefore expect it to be a theorem and indeed it is:

1	Ta → Ta	Ł$_3$∀D4, with Ta / **P**
2	(∀x)(Tx → Tx)	1, UG

On the other hand the formula *(∀x)(Tx ∨ ¬Tx)*, which is not a tautology in Ł$_3$∀, isn't a theorem of Ł$_3$∀A.

The Sorites argument

Ts_1
Es_2s_1
Es_3s_2
Es_4s_3
. . .
$Es_{193}s_{192}$
(∀x) (∀y) ((Tx ∧ Eyx) → Ty)
―――――――――――
Ts_{193}

is valid in Ł$_3\forall$A, and therefore its conclusion is derivable from the premises. We'll produce a derivation with the aid of a new derived rule (one that is also derivable in Ł$_3$A since it doesn't use the quantificational axioms or rules):

CI (Conjunction Introduction). From **P** and **Q**, infer $\mathbf{P} \wedge \mathbf{Q}$

Justification: We will derive the formula $\neg((\neg\mathbf{P} \rightarrow \neg\mathbf{Q}) \rightarrow \neg\mathbf{Q})$, in which conjunction has been rewritten using negation and the conditional, from **P** and **Q**:

1	$\mathbf{P}$	Assumption
2	$\mathbf{Q}$	Assumption
3	$\neg\neg\mathbf{P}$	1, DN
4	$\neg\neg\mathbf{P} \rightarrow (\neg\mathbf{P} \rightarrow \neg\mathbf{Q})$	Ł$_3\forall$D1, with $\neg\mathbf{P}$ / $\mathbf{P}$, $\neg\mathbf{Q}$ / $\mathbf{Q}$
5	$\neg\mathbf{P} \rightarrow \neg\mathbf{Q}$	3,4 MP
6	$(\neg\mathbf{P} \rightarrow \neg\mathbf{Q}) \rightarrow (((\neg\mathbf{P} \rightarrow \neg\mathbf{Q}) \rightarrow \neg\mathbf{Q}) \rightarrow \neg\mathbf{Q})$	Ł$_3\forall$D6, with $\neg\mathbf{P} \rightarrow \neg\mathbf{Q}$ / $\mathbf{P}$, $\neg\mathbf{Q}$ / $\mathbf{Q}$
7	$((\neg\mathbf{P} \rightarrow \neg\mathbf{Q}) \rightarrow \neg\mathbf{Q}) \rightarrow \neg\mathbf{Q}$	5,6 MP
8	$\neg\neg\mathbf{Q} \rightarrow \neg((\neg\mathbf{P} \rightarrow \neg\mathbf{Q}) \rightarrow \neg\mathbf{Q})$	7, GCON
9	$\neg\neg\mathbf{Q}$	2, DN
10	$\neg((\neg\mathbf{P} \rightarrow \neg\mathbf{Q}) \rightarrow \neg\mathbf{Q})$	8,9 MP

Here is the derivation for the Sorites argument:

1	Ts_1	Assumption
2	Es_2s_1	Assumption
3	Es_3s_2	Assumption
4	Es_4s_3	Assumption
...	...	
193	$Es_{193}s_{192}$	Assumption
194	$(\forall x)\,(\forall y)((Tx \wedge Eyx) \rightarrow Ty)$	Assumption
195	$(\forall x)\,(\forall y)\,((Tx \wedge Eyx) \rightarrow Ty) \rightarrow (\forall y)\,((Ts_1 \wedge Eys_1) \rightarrow Ty)$	Ł$_3\forall$6, with $(\forall x)\,(\forall y)\,((Tx \wedge Eyx) \rightarrow Ty)$ / $(\forall \mathbf{x})\mathbf{P}$, s_1 / **a**
196	$(\forall y)\,((Ts_1 \wedge Eys_1) \rightarrow Ty)$	194,195 MP
197	$(\forall y)\,((Ts_1 \wedge Eys_1) \rightarrow Ty) \rightarrow ((Ts_1 \wedge Es_2s_1) \rightarrow Ts_2)$	Ł$_3\forall$6, with $(\forall y)\,((Ts_1 \wedge Eys_1) \rightarrow Ty)$ / $(\forall \mathbf{x})\mathbf{P}$, s_2 / **a**
198	$(Ts_1 \wedge Es_2s_1) \rightarrow Ts_2$	196,197 MP
199	$Ts_1 \wedge Es_2s_1$	1,2 CI
200	Ts_2	198,199 MP
201	$(\forall x)\,(\forall y)\,(Tx \wedge Eyx) \rightarrow Ty) \rightarrow (\forall y)\,((Ts_2 \wedge Eys_2) \rightarrow Ty)$	Ł$_3\forall$6, with $(\forall x)\,(\forall y)\,((Tx \wedge Eyx) \rightarrow Ty)$ / $(\forall \mathbf{x})\mathbf{P}$, s_2 /**a**
202	$(\forall y)\,((Ts_2 \wedge Eys_2) \rightarrow Ty)$	194,201 MP
203	$(\forall y)\,((Ts_2 \wedge Eys_2) \rightarrow Ty) \rightarrow ((Ts_2 \wedge Es_3s_2) \rightarrow Ts_3)$	Ł$_3\forall$6, with $(\forall y)\,((Ts_2 \wedge Eys_2) \rightarrow Ty)$ / $(\forall \mathbf{x})\mathbf{P}$, s_3 / **a**
204	$(Ts_2 \wedge Es_3s_2) \rightarrow Ts_3$	202,203 MP
205	$Ts_2 \wedge Es_3s_2$	3,200 CI
206	Ts_3	204,205 MP
...	... *{repeating 195–200 with appropriate substitutions we end with}*	
1346	Ts_{193}	1344,1345 MP

Just as we symbolized formulas from Kleene's and Bochvar's three-valued propositional logics in Ł$_3$ and then used the axiomatic system Ł$_3$A to derive those formulas, we can do the same for the first-order generalizations of those systems. We'll illustrate with Kleene's system. Since no formulas of Kleene's system are tautologies, no translations of these into Ł$_3\forall$ will be theorems of Ł$_3\forall$A. But the translations of arguments that are valid in Kleene's system—and there are such arguments—will be deductively valid in Ł$_3\forall$A. Recalling that Kleene's conditional $\mathbf{P} \rightarrow_K \mathbf{Q}$ can be expressed as $\neg\mathbf{P} \vee \mathbf{Q}$ in Ł$_3\forall$, and that $\neg\mathbf{P} \vee \mathbf{Q}$ is expressible as $(\neg\mathbf{P} \rightarrow \mathbf{Q}) \rightarrow \mathbf{Q}$ using only negation and disjunction in Ł$_3\forall$, Kleene's Sorites argument is expressed in Ł$_3\forall$ as

Ts_1
Es_2s_1
Es_3s_2
Es_4s_3
. . .
$Es_{193}s_{192}$
$\underline{(\forall x)\ (\forall y)\ ((\neg(Tx \wedge Eyx) \rightarrow Ty) \rightarrow Ty)}$
Ts_{193}

Recall that Kleene's quantifiers are semantically identical to Łukasiewicz's. The conclusion of the Kleene version of the Sorites argument is derivable from the premises in Ł$_3\forall$A:

1	Ts_1	Assumption
2	Es_2s_1	Assumption
3	Es_3s_2	Assumption
4	Es_4s_3	Assumption
. . .	. . .	
193	$Es_{193}s_{192}$	Assumption
194	$(\forall x)\ (\forall y)\ ((\neg(Tx \wedge Eyx) \rightarrow Ty) \rightarrow Ty)$	Assumption
195	$(\forall x)\ (\forall y)\ ((\neg(Tx \wedge Eyx) \rightarrow Ty) \rightarrow Ty) \rightarrow$ $(\forall y)\ ((\neg(Ts_1 \wedge Eys_1) \rightarrow Ty) \rightarrow Ty)$	Ł$_3\forall$6, with $(\forall x)\ (\forall y)\ ((\neg(Tx \wedge Eyx) \rightarrow Ty) \rightarrow Ty)$ / $(\forall x)\mathbf{P}$, s_1 / $\mathbf{a}$
196	$(\forall y)\ ((\neg(Ts_1 \wedge Eys_1) \rightarrow Ty) \rightarrow Ty)$	194,195 MP
197	$(\forall y)\ ((\neg(Ts_1 \wedge Eys_1) \rightarrow Ty) \rightarrow Ty) \rightarrow$ $((\neg(Ts_1 \wedge Es_2s_1) \rightarrow Ts_2) \rightarrow Ts_2)$	Ł$_3\forall$6, with $(\forall y)\ ((\neg(Ts_1 \wedge Eys_1) \rightarrow Ty) \rightarrow Ty)$ / $(\forall x)\mathbf{P}$, s_2 / $\mathbf{a}$
198	$(\neg(Ts_1 \wedge Es_2s_1) \rightarrow Ts_2) \rightarrow Ts_2$	196,197 MP
199	$Ts_1 \wedge Es_2s_1$	1,2 CI
200	$(Ts_1 \wedge Es_2s_1) \rightarrow (\neg(Ts_1 \wedge Es_2s_1) \rightarrow Ts_2)$	Ł$_3\forall$D1, with $Ts_1 \wedge Es_2s_1$ / $\mathbf{P}$, Ts_2 / $\mathbf{Q}$
201	$\neg(Ts_1 \wedge Es_2s_1) \rightarrow Ts_2$	199,200 MP
202	Ts_2	198,201 MP
203	$(\forall x)\ (\forall y)\ ((\neg(Tx \wedge Eyx) \rightarrow Ty) \rightarrow Ty) \rightarrow$ $(\forall y)\ ((\neg(Ts_2 \wedge Eys_2) \rightarrow Ty) \rightarrow Ty)$	Ł$_3\forall$6, with $(\forall x)\ (\forall y)\ ((\neg(Tx \wedge Eyx) \rightarrow Ty) \rightarrow Ty)$ / $(\forall x)\mathbf{P}$, s_2 / $\mathbf{a}$
204	$(\forall y)\ ((\neg(Ts_2 \wedge Eys_2) \rightarrow Ty) \rightarrow Ty)$	194,203 MP
205	$(\forall y)\ ((\neg(Ts_2 \wedge Eys_2) \rightarrow Ty) \rightarrow Ty) \rightarrow$ $((\neg(Ts_2 \wedge Es_3s_2) \rightarrow Ts_3) \rightarrow Ts_3)$	Ł$_3\forall$6, with $(\forall y)\ ((\neg(Ts_2 \wedge Eys_2) \rightarrow Ty) \rightarrow Ty)$ / $(\forall x)\mathbf{P}$, s_3 / $\mathbf{a}$
206	$(\neg(Ts_2 \wedge Es_3s_2) \rightarrow Ts_3) \rightarrow Ts_3$	204,205 MP
207	$Ts_2 \wedge Es_3s_2$	3,202 CI
208	$(Ts_2 \wedge Es_3s_2) \rightarrow (\neg(Ts_2 \wedge Es_3s_2) \rightarrow Ts_3)$	Ł$_3\forall$D1, with $Ts_2 \wedge Es_3s_2$ / $\mathbf{P}$, Ts_3 / $\mathbf{Q}$
209	$\neg(Ts_2 \wedge Es_3s_2) \rightarrow Ts_3$	207,208 MP
210	Ts_3	206,209 MP
. . .	. . .*{repeating 195–202 with appropriate substitutions we end with}*	
1730	Ts_{193}	1726,1729 MP

Before closing this section we'd like to introduce a modified version of SUB. This is because although we've been using rules like GCON and DN that were justified with SUB in Chapter 6 we've been careful to use them in a restricted way and the restriction turns out to be eliminable. Consider the statement of DN in Chapter 6:

> **DN (Double Negation).** From any formula **R** that contains **P** as a constituent, infer any formula **R*** that is the result of replacing one or more occurrences of **P** in **R** with $\neg\neg$**P**, and vice versa.

We used this rule, for example, to derive $\neg\neg(\forall x)\neg Px \rightarrow \neg Pa$ from $(\forall x)\neg Px \rightarrow \neg Pa$ in the first derivation in this chapter. It looks as if DN should also allow us, for example, to derive $(\forall x)\neg\neg Px$ from $(\forall x)Px$, and vice versa. But there is, as we shall explain, a problem: in this case we would be replacing the *open* (nonclosed) subformula $\neg\neg Px$ with the *open* subformula Px—whereas in the earlier derivation we replace the *closed* subformula $\neg\neg(\forall x)\neg Px$ with the *closed* subformula $(\forall x)\neg Px$. Replacing open subformulas is a problem because the justification for DN in Chapter 6, which uses SUB, is based on the fact that both $\mathbf{P} \rightarrow \neg\neg\mathbf{P}$ and $\neg\neg\mathbf{P} \rightarrow \mathbf{P}$ are derived axiom schemata. Now, $(\forall x)\neg Px \rightarrow \neg\neg(\forall x)\neg Px$ and $\neg\neg(\forall x)\neg Px \rightarrow (\forall x)\neg Px$ are indeed derivable in Ł$_3\forall$A just as they were in Ł$_3$A, so we were justified in using DN in the earlier derivation. But for DN *as proved in Chapter* 6 also to license replacing $(\forall x)\neg\neg Px$ with $(\forall x)Px$, and vice versa, both $Px \rightarrow \neg\neg Px$ and $\neg\neg Px \rightarrow Px$ would have to be derivable in Ł$_3\forall$A. There's the problem: neither is derivable because they're not *closed* formulas, and derivations, as we've defined them, consist entirely of closed formulas.

So we introduce a new version of SUB for the cases where we want to substitute open formulas one for another. First, a definition: we define the ***universal closure*** of a formula **P** to be **P** itself if **P** is closed, and, in the case that **P** is open, to be the closed formula that results by enclosing **P** in parentheses and prefixing, in alphabetical order, a universal quantifier for each of the free variables in **P**. (The alphabetical order stipulation allows us to talk of *the* universal closure of a formula rather than merely *a* universal closure.) The universal closure of the formula $(Pa \rightarrow (\forall x)Rxz) \rightarrow (\neg Px \rightarrow Qy)$, for example, is $(\forall x)(\forall y)(\forall z)((Pa \rightarrow (\forall x)Rxz) \rightarrow (\neg Px \rightarrow Qy))$. The new version of SUB for Ł$_3\forall$A is:

> **SUB (Substitution).** From the universal closure of $\mathbf{P} \rightarrow \mathbf{Q}$, the universal closure of $\mathbf{Q} \rightarrow \mathbf{P}$, and a formula **R** that contains **P** as a subformula, infer any formula **R*** that is the result of replacing one or more occurrences of **P** in **R** with **Q**.

The version of SUB derived in Chapter 6 is a special case of this new version, since every closed formula is its own universal closure. The reader will be asked in the exercise to justify the revised SUB rule.

With this new version of SUB it's easy to prove that DN can apply in cases where a double negation is prefixed to or removed from an open subformula. For let $\mathbf{x}_1, \ldots, \mathbf{x}_n$ be the free variables in the open formula **P** that we want to replace (with $\neg\neg$**P**—or vice versa), and let $\mathbf{a}_1, \ldots, \mathbf{a}_n$ be constants that don't occur in **P** or in the derivation in which **P** occurs. We know that both $\mathbf{P(a_1/x_1)\ldots(a_n/x_n)} \rightarrow \neg\neg\mathbf{P(a_1/x_1)\ldots(a_n/x_n)}$ and $\neg\neg\mathbf{P(a_1/x_1)\ldots(a_n/x_n)} \rightarrow \mathbf{P(a_1/x_1)\ldots(a_n/x_n)}$ are derivable in Ł$_3\forall$3—they're instances of Ł$_3\forall$D3 and Ł$_3\forall$D4. Because the chosen constants $\mathbf{a_1,\ldots, a_n}$ don't occur in **P** or earlier in the derivation (and therefore not in any assumptions), UG can be applied to the conditionals $\mathbf{P(a_1/x_1)\ldots(a_n/x_n)} \rightarrow \neg\neg\mathbf{P(a_1/x_1)\ldots(a_n/x_n)}$ and $\neg\neg\mathbf{P(a_1/x_1)\ldots(a_n/x_n)} \rightarrow \mathbf{P(a_1/x_1)\ldots(a_n/x_n)}$ to obtain the universal closures $(\forall\mathbf{x_1})\ldots(\forall\mathbf{x_n})(\mathbf{P} \rightarrow \neg\neg\mathbf{P})$ and $(\forall\mathbf{x_1})\ldots(\forall\mathbf{x_n})(\neg\neg\mathbf{P} \rightarrow \mathbf{P})$. These universal closures can then be used by the new SUB to replace **P** in any formula with $\neg\neg\mathbf{P}$, and vice versa, as in

1	$(\forall x)(Px \rightarrow Qx)$	Assumption
2	$Pa \rightarrow \neg\neg Pa$	Ł$_3\forall$D3, with Pa / **P**
3	$(\forall x)(Px \rightarrow \neg\neg Px)$	2, UG
4	$\neg\neg Pa \rightarrow Pa$	Ł$_3\forall$D4, with Pa / **P**
5	$(\forall x)(\neg\neg Px \rightarrow Px)$	4, UG
6	$(\forall x)(\neg\neg Px \rightarrow Qx)$	1,3,5 SUB

Similar observations justify the use of TRAN and GCON to replace open as well as closed subformulas.

8.2 A Pavelka-Style Derivation System for Ł$_3\forall$

We extend the Pavelka-style system Ł$_3$PA to a first-order system Ł$_3\forall$PA with the first-order axiomatic system Ł$_3\forall$A (augmented with annotations) as a basis and adding the Pavelka axioms and rules from Ł$_3$PA:

Ł$_3\forall$P1. $[\mathbf{P} \rightarrow (\mathbf{Q} \rightarrow \mathbf{P}), \mathbf{T}]$
Ł$_3\forall$P2. $[(\mathbf{P} \rightarrow \mathbf{Q}) \rightarrow ((\mathbf{Q} \rightarrow \mathbf{R}) \rightarrow (\mathbf{P} \rightarrow \mathbf{R})), \mathbf{T}]$
Ł$_3\forall$P3. $[(\neg\mathbf{P} \rightarrow \neg\mathbf{Q}) \rightarrow (\mathbf{Q} \rightarrow \mathbf{P}), \mathbf{T}]$
Ł$_3\forall$P4. $[((\mathbf{P} \rightarrow \neg\mathbf{P}) \rightarrow \mathbf{P}) \rightarrow \mathbf{P}, \mathbf{T}]$
Ł$_3\forall$P5. $[(\forall\mathbf{x})(\mathbf{P} \rightarrow \mathbf{Q}) \rightarrow (\mathbf{P} \rightarrow (\forall\mathbf{x})\mathbf{Q}), \mathbf{T}]$
where **P** is a formula in which **x** does not occur free
Ł$_3\forall$P6. $[(\forall\mathbf{x})\mathbf{P} \rightarrow \mathbf{P(a/x)}, \mathbf{T}]$
where **a** is any individual constant and the expression **P(a/x)** means: *the result of substituting the constant* ***a*** *for the variable* ***x*** *wherever* ***x*** *occurs free in* ***P***
Ł$_3\forall$P7.1.1. $[(\mathbf{t} \rightarrow \mathbf{t}) \rightarrow \mathbf{t}, \mathbf{T}]$
Ł$_3\forall$P7.1.2. $[\mathbf{t} \rightarrow (\mathbf{t} \rightarrow \mathbf{t}), \mathbf{T}]$
Ł$_3\forall$P7.2.1. $[(\mathbf{t} \rightarrow \mathbf{n}) \rightarrow \mathbf{n}, \mathbf{T}]$
Ł$_3\forall$P7.2.2. $[\mathbf{n} \rightarrow (\mathbf{t} \rightarrow \mathbf{n}), \mathbf{T}]$
Ł$_3\forall$P7.3.1. $[(\mathbf{t} \rightarrow \mathbf{f}) \rightarrow \mathbf{f}, \mathbf{T}]$
Ł$_3\forall$P7.3.2. $[\mathbf{f} \rightarrow (\mathbf{t} \rightarrow \mathbf{f}), \mathbf{T}]$

Ł₃∀P7.4.1. [(n → t) → t, T]
Ł₃∀P7.4.2. [t → (n → t), T]
Ł₃∀P7.5.1. [(n → n) → t, T]
Ł₃∀P7.5.2. [t → (n → n), T]
Ł₃∀P7.6.1. [(n → f) → n, T]
Ł₃∀P7.6.2. [n → (n → f), T]
Ł₃∀P7.7.1. [(f → t) → t, T]
Ł₃∀P7.7.2. [t → (f → t), T]
Ł₃∀P7.8.1. [(f → n) → t, T]
Ł₃∀P7.8.2. [t → (f → n), T]
Ł₃∀P7.9.1. [(f → f) → t, T]
Ł₃∀P7.9.2. [t → (f → f), T]
Ł₃∀P8.1.1. [¬t → f, T]
Ł₃∀P8.1.2. [f → ¬t, T]
Ł₃∀P8.2.1. [¬n → n, T]
Ł₃∀P8.2.2. [n → ¬n, T]
Ł₃∀P8.3.1. [¬f → t, T]
Ł₃∀P8.3.2. [t → ¬f, T]
Ł₃∀P9.1. [t, T]
Ł₃∀P9.2. [n, N]
Ł₃∀P9.3. [f, F]

MP. From [**P**, v_1] and [**P** → **Q**, v_2], infer [**Q**, v_3], where v_3 is defined in terms of v_1 and v_2 as specified in the following table:

v_1	v_2	v_3
T	T	T
T	N	N
T	F	F
N	T	N
N	N	F
N	F	F
F	T	F
F	N	F
F	F	F

UG (Universal Generalization). From[**P(a/x)**, v] infer [(∀**x**)**P**, v] where **x** is any individual variable, provided that no assumption contains the constant **a** and that **P** does not contain the constant **a**

TCI (Truth-value Constant Introduction). From [**P**, v] infer [**v** → **P**, **T**] where **v** is the constant name for the value v.

All of the quantificational axiom schemata have been graded with the value **T**. The rationale for the graded values in the UG inference rule is this: if **P(a/x)** has at least the value v, where **a** is an *arbitrary* constant (not occuring in an assumption), this tells us that the value of **P** must be at least v for *any* value of **x**. If that is the case, we can infer that the value of the universal quantification must also be at least v, since the truth-value of the quantified formula is semantically defined to be the least value that **P** can have for any value of **x**.

We'll illustrate the system with some examples. The formula $(\forall x)(Tx \rightarrow Tx)$ is a tautology in Ł$_3\forall$ and so, as we would expect, the graded version of this formula is derivable with the value **T** (as in Chapter 6, we help ourselves to derived axioms with the graded value **T**):

1	[Ta $\rightarrow$ Ta, **T**]	Ł$_3\forall$PD4, with Ta / **P**
2	[$(\forall$x)(Tx $\rightarrow$ Tx), **T**]	1, UG

The Law of Excluded Middle $(\forall x)(Tx \vee \neg Tx)$ is not a tautology in Ł$_3\forall$; however, like its propositional counterpart $A \vee \neg A$, it always has at least the value **N**. We can derive $(\forall x)(Tx \vee \neg Tx)$ with graded value **N** (and with disjunction rewritten in terms of negation and the conditional) as follows:

1	[$\neg$Ta $\rightarrow$ ((Ta $\rightarrow$ $\neg$Ta) $\rightarrow$ $\neg$Ta), **T**]	Ł$_3\forall$P1, with $\neg$Ta / **P**, Ta $\rightarrow$ $\neg$Ta / **Q**
2	[(**n** $\rightarrow$ $\neg$Ta) $\rightarrow$ (($\neg$Ta $\rightarrow$ ((Ta $\rightarrow$ $\neg$Ta) $\rightarrow$ $\neg$Ta)) $\rightarrow$ (**n** $\rightarrow$ ((Ta $\rightarrow$ $\neg$Ta) $\rightarrow$ $\neg$Ta))), **T**]	Ł$_3\forall$P2, with **n** / **P**, $\neg$Ta / **Q**, (Ta $\rightarrow$ $\neg$Ta) $\rightarrow$ $\neg$Ta / **R**
3	[(**n** $\rightarrow$ $\neg$Ta) $\rightarrow$ (**n** $\rightarrow$ ((Ta $\rightarrow$ $\neg$Ta) $\rightarrow$ $\neg$Ta)), **T**]	1,2 GMP
4	[Ta $\rightarrow$ ((Ta $\rightarrow$ $\neg$Ta) $\rightarrow$ $\neg$Ta), **T**]	Ł$_3\forall$PD6, with Ta / **P**, $\neg$Ta / **Q**
5	[(**n** $\rightarrow$ Ta) $\rightarrow$ ((Ta $\rightarrow$ ((Ta $\rightarrow$ $\neg$Ta) $\rightarrow$ $\neg$Ta)) $\rightarrow$ (**n** $\rightarrow$ ((Ta $\rightarrow$ $\neg$Ta) $\rightarrow$ $\neg$Ta))), **T**]	Ł$_3\forall$P2, with **n** / **P**, Ta / **Q**, (Ta $\rightarrow$ $\neg$Ta) $\rightarrow$ $\neg$Ta / **R**
6	[(**n** $\rightarrow$ Ta) $\rightarrow$ (**n** $\rightarrow$ ((Ta $\rightarrow$ $\neg$Ta) $\rightarrow$ $\neg$Ta)), **T**]	4,5 GMP
7	[(**n** $\rightarrow$ Ta) $\vee$ (Ta $\rightarrow$ **n**), **T**]	Ł$_3\forall$PD10, with **n** / **P**, Ta / **Q**
8	[(**n** $\rightarrow$ Ta) $\vee$ ($\neg$**n** $\rightarrow$ $\neg$Ta), **T**]	7, GCON
9	[$\neg$**n** $\rightarrow$ **n**, **T**]	Ł$_3\forall$P6.2.1
10	[**n** $\rightarrow$ $\neg$**n**, **T**]	Ł$_3\forall$P6.2.2
11	[(**n** $\rightarrow$ Ta) $\vee$ (**n** $\rightarrow$ $\neg$Ta), **T**]	8,9,10 SUB
12	[**n** $\rightarrow$ ((Ta $\rightarrow$ $\neg$Ta) $\rightarrow$ $\neg$Ta), **T**]	3,6,11 DS
13	[**n**, **N**]	Ł$_3\forall$P9.2
14	[(Ta $\rightarrow$ $\neg$Ta) $\rightarrow$ $\neg$Ta, **N**]	12,13 MP
15	[$(\forall$x)((Tx $\rightarrow$ $\neg$Tx) $\rightarrow$ $\neg$Tx), **N**]	14, UG

In Chapter 7 we showed that the formula $(\exists x)(\neg Tx \wedge Tx)$, figuring in Black's Problem of the Fringe, can have the value **N** in Ł$_3\forall$—it is a quasi-contradiction. The formula's negation $\neg(\exists x)\ (\neg Tx \wedge Tx)$ is a quasi-tautology. In Ł$_3\forall$PA we can produce a derivation establishing that this latter formula always has at least the value **N**. Using only negation, the conditional, and the universal quantifier we rewrite the formula as $\neg\neg(\forall x)\neg\neg((\neg\neg Tx \rightarrow \neg Tx) \rightarrow \neg Tx)$. Here we have replaced the existential quantifier $(\exists x)$ with $\neg(\forall x)\neg$, and if we replace the conjunction $\neg Tx \wedge Tx$ with $\neg(\neg\neg Tx \vee \neg Tx)$, this latter formula becomes $\neg((\neg\neg Tx \rightarrow \neg Tx) \rightarrow \neg Tx)$ when we rewrite the disjunction. We begin with the previous derivation of the Ław of Excluded Middle and add the following line:

16	$[\neg\neg(\forall x)\neg\neg((\neg\neg Tx \rightarrow \neg Tx) \rightarrow \neg Tx), \mathbf{N}]$	17, DN (three times)

Our next derivation establishes that the formula $(\forall x)(Px \rightarrow Qx) \rightarrow ((\mathbf{n} \rightarrow (\exists x)Px) \rightarrow (\mathbf{n} \rightarrow (\exists x)Qx))$ always has the value **T**. We rewrite this formula as $(\forall x)(Px \rightarrow Qx) \rightarrow ((\mathbf{n} \rightarrow \neg(\forall x)\neg Px) \rightarrow (\mathbf{n} \rightarrow \neg(\forall x)\neg Qx))$:

1	$[(\forall x)(Px \rightarrow Qx) \rightarrow (\forall x)(Px \rightarrow Qx), \mathbf{T}]$	$Ł_3\forall$GD4, with $(\forall x)(Px \rightarrow Qx)$ / **P**
2	$[(\forall x)(Px \rightarrow Qx) \rightarrow (\forall x)(\neg Qx \rightarrow \neg Px), \mathbf{T}]$	1,GCON
3	$[(\forall x)(\neg Qx \rightarrow \neg Px) \rightarrow ((\forall x)\neg Qx \rightarrow (\forall x)\neg Px), \mathbf{T}]$	$Ł_3\forall$PD12, with $(\forall x)(\neg Qx \rightarrow \neg Px)$ / $(\forall \mathbf{x})(\mathbf{P} \rightarrow \mathbf{Q})$
4	$[(\forall x)(Px \rightarrow Qx) \rightarrow ((\forall x)\neg Qx \rightarrow (\forall x)\neg Px), \mathbf{T}]$	2,3 HS
5	$[(\forall x)(Px \rightarrow Qx) \rightarrow (\neg(\forall x)\neg Px \rightarrow \neg(\forall x)\neg Qx), \mathbf{T}]$	4, GCON
6	$[(\mathbf{n} \rightarrow \neg(\forall x)\neg Px) \rightarrow ((\neg(\forall x)\neg Px \rightarrow \neg(\forall x)\neg Qx) \rightarrow (\mathbf{n} \rightarrow \neg(\forall x)\neg Qx)), \mathbf{T}]$	$Ł_3\forall$P2, with **n** / **P**, $\neg(\forall x)\neg Px$ / **Q**, $\neg(\forall x)\neg Qx$ / **R**
7	$[(\neg(\forall x)\neg Px \rightarrow \neg(\forall x)\neg Qx) \rightarrow ((\mathbf{n} \rightarrow \neg(\forall x)\neg Px) \rightarrow (\mathbf{n} \rightarrow \neg(\forall x)\neg Qx)), \mathbf{T}]$	6, TRAN
8	$[((\forall x)(Px \rightarrow Qx) \rightarrow ((\mathbf{n} \rightarrow \neg(\forall x)\neg Px) \rightarrow (\mathbf{n} \rightarrow \neg(\forall x)\neg Qx)), \mathbf{T}]$	5,7 HS

Finally, we showed in Chapter 7 that although the Sorites argument is valid in $Ł_3\forall$, it is *not quasi-valid* in $Ł_3\forall$. So (given the soundness of our system) we cannot produce a derivation that has the Principle of Charity premise graded with the value **N**, the other premises graded with the value **T**, and the conclusion graded with either the value **T** or the value **N**. For example, if we assign the value **N** to the Principle of Charity premise and the value **T** to all of the other premises, the first Sorites derivation in Section 8.1 ends up with the value **F** for the conclusion.

To show this we need a graded version of Conjunction Introduction. If we simply add grades to the derivation of CI in Section 8.1, we'll end up with a weaker grade for the conjunction than is possible; so we'll justify CI in a different way here to give us the strongest grade for the inferred conjunction. (Recall that we had to do something similar to justify the graded version of Modified Constructive Dilemma in Section 6.2 of Chapter 6.)

CI (Conjunction Introduction). From $[\mathbf{P}, v_1]$ and $[\mathbf{Q}, v_2]$, infer $[\mathbf{P} \wedge \mathbf{Q}, \min(v_1, v_2)]$.

Justification: We rewrite $\mathbf{P} \wedge \mathbf{Q}$ as $\neg((\neg\mathbf{P} \rightarrow \neg\mathbf{Q}) \rightarrow \neg\mathbf{Q})$. Note first that either $v_1 \leq v_2$ or $v_2 < v_1$. In the case where $v_1 \leq v_2$, we begin the derivation with

1	$[\mathbf{P}, v_1]$	Assumption
2	$[\mathbf{Q}, v_2]$	Assumption
3	$[\mathbf{v_1} \to \mathbf{P}, \mathbf{T}]$	1, TCI
4	$[\mathbf{v_2} \to \mathbf{Q}, \mathbf{T}]$	2, TCI
5	$[\mathbf{t} \to (\mathbf{v_1} \to \mathbf{v_2}), \mathbf{T}]$	Ł$_3$∀P7.x.2
	{This is an instance of Ł$_3$∀P7.x.2 because of the assumption in this case that $v1 \leq v2$}	
6	$[\mathbf{t}, \mathbf{T}]$	Ł$_3$∀P9.1
7	$[\mathbf{v_1} \to \mathbf{v_2}, \mathbf{T}]$	5,6 MP
8	$[\mathbf{v_1} \to \mathbf{Q}, \mathbf{T}]$	4,7 HS

where $\mathbf{v_1}$ stands for the atomic formula that denotes the value v_1 and $\mathbf{v_2}$ stands for the atomic formula that denotes the value v_2. Because $v_1 \leq v_2$ in this case, $v_1 = \min(v_1, v_2)$ and so we can rewrite the preceding as

1	$[\mathbf{P}, v_1]$	Assumption
2	$[\mathbf{Q}, v_2]$	Assumption
3	$[\mathbf{min}\ (\mathbf{v_1}, \mathbf{v_2}) \to \mathbf{P}, \mathbf{T}]$	1, TCI
4	$[\mathbf{v_2} \to \mathbf{Q}, \mathbf{T}]$	2, TCI
5	$[\mathbf{t} \to (\mathbf{min}\ (\mathbf{v_1}, \mathbf{v_2}) \to \mathbf{v_2}), \mathbf{T}]$	Ł$_3$∀P7.x.2
6	$[\mathbf{t}, \mathbf{T}]$	Ł$_3$∀P9.1
7	$[\mathbf{min}\ (\mathbf{v_1}, \mathbf{v_2}) \to \mathbf{v_2}, \mathbf{T}]$	5,6 MP
8	$[\mathbf{min}\ (\mathbf{v_1}, \mathbf{v_2}) \to \mathbf{Q}, \mathbf{T}]$	4,7 HS

If $v_2 < v_1$ we begin the derivation with

1	$[\mathbf{P}, v_1]$	Assumption
2	$[\mathbf{Q}, v_2]$	Assumption
3	$[\mathbf{min}\ (\mathbf{v_1}, \mathbf{v_2}) \to \mathbf{Q}, \mathbf{T}]$	2, TCI
4	$[\mathbf{v_1} \to \mathbf{P}, \mathbf{T}]$	1, TCI
5	$[\mathbf{t} \to (\mathbf{min}\ (\mathbf{v_1}, \mathbf{v_2}) \to \mathbf{v_1}), \mathbf{T}]$	Ł$_3$∀P7.x.2
6	$[\mathbf{t}, \mathbf{T}]$	Ł$_3$∀P9.1
7	$[\mathbf{min}\ (\mathbf{v_1}, \mathbf{v_2}) \to \mathbf{v_1}, \mathbf{T}]$	5,6 MP
8	$[\mathbf{min}\ (\mathbf{v_1}, \mathbf{v_2}) \to \mathbf{P}, \mathbf{T}]$	4,7 HS

In either case the derivation continues as follows (with $\{x, y\}$ meaning line x if we began the first way, and line y if we began the second way):

9	$[\neg\mathbf{P} \to \neg\mathbf{min}\ (\mathbf{v_1}, \mathbf{v_2}), \mathbf{T}]$	$\{3, 8\}$ GCON
10	$[\neg\mathbf{Q} \to \neg\mathbf{min}\ (\mathbf{v_1}, \mathbf{v_2}), \mathbf{T}]$	$\{8, 3\}$ GCON
11	$[((\neg\mathbf{P} \to \neg\mathbf{Q}) \to \neg\mathbf{Q}) \to \neg\mathbf{min}\ (\mathbf{v_1}, \mathbf{v_2}), \mathbf{T}]$	9,10 DC
12	$[\neg\neg((\neg\mathbf{P} \to \neg\mathbf{Q}) \to \neg\mathbf{Q}) \to \neg\mathbf{min}\ (\mathbf{v_1}, \mathbf{v_2}), \mathbf{T}]$	11, DN
13	$[\mathbf{min}\ (\mathbf{v_1}, \mathbf{v_2}) \to \neg((\neg\mathbf{P} \to \neg\mathbf{Q}) \to \neg\mathbf{Q}), \mathbf{T}]$	12, GCON
14	$[\mathbf{min}\ (\mathbf{v_1}, \mathbf{v_2}), \min(v_1, v_2)]$	Ł$_3$∀P9.x
15	$[\neg((\neg\mathbf{P} \to \neg\mathbf{Q}) \to \neg\mathbf{Q}), \min(v_1, v_2)]$	13,14 MP

Here's the graded Sorites derivation:

1	$[Ts_1, \mathbf{T}]$	Assumption
2	$[Es_2s_1, \mathbf{T}]$	Assumption
3	$[Es_3s_2, \mathbf{T}]$	Assumption
4	$[Es_4s_3, \mathbf{T}]$	Assumption
...	...	
193	$[Es_{193}s_{192}, \mathbf{T}]$	Assumption
194	$[(\forall x)\ (\forall y)((Tx \wedge Eyx) \rightarrow Ty), \mathbf{N}]$	Assumption
195	$[(\forall x)\ (\forall y)\ ((Tx \wedge Eyx) \rightarrow Ty) \rightarrow$	Ł$_3$∀P6, with $(\forall x)\ (\forall y)\ ((Tx \wedge Eyx) \rightarrow Ty)$ /
	$(\forall y)\ ((Ts_1 \wedge Eys_1) \rightarrow Ty), \mathbf{T}]$	$(\forall \mathbf{x})\mathbf{P}$, s_1 / $\mathbf{a}$
196	$[(\forall y)\ ((Ts_1 \wedge Eys_1) \rightarrow Ty), \mathbf{N}]$	194,195 MP
197	$[(\forall y)\ ((Ts_1 \wedge Eys_1) \rightarrow Ty) \rightarrow$	Ł$_3$∀P6, with $(\forall y)\ ((Ts_1 \wedge Eys_1) \rightarrow Ty)$ /
	$((Ts_1 \wedge Es_2s_1) \rightarrow Ts_2), \mathbf{T}]$	$(\forall \mathbf{x})\mathbf{P}$, s_2 / $\mathbf{a}$
198	$[(Ts_1 \wedge Es_2s_1) \rightarrow Ts_2, \mathbf{N}]$	196,197 MP
199	$[Ts_1 \wedge Es_2s_1, \mathbf{T}]$	1,2 CI
200	$[Ts_2, \mathbf{N}]$	198,199 MP
201	$[(\forall x)\ (\forall y)\ (Tx \wedge Eyx) \rightarrow Ty) \rightarrow$	Ł$_3$∀6, with $(\forall x)\ (\forall y)\ ((Tx \wedge Eyx) \rightarrow Ty)$ /
	$(\forall y)\ ((Ts_2 \wedge Eys_2) \rightarrow Ty), \mathbf{T}]$	$(\forall \mathbf{x})\mathbf{P}$, s_2 / $\mathbf{a}$
202	$[(\forall y)\ ((Ts_2 \wedge Eys_2) \rightarrow Ty), \mathbf{N}]$	194,201 MP
203	$[(\forall y)\ ((Ts_2 \wedge Eys_2) \rightarrow Ty) \rightarrow$	Ł$_3$∀6, with $(\forall y)\ ((Ts_2 \wedge Eys_2) \rightarrow Ty)$ /
	$((Ts_2 \wedge Es_3s_2) \rightarrow Ts_3), \mathbf{T}]$	$(\forall \mathbf{x})\mathbf{P}$, s_3 / $\mathbf{a}$
204	$[(Ts_2 \wedge Es_3s_2) \rightarrow Ts_3, \mathbf{N}]$	202,203 MP
205	$[Ts_2 \wedge Es_3s_2, \mathbf{N}]$	3,200 CI
206	$[Ts_3, \mathbf{F}]$	204,205 MP
...	... *{repeating 195–200 with appropriate substitutions we end with}*	
1346	$[Ts_{193}, \mathbf{F}]$	1344,1345 MP

By the time we get to the intermediate inference of Ts_3 the resulting grade says only that this sentence is at least **F**—and that will be the case for each inference of a formula Ts_i from that point on. The truth of Ts_1 at the outset ensures that we derive Ts_2 with the graded value **N**, given grade **N** for the Principle of Charity premise. But that **N** subsequently causes the application of Modus Ponens on line 206 to produce the grade **F**—again, given the value **N** for the Principle of Charity premise. This **F** together with the **N**-value of the Principle of Charity premise will continue to cause **F**-grades for all of the following inferred formulas Ts_i.

Note that this derivation alone is not sufficient to prove that the argument fails to be quasi-valid in Ł$_3$∀, for as we noted in Chapter 6 an argument from $\mathbf{P}_1, \ldots, \mathbf{P}_n$ to $\mathbf{Q}$ is quasi-valid if there is *at least one derivation* of the graded formula [**Q**, **N**] from the graded formulas $[\mathbf{P}_i, \mathbf{N}]$. The single derivation above doesn't establish that there is no *other* derivation in which each of the Sorites premises is graded with the value **N** and the conclusion is graded with the value **N** as well. However, the system Ł$_3$∀PA is *sound* for Ł$_3$∀A, and so we can conclude that in fact there is no such derivation.

8.3 Exercises

SECTION 8.1

1 Construct derivations that show that the following can be introduced as derived axiom schemata in $Ł_3∀A$:

a. **P(a/x) → (∃x)P**

b. **(∀x)(P → Q) → ((∃x)P → (∃x)Q)**

2 Show how to derive the following rules in $Ł_3∀A$:

a. **Universal Generalization in the Antecedent**. From **P(a/x) → Q** infer **(∀x)P → Q**

where **x** is any individual variable.

b. **Existential Generalization in the Consequent**. From **P → Q(a/x)** infer **P → (∃x)Q**

where **x** is any individual variable.

c. **Existential Generalization in the Antecedent**. From **P(a/x) → Q** infer **(∃x)P → Q**

where **x** is any individual variable, provided that no assumption contains the constant **a** and that **P** does not contain the constant **a**.

3 Construct derivations that show that the following formulas are theorems of $Ł_3∀A$:

a. (∃x)Qx → (∃x)(Pa → Qx)

b. (∃x)Qx → (∃x)(Px → Qx)

c. (∃x)(Pa → Qx) → ((∀x)Px → (∃x)Qx)

d. (Pa ∨ (∃x)Qx) → (∃x)(Pa ∨ Qx)

4 Show that the new SUB:

SUB (Substitution). From the universal closure of **P → Q**, the universal closure of **Q → P**, and a formula **R** that contains **P** as a subformula, infer any formula **R*** that is the result of replacing one or more occurrences of **P** in **R** with **Q**.

is indeed derivable in $Ł_3∀A$ by proving the following:

If formulas **P(a/x) → Q(a/x)** and **Q(a/x) → P(a/x)** occur in a derivation, where **a** is a constant that does not occur in **P**, **Q**, or any assumption, then both **(∀x)P → (∀x)Q and (∀x)Q → (∀x)P** can also be derived (this would be *Case 4* of the new proof).

This will do the trick because the new SUB applies when we have already derived the universal closures of **P → Q** and **Q → P**, say, $(\forall \mathbf{x}_1)\ldots(\forall \mathbf{x}_n)(\mathbf{P} \rightarrow \mathbf{Q})$ and $(\forall \mathbf{x}_1)\ldots(\forall \mathbf{x}_n)(\mathbf{Q} \rightarrow \mathbf{P})$. From these universal closures we can then derive specific instances $(\mathbf{P} \rightarrow \mathbf{Q})(\mathbf{a}_1/\mathbf{x}_1)\ldots(\mathbf{a}_n/\mathbf{x}_n)$ and $(\mathbf{Q} \rightarrow \mathbf{P})(\mathbf{a}_1/\mathbf{x}_1)\ldots(\mathbf{a}_n/\mathbf{x}_n)$ using $Ł_3∀6$ and MP. Thus as we derive the series of conditionals building up the formulas described in Chapter 6's justification of SUB we can use constants $\mathbf{a}_1, \mathbf{a}_2, \ldots, \mathbf{a}_n$ that do not occur in the derivation in place of the free variables $\mathbf{x}_1, \mathbf{x}_2, \ldots, \mathbf{x}_n$ in **P** and **Q**, reintroducing the free variables along with their universal quantifiers at the relevant point as the formulas are built up. Case 4 licenses the necessary universal quantification.

SECTION 8.2

5 Derive graded versions of the following rules for $Ł_3\forall PA$ (the first was derived in Section 8.1, and the rest were derived in Exercise 2):

a. Universal Generalization in the Consequent

b. Universal Generalization in the Antecedent

c. Existential Generalization in the Consequent

d. Existential Generalization in the Antecedent

6 Construct derivations (without assumptions) of the following graded formulas in $Ł_3\forall PA$:

a. $[(\mathbf{n} \to Pa) \to (\mathbf{n} \to (\exists x)Px), \mathbf{T}]$

b. $[(\mathbf{n} \to Pa) \to (\exists x)Px, \mathbf{N}]$

c. $[(\forall x)(Px \to Qx) \to ((\mathbf{n} \to Pa) \to (\mathbf{n} \to Qa)), \mathbf{T}]$

d. $[(\forall x)(Px \to Qx) \to (((\forall x)\neg Px \to \mathbf{n}) \to (\exists x)Qx), \mathbf{N}]$

e. $[(\mathbf{n} \to (\forall x)(Px \to Qx)) \to ((\mathbf{t} \to (\forall x)Px) \to (\mathbf{n} \to (\exists x)Qx)), \mathbf{T}]$

7 Construct derivations of the conclusions of the following graded arguments from the graded premises:

a. $\dfrac{[\mathbf{n} \to (\forall x)Qx, \mathbf{T}]}{[(\forall x)(\mathbf{n} \to Qx), \mathbf{T}]}$

b. $\dfrac{[(\forall x)(\mathbf{n} \to Qx), \mathbf{T}]}{[\mathbf{n} \to (\forall x)Qx, \mathbf{T}]}$

c. $[(\forall x)(\forall y)Rxy, \mathbf{N}]$

$\dfrac{[(\forall x)(Rxx \to Sx), \mathbf{T}]}{[(\forall x)Sx, \mathbf{N}]}$

d. $\dfrac{[(\forall x)(Px \to (Qx \to \mathbf{n})), \mathbf{T}]}{[(\forall x)(Px \to (Qx \to \mathbf{f})), \mathbf{N}]}$

9 Alternative Semantics for Three-Valued Logic

9.1 Numeric Truth-Values for Three-Valued Logic

In three-valued logic we can use the numeric values 1 and 0 for *true* and *false*, as we did for classical logic, and the value $\frac{1}{2}$ for *neither true nor false*. Chapter 4's numeric definitions of truth-values for complex formulas in classical propositional logic, which we repeat here, produce exactly the truth-conditions for the $K^S{}_3$ connectives:

1. $V(\neg \mathbf{P}) = 1 - V(\mathbf{P})$
2. $V(\mathbf{P} \wedge \mathbf{Q}) = \min(V(\mathbf{P}), V(\mathbf{Q}))$
3. $V(\mathbf{P} \vee \mathbf{Q}) = \max(V(\mathbf{P}), V(\mathbf{Q}))$
4. $V(\mathbf{P} \rightarrow \mathbf{Q}) = \max(1-V(\mathbf{P}), V(\mathbf{Q}))$
5. $V(\mathbf{P} \leftrightarrow \mathbf{Q}) = \min(\max(1-V(\mathbf{P}), V(\mathbf{Q})), \max(1-V(\mathbf{Q}), V(\mathbf{P})))$

The reader can easily verify that these clauses yield the following truth-tables:

P	$\neg_K$P
1	0
0	1
½	½

P $\wedge_K$ Q

P \ Q	1	½	0
1	1	½	0
½	½	½	0
0	0	0	0

P $\vee_K$ Q

P \ Q	1	½	0
1	1	1	1
½	1	½	½
0	1	½	0

P $\rightarrow_K$ Q

P \ Q	1	½	0
1	1	½	0
½	1	½	½
0	1	1	1

P $\leftrightarrow_K$ Q

P \ Q	1	½	0
1	1	½	0
½	½	½	½
0	0	½	1

The fact that the classical numeric clauses from Chapter 4 generate Kleene's strong three-valued system raises an important question, namely, Does this show that $K^S{}_3$ is the "true" generalization of classical propositional logic, whereas the other three-valued systems we have studied are not? The answer is *no*. All that we have shown is that if we use exactly the clauses for classical logic presented in Section 4.1 and add the third value $\frac{1}{2}$ to the classical values, we get Kleene's system. We have not shown anything stronger, because there are *alternative* numeric clauses for generating classical logic and these alternatives generalize to *different* three-valued systems.

For example, we can replace the classical clauses 4 and 5 with

4. $V(\mathbf{P} \rightarrow \mathbf{Q}) = \min(1, 1 - V(\mathbf{P}) + V(\mathbf{Q}))$
5. $V(\mathbf{P} \leftrightarrow \mathbf{Q}) = \min(1, 1 - V(\mathbf{P}) + V(\mathbf{Q}), 1 - V(\mathbf{Q}) + V(\mathbf{P}))$

to obtain the propositional system Ł3 with the following truth-tables for the conditional and biconditional:

$\mathbf{P} \rightarrow_L \mathbf{Q}$

P \ Q	1	½	0
1	1	½	0
½	1	1	½
0	1	1	1

$\mathbf{P} \leftrightarrow_L \mathbf{Q}$

P \ Q	1	½	0
1	1	½	0
½	½	1	½
0	0	½	1

Moreover, these clauses also work for classical propositional logic—as they should, since we know that Ł3 is a *normal* system.

To understand the formulas used for the Ł3 conditional and biconditional recall our comment in Section 4.2 that in classical logic disjunction is almost like addition: $V(\mathbf{P} \vee \mathbf{Q}) = V(\mathbf{P}) + V(\mathbf{Q})$, *except when* both $V(\mathbf{P})$ and $V(\mathbf{Q})$ are 1, so that classical disjunction can be defined as $\min(1, V(\mathbf{P}) + V(\mathbf{Q}))$.[1] Now, this formula can't work for weak disjunction in Ł3: $\min(1, V(\mathbf{P}) + V(\mathbf{Q}))$ is 1 when $V(\mathbf{P})$ and $V(\mathbf{Q})$ are both ½, but Ł3's weak disjunction has the value ½ in this case. However, the formula *does* characterize *bold* disjunction, which produces a true formula when both disjuncts are neither true nor false. But then, if we think of $\mathbf{P} \rightarrow_L \mathbf{Q}$ as $\neg_L \mathbf{P} \nabla_L \mathbf{Q}$ (we showed in Chapter 5 that Łukasiewicz's conditional could be so defined in terms of negation and bold disjunction), we get clause 4 for the conditional. Clause 5 results from defining $\mathbf{P} \leftrightarrow_L \mathbf{Q}$ as $(\mathbf{P} \rightarrow_L \mathbf{Q}) \wedge_L (\mathbf{Q} \rightarrow_L \mathbf{P})$.

There's another way to look at clause 4 for Łukasiewicz's conditional. The formula $1 - V(\mathbf{P}) + V(\mathbf{Q})$ is equivalent to $1 - (V(\mathbf{P}) - V(\mathbf{Q}))$, and $V(\mathbf{P}) - V(\mathbf{Q})$ is a kind of measure of how much "truer" **P** is than **Q**. If **P** is indeed truer than **Q**, then we subtract this value from 1, so that the truer **P** is than **Q**, the less true the conditional is. If **P** is not truer than **Q**, that is, if it is the same or less true than **Q**, then the conditional is simply true.

For Bochvar's internal connectives we have the numeric clauses

2. $V(\mathbf{P} \wedge_{BI} \mathbf{Q}) = ½$ if $V(\mathbf{P}) = ½$ or $V(\mathbf{Q}) = ½$
   $\min(V(\mathbf{P}), V(\mathbf{Q}))$ otherwise
3. $V(\mathbf{P} \vee_{BI} \mathbf{Q}) = ½$ if $V(\mathbf{P}) = ½$ or $V(\mathbf{Q}) = ½$
   $\max(V(\mathbf{P}), V(\mathbf{Q}))$ otherwise

[1] Recall that for classical logic we could also define conjunction in terms of multiplication: $V(\mathbf{P} \wedge \mathbf{Q}) = V(\mathbf{P}) \cdot V(\mathbf{Q})$. However, we can't define conjunction this way in three-valued logical systems because it would be ill-defined—when **P** and **Q** both have the value ½, the product of their values is ¼, which is not a numeric truth-value in our three-valued systems. We'll revisit this definition of conjunction, however, when we turn to fuzzy logics.

for conjunction and disjunction. The "otherwise" cases give us a normal system—Bochvar's system, like the others, behaves classically when only classical truth-values are involved. We leave it as an exercise to develop numeric clauses for Bochvar's internal conditional and biconditional as well as for his external connectives.

Turning to three-valued first-order logics, we revise numeric clause 1 for classical first-order logic, assigning $\frac{1}{2}$ to formulas that are undetermined:

1. $I_v(\mathbf{P}\mathbf{t}_1 \dots \mathbf{t}_n) = 1$ if $<I^*(\mathbf{t}_1), \dots, I^*(\mathbf{t}_n)> \in \text{ext}(\mathbf{P})$, where $I^*(\mathbf{t}_i)$ is $I(\mathbf{t}_i)$ if $\mathbf{t}_i$ is a constant and is $v(\mathbf{t}_i)$ if $\mathbf{t}_i$ is a variable,
   0 if $<I^*(\mathbf{t}_1), \dots, I^*(\mathbf{t}_n)> \in \text{cxt}(\mathbf{P})$, and
   $\frac{1}{2}$ if $<I^*(\mathbf{t}_1), \dots, I^*(\mathbf{t}_n)> \in \text{fge}(\mathbf{P})$.

If we then use the remaining numeric satisfaction clauses 2–8 for classical first-order logic, we get $K^S{}_3\forall$. Here are the quantifier clauses:

7. $I_v((\forall \mathbf{x})\mathbf{P}) = \min\{I_{v'}(\mathbf{P})\text{: } v' \text{ is an } \mathbf{x}\text{-variant of } v\}$
8. $I_v((\exists \mathbf{x})\mathbf{P}) = \max\{I_{v'}(\mathbf{P})\text{: } v' \text{ is an } \mathbf{x}\text{-variant of } v\}$

Clause 7 gives a universal quantification the value 1 if the quantified formula is satisfied by everything, 0 if it is *dis*satisfied by at least one thing, and $\frac{1}{2}$ otherwise. Clause 8 gives an existential quantification the value 1 if the quantified formula is satisfied by at least one thing, 0 if it is dissatisfied by *everything*, and $\frac{1}{2}$ otherwise. Since the quantifiers have the same definitions in both $K^S{}_3$ and $Ł_3$, the revised clauses 1, 7, and 8 will generate $Ł_3\forall$ if we add them to the numeric propositional clauses for $Ł_3$, revised to talk of interpretations and variable assignments.

The following clause suffices for the universal quantifier in the first-order generalization of $B^I{}_3$:

7. $I_v((\forall_{BI}\, \mathbf{x})\mathbf{P}) = \frac{1}{2}$ if $I_{v'}(\mathbf{P}) = \frac{1}{2}$ for any $\mathbf{x}$-variant v' of v
   $\min\{I_{v'}(\mathbf{P})\text{: } v' \text{ is an } \mathbf{x}\text{-variant of } v\}$otherwise.

Again, the "otherwise" cases give us classically defined values when the third truth-value is ignored. We leave it as an exercise to develop numeric clauses for Bochvar's internal existential quantifier and for his external universal and existential quantifiers.

9.2 Abstract Algebras for $Ł_3$, $K^S{}_3$, $B^I{}_3$, and $B^E{}_3$

We'll now explore the abstract algebraic structures characterizing the operations of three-valued logical systems, whether those operations are defined for **T**, **N**, and **F** or $\{1, {}^1/_2, 0\}$. We begin with the algebra induced by Łukasiewicz's weak disjunction and weak conjunction (and hence by Kleene's strong operations) as defined on the values $\{1, {}^1/_2, 0\}$: this algebra is a distributed lattice with domain $\{1, {}^1/_2, 0\}$ in which

1 and 0 serve, respectively, as *unit* and *zero* elements. Recall that conditions i–v from the definition of Boolean algebras in Chapter 4 define distributed lattices and that condition vi defines *unit* and *zero* elements for a lattice:

i. $x \cup y = y \cup x$, and $x \cap y = y \cap x$ (*commutation*)
ii. $x \cup (y \cup z) = (x \cup y) \cup z$, and $x \cap (y \cap z) = (x \cap y) \cap z$ (*association*)
iii. $x \cup x = x$, and $x \cap x = x$ (*idempotence*)
iv. $x \cup (x \cap y) = x$, and $x \cap (x \cup y) = x$ (*absorption*)
v. $x \cup (y \cap z) = (x \cup y) \cap (x \cup z)$, and $x \cap (y \cup z) = (x \cap y) \cup (x \cap z)$ (*distribution*)
vi. $x \cup \textit{zero} = x$, and $x \cap \textit{unit} = x$ (*identity for join and meet*)

The reader will be asked in the exercises to verify that $Ł_3$ weak (or $K^S{}_3$) disjunction and conjunction satisfy these join and meet conditions for the domain $\{1, {}^1/_2, 0\}$.

We'll call the lattice arising from these operations *ŁKL*. If we try to add Łukasiewicz-Kleene negation to ŁKL as its complementation operation the result is *not* a Boolean algebra because the complementation condition vii, $x \cup x' = \textit{unit}$ and $x \cap x' = \textit{zero}$, doesn't hold. When $x = {}^1\!/_2$, $x \cup x'$ is ${}^1\!/_2$ rather than the *unit* element 1 and $x \cap x'$ is ${}^1\!/_2$ rather than the *zero* element 0. It should be no surprise that $x \cup x'$ fails to be the *unit* element, because $x \cup x'$ is the algebraic counterpart to the Law of Excluded Middle, which fails in $Ł_3$ (using weak disjunction $\vee$) and in $K^S{}_3$. In fact, for the condition $x \cup x' = \textit{unit}$ to hold for a complementation operation in ŁKL we would have to define both ${}^1\!/_2{}'$ and $0'$ to be 1 since this condition would require that $\max({}^1/_2, {}^1\!/_2{}') = 1$ and that $\max(0, 0') = 1$, and for the condition $x \cap x' = \textit{zero}$ to hold as well we would have to define both ${}^1\!/_2{}'$ and $1'$ to be 0. But we can't have ${}^1\!/_2{}'$ be both 1 and 0—so it is impossible to define a Boolean complementation operation for ŁKL. We'll revisit this point in Section 9.3.

Although it is not full complementation, Łukasiewicz-Kleene negation is an example of an *orthocomplementation* operation. An algebraic **orthocomplement** is a unary operation′ that satisfies the conditions

vii. $\textit{zero}' = \textit{unit}$ and $\textit{unit}' = \textit{zero}$

and

viii. $x = x''$

for all x in the domain.[2] The normality of Łukasiewicz-Kleene negation clinches condition vii, while condition viii is the algebraic counterpart to the Law of Double Negation. A distributive lattice that has *unit* and *zero* elements and an orthocomplement satisfying conditions vii and viii is called a ***DeMorgan algebra*** if it also satisfies the DeMorgan condition ix:

ix. $(x \cup y)' = x' \cap y'$ and $(x \cap y)' = x' \cup y'$

[2] Definitions of *orthocomplementation* vary, depending on the type of lattice for which the operation is being defined. We've chosen a definition that's appropriate for DeMorgan and Kleene algebras.

The reader will verify in the exercises that the algebra ŁKL′ in which Łukasiewicz-Kleene negation has been added to ŁKL is a DeMorgan algebra.

More specifically, ŁKL′ is a *Kleene algebra*. To define Kleene algebras we recall the natural lattice ordering relation $\leq$ defined as

$$x \leq y =_{def} x \cap y = x.$$

This produces the natural numeric ordering $0 \leq \frac{1}{2} \leq 1$ in ŁKL′ because min $(0, {}^1/_2) = 0$ and min $({}^1/_2, 1) = 1$. A **Kleene algebra**[3] is a DeMorgan algebra that satisfies the additional condition

x. $x \cap x' \leq y \cup y'$

for all x, y in the domain. For the algebra ŁKL′ we have min $(x, 1-x) \leq$ max $(y, 1-y)$ because for any x in $\{1, \frac{1}{2}, 0\}$, min $(x, 1-x) \leq \frac{1}{2}$ while for any y, $\frac{1}{2} \leq$ max $(y, 1-y)$.

In Section 4.3 of Chapter 4 we used the lattice ordering relation to define a Boolean algebra conditional operation $\Rightarrow$ as

$$x \Rightarrow y = \textit{unit} \text{ if and only if } x \leq y.$$

So the question naturally arises, Does Łukasiewicz's or Kleene's strong conditional operation, or neither, meet this condition? Well, putting this condition together with the definition of $\leq$ we have

$$x \Rightarrow y = \textit{unit} \text{ if and only if } x \cap y = x$$

and in the case of ŁKL′:

$$x \Rightarrow y = \textit{unit} \text{ in ŁKL}' \text{ if and only if } \min(x, y) = x.$$

This is exactly the condition under which a Łukasiewicz conditional is true, but it isn't satisfied by Kleene's conditional. Given a domain containing more than two elements, though, the condition doesn't specify the conditions under which $x \Rightarrow y$ evaluates to *zero*, and so it doesn't define when a Łukasiewicz conditional has the value $\frac{1}{2}$ and when it has the value 0. In Section 9.3 we'll introduce another type of algebra, which more fully captures the Łukasiewicz conditional. On the other hand, a Kleenean conditional operation can be defined for Kleene algebras in the obvious way, based on $K^S{}_3$'s definition: $x \Rightarrow_K y =_{def} x' \cup y$.

We have a result relating $K^S{}_3$ and Kleene algebras analogous to a result in Chapter 4 relating classical propositional logic and Boolean algebras. When we use a three-valued Kleene algebra to interpret formulas of propositional logic by assigning either *unit*, *zero*, or the third element to each atomic formula and using the algebra's join, meet, and orthocomplement operations to define the respective values of disjunctions, conjunctions, and negations, we call this an ***algebraic interpretation*** based on that Kleene algebra and the set of all such interpretations a ***semantics***

[3] Balbes and Dwinger (1974, p. 215).

based on the Kleene algebra. The following result tells us that the semantics for propositional logic based on any three-valued Kleene algebra give us $K^S{}_3$'s truth-functions for disjunction, conjunction, and negation, with *unit*, *zero*, and the third element, respectively, in place of 1, 0, and ½:

Result 9.1: Every three-valued Kleene algebra KA = <{*unit*, *zero*, *other*}, ∪, ∩, ′, *unit*, *zero*> generates the following truth-tables for assignments of *unit*, *zero*, or *other* to each atomic formula of propositional logic when ∪, ∩, and ′ respectively define the disjunction, conjunction, and negation operations (we use *u*, *z*, and *o* to stand for *unit*, *zero*, and *other*):

P	**¬P**
u	*z*
o	*o*
z	*u*

P ∨ Q

P \ Q	*u*	*o*	*z*
u	*u*	*u*	*u*
o	*u*	*o*	*o*
z	*u*	*o*	*z*

P ∧ Q

P \ Q	*u*	*o*	*z*
u	*u*	*o*	*z*
o	*o*	*o*	*z*
z	*z*	*z*	*z*

Proof: Left as an exercise.

Because there are no $K^S{}_3$ tautologies, it follows that no formulas of propositional logic are tautologies of any three-valued Kleene algebra (where the algebraic definition of a tautology is analogous to the definition of BA-tautologies in Chapter 4).

The algebra B^IA induced by Bochvar's internal disjunction and conjunction isn't a lattice because lattice condition iv, absorption, doesn't hold—when x is 1 or 0 the expressions x ∪ (x ∩ ½) and x ∩ (x ∪ ½) both evaluate to ½ in B^IA, not x. Conditions i–iii do hold for B^IA, as do conditions v and vi earlier specifying distributivity and unit and zero elements. The equations involving the join operation in conditions i–iii define a structure known as a ***semi-lattice***, and so do the equations involving the meet operation. (It is idempotency condition iv that unifies join and meet semi-lattices into a single lattice.) Thus we may simply describe B^IA as a distributive dual system of semi-lattices with unit and zero elements. The addition of Bochvar's internal negation would give us an orthocomplement for the dual system. Finally, the algebra B^EA defined by Bochvar's external disjunction and conjunction is a distributed lattice, but the elements 0 and 1 don't serve as identity elements for join and meet as specified by condition vi (½ ∪ 0 = 0 in B^EA, not ½, and ½ ∩ 1 = 0, not ½). The reader will be asked to discuss these systems, as well as the algebraic operation corresponding to Bochvar's external negation, in the exercises.

9.3 MV-Algebras

In Section 9.1 we gave Łukasiewicz's bold disjunction the numeric truth-condition V(**P** ∇ **Q**) = min(1, V(**P**) + V(**Q**)). Bold conjunction can be defined as **P & Q** $=_{def}$ ¬(¬**P** ∇ ¬**Q**), from which it follows that V(**P & Q**) = max (0, V(**P**) + V(**Q**) – 1). The algebraic structure $Ł_3MV$ induced by $Ł_3$ bold disjunction, bold conjunction, and negation, that is, the structure <{1, ½, 0}, $\oplus_L$, $\otimes_L$, 1−, 1, 0> where $\oplus_L$, $\otimes_L$, and 1− are $Ł_3$ bold disjunction, bold conjunction, and negation, is an *MV-algebra.* An MV-algebra is a an algebra <M, ⊕, ⊗, ′, *unit, zero*> that meets the following conditions for all x, y, and z in M:[4]

i.	$x \oplus y = y \oplus x$, and $x \otimes y = y \otimes x$	*(commutation)*
ii.	$x \oplus (y \oplus z) = (x \oplus y) \oplus z$, and $x \otimes (y \otimes z) = (x \otimes y) \otimes z$	*(association)*
iii.	$x \oplus zero = x$, and $x \otimes unit = x$	*(identity for join and meet)*
iv.	$x \oplus unit = unit$, and $x \otimes zero = zero$	*(unit and zero consumption)*
v.	$x \oplus x' = unit$, and $x \otimes x' = zero$	*(complementation)*
vi.	$(x \oplus y)' = x' \otimes y'$, and $(x \otimes y)' = x' \oplus y'$	*(DeMorgan's Laws)*
vii.	$x = x''$	*(Double Negation)*
viii.	$zero' = unit$	*(duality of zero and unit)*
ix.	$(x' \oplus y)' \oplus y = (y' \oplus x)' \oplus x$	*(lattice meet commutation)*

(The reason for calling condition ix *lattice meet commutation* will be explained later.) We'll show that bold disjunction and bold conjunction are both commutative and associative operations on {1, ½, 0}, thus meeting the first two conditions on MV-algebras:

Commutation: $x \oplus_L y = x \oplus_L y$, and $x \otimes_L y = x \otimes_L y$

Proof of first equation: For all x, y ∈ {1, ½, 0}, min (1, x + y) = min (1, y + x) since x + y = y + x.

Proof of second equation: Left as an exercise.

Association: $x \oplus_L (y \oplus_L z) = (x \oplus_L y) \oplus_L z$, and $x \otimes_L (y \otimes_L z) = (x \otimes_L y) \otimes_L z$

Proof of first equation: For all x, y, z ∈ {1, ½, 0}, min (1, x + min (1, y + z)) = min (1, min (x + 1, x + y + z)) = min (1, x + 1, x + y + z), and since x ≥ 0 it follows that x + 1 ≥ 1, so min (1, x + 1, x + y + z) = min (1, x + y + z). Similarly, min (1, min (1, x + y) + z) = min (1, min (1 + z, x + y + z)) = min (1, 1 + z, x + y + z) = min (1, x + y + z).

Proof of second equation: Left as an exercise.

[4] MV-algebras were first developed in Chang (1958b, 1959). *MV* is short for *many-valued.* These are not Chang's original conditions but are a variation of an equivalent formulation given in Mangani (1973). An excellent monograph that explores various formulations of MV-algebras is Cignoli, D'Ottaviano, and Mundici (2000).

It is also left as an exercise to show that the remaining conditions on MV-algebras hold for Ł3MV.

We noted in Section 9.2 that Łukasiewicz/Kleene negation cannot serve as a complementation operation for ŁKL because it doesn't satisfy the complementation conditions with respect to ŁKL's lattice join and meet. However, as will be verified in the exercises, Łukasiewicz/Kleene negation *does* serve as complementation with respect to the MV-algebraic join and meet operations $\oplus_L$ and $\otimes_L$. This isn't surprising since we already knew that the *bold* disjunction Law of Excluded Middle $A \triangledown \neg A$ is an Ł3-tautology, while the bold conjunction $A \,\&\, \neg A$ is an Ł3-contradiction.

The first complementation condition for MV-algebras isn't required as a separate condition since it can be derived using the other conditions:

$\mathrm{x} \oplus \mathrm{x}' = \mathrm{x}' \oplus \mathrm{x}$	(commutation)
$= (\mathrm{x} \oplus zero)' \oplus \mathrm{x}$	(identity for join)
$= (zero \oplus \mathrm{x})' \oplus \mathrm{x}$	(commutation)
$= (zero'' \oplus \mathrm{x})' \oplus \mathrm{x}$	(Double Negation)
$= (\mathrm{x}' \oplus zero')' \oplus zero'$	(lattice meet commutation)
$= (\mathrm{x}' \oplus zero')' \oplus unit$	(duality of *zero* and *unit*)
$= unit$	(*unit* consumption)

But this equality is so important that we have included it as a separate condition.

We know that we can also derive the second complementation condition from the other conditions because there is a general principle of duality for MV-algebras similar to that for Boolean algebras. Each of the conditions i–vi consists of a pair of equations such that join and meet have been interchanged and so have *unit* and *zero*. We can derive $unit' = zero$, the dual to condition viii, from conditions vii and viii (see Exercise 7 for Section 9.2), and the dual to lattice meet commutation (which is *lattice join commutation*) is also derivable from other conditions—this is left as an exercise.

Every Boolean algebra is an MV-algebra, where the Boolean join $\cup$ serves as the MV-algebra join $\oplus$ and the Boolean meet $\cap$ serves as the MV-algebra meet $\otimes$. We have already shown (in Chapter 4) that most of the conditions for MV-algebras hold for Boolean algebras; we leave it as an exercise to show that the duality of *zero* and *unit* and lattice meet commutation both hold for Boolean algebras as well. On the other hand, not all MV-algebras are Boolean algebras. For example, Ł3MV is an MV-algebra but it isn't a Boolean algebra. In particular, idempotence fails since $\frac{1}{2} \oplus \frac{1}{2} = 1$, not $\frac{1}{2}$ as would be required by idempotence, and $\frac{1}{2} \otimes \frac{1}{2} = 0$. Distribution also fails—for example, $\frac{1}{2} \otimes (1 \oplus 1) = \frac{1}{2}$, but $(\frac{1}{2} \otimes 1) \oplus (\frac{1}{2} \otimes 1) = 1$. (If distribution held then idempotence would hold as well—we showed in Chapter 4 how to derive idempotence from distribution, complementation, and identity for meet and join, and the latter conditions both hold in MV-algebras). In fact, an MV-algebra is a Boolean algebra if and only if idempotence and/or distribution holds for the MV-algebra's meet and join.

There is a natural ordering on MV-algebras defined as

$x \leq y$ if and only if $x' \oplus y = \textit{unit}$

Like its lattice counterpart, this ordering is reflexive, antisymmetric, and transitive. Alternatively, we can define an MV-algebra operation $\cap$ that satisfies the conditions defining lattice meet operations:

$$x \cap y =_{\text{def}} x \otimes (x' \oplus y)$$

and then define the ordering as we did for lattices:

$x \leq y$ if and only if $x \cap y = x$

It is left as an exercise to prove that when the lattice meet operation $\cap$ is defined as $x \otimes (x' \oplus y)$, $x \cap y = x$ if and only if $x' \oplus y = \textit{unit}$, so that the two definitions for the MV-relation $\leq$ coincide. An operation $\cup$ satisfying the conditions defining lattice join operations can also be defined within an MV-algebra, for example, as

$$x \cup y = \text{def } (x' \cap y')'$$

MV-algebras are said to contain lattices as *substructures* because these meet and join operations $\cap$ and $\cup$ together with the MV-algebraic operation $'$ form a lattice over the MV-algebra's domain. And because of these substructures, the MV-algebra meet and join operations $\otimes$ and $\oplus$ are sometimes called *bold* meet and join to distinguish them from the lattice operations.

To prove that the operations $\cap$ and $\cup$ as defined in the previous paragraph do indeed form a lattice, we need to show that lattice conditions i–iv hold, that is, that these meet and join operations are commutative, associative, and idempotent, and that the absorption conditions hold. The name for condition ix in the definition of MV-algebras—*lattice meet commutation*—refers to the fact that this condition guarantees that the meet operation $\cap$ as we have just defined it is a commutative operation:

$$\begin{aligned}
x \cap y &= x \otimes (x' \oplus y) && \text{(definition)}\\
&= (x \otimes (x' \oplus y))'' && \text{(Double Negation)}\\
&= (x' \oplus (x' \oplus y)')' && \text{(DeMorgan's Law)}\\
&= ((y \oplus x')' \oplus x')' && \text{(commutation, twice)}\\
&= ((y'' \oplus x')' \oplus x')' && \text{(Double Negation)}\\
&= ((x'' \oplus y')' \oplus y')' && \text{(lattice meet commutation)}\\
&= ((x \oplus y')' \oplus y')' && \text{(Double Negation)}\\
&= (y' \oplus (y' \oplus x)')' && \text{(commutation, twice)}\\
&= (y \otimes (y' \oplus x))'' && \text{(DeMorgan's Law)}\\
&= y \otimes (y' \oplus x) && \text{(Double Negation)}\\
&= y \cap x && \text{(definition)}
\end{aligned}$$

(Note the second half of this derivation reverses the steps of the first half, but on different operands.) An equivalent derivation without lattice meet commutation cannot be constructed.

The associativity of $\cap$ can be derived as follows (we use lattice join commutation, which is the dual of lattice meet commutation, the derivation of which is left as an exercise). This derivation is tricky to follow, and it may help the reader to note that once again the second half of the derivation reverses the steps of the first half:

$$
\begin{aligned}
x \cap (y \cap z) &= (y \cap z) \cap x && \text{(lattice commutation, just established)}\\
&= (y \otimes (y' \oplus z)) \otimes ((y \otimes (y' \oplus z))' \oplus x) && \text{(definition)}\\
&= y \otimes ((y' \oplus z) \otimes ((y \otimes (y' \oplus z))' \oplus x)) && \text{(association)}\\
&= y \otimes ((y' \oplus z) \otimes ((y' \oplus (y' \oplus z)') \oplus x)) && \text{(DeMorgan's Law)}\\
&= y \otimes ((y' \oplus z) \otimes (((y' \oplus z)' \oplus y') \oplus x)) && \text{(commutation)}\\
&= y \otimes ((y' \oplus z) \otimes ((y' \oplus z)' \oplus (y' \oplus x))) && \text{(association)}\\
&= y \otimes ((y' \oplus z) \otimes ((y' \oplus z)' \oplus (y' \oplus x))'') && \text{(Double Negation)}\\
&= y \otimes ((y' \oplus z) \otimes ((y' \oplus z)'' \otimes (y' \oplus x)')') && \text{(DeMorgan's Law)}\\
&= y \otimes ((y' \oplus z) \otimes ((y' \oplus z) \otimes (y' \oplus x)')') && \text{(Double Negation)}\\
&= y \otimes (((y' \oplus x)' \otimes (y' \oplus z))' \otimes (y' \oplus z)) && \text{(commutation, twice)}\\
&= y \otimes (((y' \oplus z)' \otimes (y' \oplus x))' \otimes (y' \oplus x)) && \text{(lattice join commutation)}\\
&= y \otimes ((y' \oplus x) \otimes ((y' \oplus x) \otimes (y' \oplus z)')') && \text{(commutation, twice)}\\
&= y \otimes ((y' \oplus x) \otimes ((y' \oplus x)'' \otimes (y' \oplus z)')') && \text{(Double Negation)}\\
&= y \otimes ((y' \oplus x) \otimes ((y' \oplus x)' \oplus (y' \oplus z))'') && \text{(DeMorgan's Law)}\\
&= y \otimes ((y' \oplus x) \otimes ((y' \oplus x)' \oplus (y' \oplus z))) && \text{(Double Negation)}\\
&= y \otimes ((y' \oplus x) \otimes (((y' \oplus x)' \oplus y') \oplus z)) && \text{(association)}\\
&= y \otimes ((y' \oplus x) \otimes ((y' \oplus (y' \oplus x)') \oplus z)) && \text{(commutation)}\\
&= y \otimes ((y' \oplus x) \otimes ((y \otimes (y' \oplus x))' \oplus z)) && \text{(DeMorgan's Law)}\\
&= (y \otimes (y' \oplus x)) \otimes ((y \otimes (y' \oplus x))' \oplus z) && \text{(association)}\\
&= (y \cap x) \cap z && \text{(definition)}\\
&= (x \cap y) \cap z && \text{(commutation)}
\end{aligned}
$$

We leave it as an exercise to establish the remaining lattice properties for MV-algebra lattice meet and join operations.

Given the lattice substructure definable in an MV-algebra, it should come as no surprise that Ł$_3$'s weak conjunction and disjunction connectives, corresponding to the meet and join operations of this lattice substructure, are definable using negation and the Ł$_3$ bold connectives. Of course, we already knew this because we showed how to use the Ł$_3$ bold connectives and negation to define the Ł$_3$ conditional, and how to use this conditional and negation to define Ł$_3$ weak disjunction and conjunction, but this shows how algebraic structures can also be used to support claims about logical systems. Using the definition of lattice meet in MV-algebras, for example, we conclude that $\mathbf{P} \wedge \mathbf{Q}$ may be defined as $\mathbf{P}$ **&** $(\neg\mathbf{P} \triangledown \mathbf{Q})$.

We can also use bold join to define $\Rightarrow_L$, the algebraic counterpart of the Łukasiewicz conditional, completely for any MV-algebra as

$$x \Rightarrow_L y = x' \oplus y$$

for all x and y, for recall that in $Ł_3$ we have $V(\mathbf{P} \rightarrow \mathbf{Q}) = V(\neg\mathbf{P} \nabla \mathbf{Q})$. In $Ł_3$MV this gives us exactly the truth-conditions for a Łukasiewicz conditional.

We can now relate $Ł_3$ tautologies and MV-algebra tautologies. We will say that a formula of propositional logic is a ***tautology of an MV-algebra*** if the formula evaluates to *unit* under every algebraic interpretation based on that algebra. First, we have

Result 9.2: Every three-valued MV-algebra MV = <{*unit, zero, other*}, ⊕, ⊗, ′, *unit, zero*> generates the following truth-tables for assignments of *unit, zero,* or *other* to each atomic formula of propositional logic when ⊕, ⊗, and ′ respectively define the bold disjunction, bold conjunction, and negation operations (we use *u, z,* and *o* to stand, respectively, for *unit, zero,* and *other*):

P	**¬P**
u	*z*
o	*o*
z	*u*

P ∇ Q

P \ Q	*u*	*o*	*z*
u	*u*	*u*	*u*
o	*u*	*u*	*o*
z	*u*	*o*	*z*

P & Q

P \ Q	*u*	*o*	*z*
u	*u*	*o*	*z*
o	*o*	*z*	*z*
z	*z*	*z*	*z*

Proof: Left as an exercise.

Given definitions of the $Ł_3$ conditional and biconditional in terms of negation, bold disjunction, and bold conjunction we therefore have

Result 9.3: For any three-valued MV-algebra BA, the set of formulas of propositional logic that are MV-tautologies is exactly the set of $Ł_3$ tautologies under the standard semantics $\{1, \frac{1}{2}, 0\}$.

In Chapter 4 we proved a stronger result relating Boolean algebras and classical propositional logic: that for any Boolean algebra BA, the set of formulas that are BA-tautologies is exactly the set of classical tautologies. There is *not* an analogous result for relating $Ł_3$ and MV-algebras, for some $Ł_3$ tautologies are not tautologies in every MV-algebra. This will become clear when we study fuzzy propositional logics and their algebras in Chapters 11 and 12.

9.4 Exercises

SECTION 9.1

1 Develop numeric truth-condition clauses for Bochvar's internal conditional and biconditional.

2 Develop numeric truth-condition clauses for Bochvar's external connectives.

3 Develop numeric truth-value clauses for Bochvar's internal existential quantifier.

4 Develop numeric truth-value clauses for Bochvar's external quantifiers.

SECTION 9.2

5 Prove that Ł3 weak (or $K^S{}_3$) conjunction and disjunction satisfy distributed lattice meet and join conditions for the domain $\{1, {}^1/_2, 0\}$, with 1 and 0 serving respectively as *unit* and *zero* elements.

6 Do the Ł3 *bold* conjunction and disjunction operations produce a distributed lattice structure over $\{1, {}^1\!/_2, 0\}$? Defend your answer.

7 Show that the second part of the orthocomplement condition vii, $unit' = zero$, follows from the first part of condition vii ($zero' = unit$) and condition viii, $x'' = x$.

8 Prove that ŁKL′ is a DeMorgan algebra.

9 Prove that every Boolean algebra is a DeMorgan algebra, where the Boolean complement serves as the DeMorgan algebra's orthocomplement.

10 Prove Result 9.1.

11 Prove that B^IA meets conditions v and vi specifying distributivity and *unit* and *zero* elements:

v. $x \cup (y \cap z) = (x \cup y) \cap (x \cup z)$, and $x \cap (y \cup z) = (x \cap y) \cup (x \cap z)$

vi. $x \cup zero = x$, and $x \cap unit = x$.

12 Prove that B^EA is a distributed lattice.

13 Is Bochvar's external negation an orthocomplement for the algebra B^EA? Explain. Is it a complementation operation (such that $x \cup x' = unit$ and $x \cap x' = zero$)? Explain.

14 There does not exist a result analogous to Result 9.1 that relates $B^I{}_3$'s truth-tables and three-valued distributive dual systems of semi-lattices with *unit* and *zero* elements and an orthocomplement because every Kleene algebra satisfies the conditions of these structures and we know that three-valued Kleene algebras produce the $K^S{}_3$ truth-tables. Here's your task: find one or more conditions that can be added to i–iii, v, vi, vii, and viii that will force a three-valued algebra to produce the truth-tables for $B^I{}_3$ disjunction, conjunction, and negation.

SECTION 9.3

15 Prove that the equation $V(\mathbf{P} \,\&\, \mathbf{Q}) = \max(0, V(\mathbf{P}) + V(\mathbf{Q}) - 1)$ correctly defines the truth-conditions for Łukasiewicz bold conjunction.

16 Complete the proofs that the bold disjunction and bold conjunction operations $\oplus_L$ and $\otimes_L$ are both commutative and associative.

17 Prove that conditions iii–ix for MV-algebras hold for Ł$_3$MV.

18 Derive lattice join commutation, the dual of lattice meet commutation, from the conditions defining MV-algebras.

19 Derive the second complementation condition for MV-algebras using the other conditions defining MV-algebras.

20 Show that the second DeMorgan Law for MV-algebras follows from the first DeMorgan Law and conditions i–v and vii–ix for MV-algebras.

21 Show that the duality of *zero* and *unit*, and lattice meet commutation also hold for all Boolean algebras.

22 Prove that in an MV-algebra, $x \cap y = x$ if and only if $x' \oplus y =$ *unit*, where $x \cap y =_{def} x \otimes (x' \oplus y)$.

23 Prove that

a. the MV-algebra lattice meet operation $\cap$ defined as $x \cap y =_{def} x \otimes (x' \oplus y)$ is idempotent

b. the MV-algebra lattice join operation $\cup$ defined as $x \cup y =_{def} (x' \cap y')'$ is commutative, associative, and idempotent

c. these two operations satisfy the lattice absorption conditions $x \cup (x \cap y) = x$ and $x \cap (x \cup y) = x$.

24 Prove Result 9.2.

10 The Principle of Charity Reconsidered and a New Problem of the Fringe

It's time to face two problems that we sidestepped while exploring three-valued logical systems for vagueness.

Although the Sorites argument is valid in all of the systems we've presented, we claimed that the paradox can nevertheless be dissolved in three-valued logic because the Principle of Charity premise is not true on any reasonable interpretation. The first problem concerns the exact nature of the principle's nontruth. Our sample interpretations rendered the premise false in Bochvar's external system, which didn't sound right because its negation—which states that $^1/_8''$ *does* make a difference—must then be true. However, the situation looked more promising in the other three systems, where the Principle of Charity and its negation were neither true nor false. But now let us recall that the Principle of Charity is so called by virtue of the colloquial reading, *One-eighth of an inch doesn't make a difference*. Put that way, the Principle of Charity seems true, or close to it, doesn't it? If you shrink a tall person by $^1/_8''$, surely that person will still be tall. (If you disagree, change the shrinking to $^1/_{100}''$—we'll still get the paradox, but surely $^1/_{100}''$ doesn't make a difference.) Three-valued accounts can avoid the paradox by claiming that the Principle of Charity is either false or neither true nor false, but that leaves another puzzle: why does the principle seem to be true?

We'll see that in fuzzy logical systems we can do better: we'll be able to say that the Principle of Charity, although not exactly true, is very close to true, because we'll allow sentences to have one of *infinitely many* truth-values—ranging from true at one end to false at the other, and reflecting various degrees of truth in between. The alternative, which three-valued theorists can embrace, is to supplement the logical solution to the paradox with an explanation of why the apparent truth of the Principle of Charity is in fact illusory.[1] So we might say that fuzzy logic takes the apparent truth *at face value*.

The second and more serious problem, which we mentioned in footnote 2 to Chapter 7 and which we will call the *New Problem of the Fringe*, is that our three-valued interpretations for vague predicates assume clear cutoff points between the extension of a predicate and its fringe and between the fringe and the counterextension. This can't be right. Where is the cutoff between the extension of *tall* and

[1] Kit Fine (1975), for example, takes this route.

the fringe of *tall*, for example? Is it 5′ 11″, as we assumed in our interpretations in Chapter 7? Or is it 5′ 10″, or 5′ 9″, or perhaps 6′? And even if we can agree on where to make the cutoff, its mere existence shows that an eighth of an inch *can* make a difference, contrary to the Principle of Charity. That is, you can go from being tall to being neither tall nor not tall by shrinking $^1/_8$″, and that seems plain wrong.

Indeed, Bertrand Russell recognized that the existence of a fringe seems to require **"higher-order" fringes** between a vague predicate's extension and fringe and also between the predicate's fringe and counterextension—to replace the objectionable cutoff points.[2] But now if we countenance these higher-order fringes, we may find that we need yet higher-order fringes to set them off—lest we posit an exact cutoff between, say, the extension of a predicate and the higher-order fringe that separates it from the predicate's fringe. We thus find ourselves going from three values to five values to nine values to . . . an infinite number of truth-values. And that is exactly what we have in fuzzy logic, which is a type of infinite-valued logic.

So let's get fuzzy!

2 Bertrand Russell (1923, p. 87). The term *fringe* is from Black (1937), not from Russell.

11 Fuzzy Propositional Logics: Semantics

11.1 Fuzzy Sets and Degrees of Truth

Chapter 10 noted two problems that crop up for three-valued approaches to vagueness. The first problem is that the Principle of Charity seems to be, if not completely true, then at least very close to true: 1/8″ doesn't make a significant difference where tallness is concerned. Three-valued logic has no obvious way to capture "very close to true."[1] The other problem is that the three sets used to interpret predicates—the extension, counterextension, and fringe—require clear cutoff points. Heights of 5′ 11″ and greater might be clearly in the extension of *tall*, heights of 5′ 3″ or less might be clearly in the counterextension of *tall*, and 5′ 7″ might be clearly in the fringe—but can we classify all heights in this way? If so, there is a sharp cutoff point between being *tall* (the extension of *tall*) and being *neither tall nor not tall* (the fringe of *tall*), and one between being *neither tall nor not tall* and being *not tall* (the counterextension of *tall*), something like:

Height

4′ 7″ | 5′ 3″ | 5′ 11″ | 6′ 7″

|————————|————————|————————|

←------- *not tall* -------→ ←*neither tall nor not tall*→ ← -------- *tall* -------- →

Or may be the cutoff points are at 5′ 2″ and 5′ 10″, or . . . ? But wherever we draw them, it doesn't seem true to the facts, to wit: *there are no sharp cutoff points* between an extension, fringe, and counterextension for our ordinary concept *tall* (or for any other vague concepts).[2]

Rather, there are infinitely many ***degrees*** of tallness, which we may indicate with values between 0 and 1 inclusive. Gina Biggerly, at 6′ 7″, is tall to degree 1 (i.e., clearly *tall*), while Tina Littleton, at 4′ 7″, is tall to degree 0 (clearly *not tall*). Mary Middleford, at 5′ 7″, is perhaps tall to degree .5—smack in the middle between being tall and being not tall; Anne, at 5′ 8″, is perhaps tall to degree .6: somewhat closer to

[1] Unless we use *N* to stand for *very close to true*—but then we will have no way also to capture *very close to false*, and so forth.

[2] The claim here is independent of intersubjective agreement or relativity of concepts. Just consider your own idea of tallness for some class of people: you'll find that there are not sharp cutoff points for the application of that concept; any attempt to fix such points seems arbitrary.

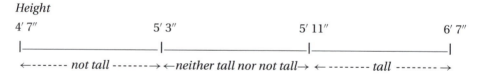

tall than to *not tall.* Crystal, at 5′ 2″, is perhaps tall to degree .1—not as clearly *not* tall as Tina, but almost there.

Deciding exactly how to assign degrees of tallness is interesting, and we'll return to this issue in Chapter 17. But our logic doesn't depend on any specific way of assigning these degrees, so for now we'll introduce *one* way this might be done in order to explore the logic that unfolds. Let's consider only the heights between 4′ 7″ and 6′ 7″ inclusive. We first express the heights on our previous scale as heights in excess of 4′ 7″:

Height

4′ 7″	5′ 3″	5′ 11″	6′ 7″

Height in excess of 4′ 7″

0″	8″	16″	24″

That is, 4′ 7″ is 0 inch greater than 4′ 7″, while 6′ 7″ is 24 inches greater than 4′ 7″. This gives us a scale ranging from 0 to 24. But we want a scale ranging from 0 to 1, so we'll divide these values by 24 to arrive at the degree of tallness for each height:

Height

4′ 7″	5′ 3″	5′ 11″	6′ 7″

Height in excess of 4′ 7″

0″	8″	16″	24″

Degree of tallness

0 (= $^0/_{24}$)	0.333 . . . (= $^8/_{24}$)	0.666 . . . (= $^{16}/_{24}$)	1 (= $^{24}/_{24}$)

Thus 4′ 7″ is *tall to degree 0,* 5′ 3″ is *tall to degree .333 . . . ,* 5′ 11″ is *tall to degree .666 . . . ,* and 6′ 7″ is *tall to degree 1.*

We've just described what is known as a *fuzzy set*: a fuzzy set of heights between 4′ 7″ and 6′ 7″. A non-fuzzy (crisp)[3] set is a collection of entities, such that each entity either is or isn't a member of the set. But with fuzzy sets we don't have entities either being or not being a member of the set; rather, we have entities being members of the set to some degree. More technically, a **fuzzy set** is defined by a function that assigns to each entity in its domain a value between 0 and 1 inclusive, representing the entity's **degree of membership** in the set. 4′ 7″ is a member of our fuzzy set of heights to degree 0; 5′ 3″ is a member to degree .333 . . . ; and so on.

We can in fact describe a crisp set in fuzzy terms: it is a set for which the degree of membership of any entity is either 1 (the entity *is* in the set) or 0 (the entity is *not* in the set). Our three-valued interpretations for vague predicates in earlier chapters may also be characterized as fuzzy sets: the entities in the extension of the predicate

[3] *Crisp* is the fuzzy community's term for classical sets, *crisp* contrasting with *fuzzy*.

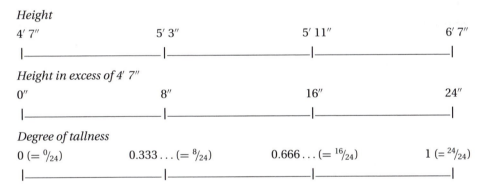

are members of the fuzzy set corresponding to the predicate to degree 1; the entities in the counterextension are members to degree 0; and the entities in the fringe are members to degree .5. So what we have done in this text is to move from two distinct degrees of set membership (classical logic) to three degrees of membership (three-valued logic) to an infinite number of degrees of membership in the sets that are used to interpret predicates.

Just as we used the extension, counterextension, and fringe assigned to predicates to determine the truth-value of simple subject-predicate sentences, we may now use fuzzy sets assigned to predicates to determine truth-values. Our truth-values will be values between 0 and 1 inclusive, and in the case of simple sentences will correspond directly to degrees of membership. If Anne's height is tall to degree 1, for example, we will assign the value 1 to the sentence *Anne is tall*. If her height is tall to degree .2, then we will assign the value .2 to the sentence *Anne is tall*, and so on. We call these values ***degrees of truth***. A logical system in which sentences may have any of an infinite number of degrees of truth (e.g., values between 0 and 1) is an **infinite-valued logical system**. When the bases for assigning the degrees of truth are fuzzy sets, we call the system a ***fuzzy logic***.[4]

The move to fuzzy logic does more than just settle—or eliminate—the problem of exact boundaries for the fringe of a vague predicate. With a variety of degrees of truth we can also address the other major problem that hounds three-valued accounts of vagueness: we can accommodate the intuition that the Principle of Charity is very close to true by assigning it a high degree of truth.[5]

11.2 Łukasiewicz Fuzzy Propositional Logic

We'll use fuzzy sets explicitly in fuzzy first-order logics in Chapter 14, but first we present propositional systems, just as we did for three-valued logic, so that we can examine some of the logical principles in a simpler setting.

To specify a full fuzzy propositional logic, we begin with an assignment V of fuzzy truth-values, between 0 and 1 inclusive, to the atomic formulas of the language. We call the set of real numbers between 0 and 1 inclusive the ***unit interval***, and we use the notation *[0. .1]* to designate the unit interval. So we may say that for each atomic formula **P**, V(**P**) is a member of the unit interval, or in more concise notation, V(**P**) ∈ [0. .1]. We can then use the same numeric clauses that we presented for

4 There is disagreement in the literature about whether this is enough to call a logic *fuzzy* (as opposed to an *infinite-valued logic*), or whether the logic also needs to include fuzzy semantic and syntactic concepts like *n*-degree-validity that will be introduced later in this chapter. Our logics will be fuzzy in any case since we are introducing these latter concepts.

5 Graham Priest (2001) notes that many Sorites paradoxes involve discrete steps (like the tallness version in which we subtract ⅛″ each time) rather than continuous ones (such as a version that said that any reduction of height of ⅛″ or less can't take us from a tall person to one who's not tall) and that in these cases a finite-valued logic would suffice for a solution to the paradox. But he adds (and we agree) that "the continuum-valued semantics [the semantics that includes all real numbers between 0 and 1 as truth-values] is more general, and can be applied to all [S]orites paradoxes, giving, what is clearly desirable, a uniform account" (p. 214).

Łukasiewicz three-valued logic to obtain the Łukasiewicz fuzzy system $\text{Fuzzy}_{\text{Ł}}$:[6]

1. $V(\neg\mathbf{P}) = 1 - V(\mathbf{P})$
2. $V(\mathbf{P} \wedge \mathbf{Q}) = \min(V(\mathbf{P}), V(\mathbf{Q}))$
3. $V(\mathbf{P} \vee \mathbf{Q}) = \max(V(\mathbf{P}), V(\mathbf{Q}))$
4. $V(\mathbf{P} \rightarrow \mathbf{Q}) = \min(1, 1 - V(\mathbf{P}) + V(\mathbf{Q}))$
5. $V(\mathbf{P} \leftrightarrow \mathbf{Q}) = \min(1, 1 - V(\mathbf{P}) + V(\mathbf{Q}), 1 - V(\mathbf{Q}) + V(\mathbf{P}))$
6. $V(\mathbf{P} \,\&\, \mathbf{Q}) = \max(0, V(\mathbf{P}) + V(\mathbf{Q}) - 1)$
7. $V(\mathbf{P} \nabla \mathbf{Q}) = \min(1, V(\mathbf{P}) + V(\mathbf{Q}))$

Some of these connectives can be defined using others, as we did in $Ł_3$. For example, in Section 9.3 of Chapter 9 we noted that $\mathbf{P} \wedge \mathbf{Q}$ is definable as $\mathbf{P} \,\&\, (\neg\mathbf{P} \nabla \mathbf{Q})$ in $Ł_3$, and $V(\mathbf{P} \,\&\, (\neg\mathbf{P} \nabla \mathbf{Q})) = \max(0, V(\mathbf{P}) + \min(1, 1 - V(\mathbf{P}) + V(\mathbf{Q})) - 1) = \max(0, \min(V(\mathbf{P}) + 1 - 1, V(\mathbf{P}) + 1 - V(\mathbf{P}) + V(\mathbf{Q}) - 1)) = \max(0, \min(V(\mathbf{P}), V(\mathbf{Q}))) = \min(V(\mathbf{P}), V(\mathbf{Q}))$.

Assuming that $V(P) = 1$, $V(Q) = .75$, and $V(R) = .5$, here are the values of various compound formulas:

Formula	*Value*
$\neg P$	0
$\neg Q$	.25
$\neg R$	.5
$Q \wedge P$	.75
$Q \wedge R$	.5
$P \wedge \neg P$	0
$Q \wedge \neg Q$	.25
$Q \vee R$	.75
$P \vee \neg P$	1
$R \vee \neg R$	.5
$P \rightarrow Q$	.75
$P \rightarrow R$	.5
$Q \rightarrow R$	.75
$Q \rightarrow Q$	1
$Q \rightarrow \neg Q$	.5
$P \,\&\, Q$	.75
$Q \,\&\, R$	.25
$Q \nabla \neg Q$	0
$P \nabla R$	1
$R \nabla \neg R$	1

[6] Apropos of footnote 4, Łukasiewicz developed an infinite-valued logic but his work predated by a good 40 years the introductions of fuzzy sets and of fuzzy semantic concepts. Łukasiewicz's work on infinite-valued logic is discussed in Łukasiewicz and Tarski (1930). We also note that Łukasiewicz assumed a range of truth-values consisting of the rational numbers, rather than all of the real numbers, in the unit interval. However, it turns out that the set of tautologies obtained when the truth-values are restricted to the rationals is identical to the set of tautologies when all of the real numbers in the unit interval are used as truth-values. See Gottwald (2001, pp. 191–192), for a concise proof of this equivalence.

Note that every formula of the form **P** → **P** will still (as in $Ł_3$) have the value 1, since min (1, 1−V(**P**) + V(**P**)) = 1, no matter what the value V(**P**) is. Every formula of the form **P** ∇ ¬**P** will have the value 1, since V(**P** ∇ ¬**P**) = min (1, V(**P**) + 1 −V(**P**)) = min (1,1). Similarly, every formula of the form ¬(**P** & ¬**P**) will always have the value 1. On the other hand, a formula of the form **P** ∨ ¬**P** may have a value as low as .5 (but no lower)—it has that value when V(**P**) = .5—and the same holds for formulas of the form ¬(**P** ∧ ¬**P**). Thus the Laws of Excluded Middle and Noncontradiction hold in $Fuzzy_Ł$ when expressed using the bold connectives but fail when expressed using the weak connectives.

We remind the reader of the rationale for the truth-conditions of the Lukasiewicz conditional. By the definition in clause 4, whenever V(**P**) is less than or equal to V(**Q**), V(**P** → **Q**) = 1. This is obviously as it should be if **Q** is "truer" than **P**. Whenever V(**P**) is *greater* than V(**Q**), V(**P** → **Q**) = 1 − V(**P**) + V(**Q**), which is 1 − (V(**P**) − V(**Q**)). Thus the value of the conditional is 1 minus the value representing how much "truer" **P** is than **Q**. As Graham Priest explains it: "If A is more true than B, then there is something faulty about the conditional: its truth value must be less than 1. How much less? The amount that the truth value falls in going from A to B. In particular, if it falls all the way from 1 to 0, then the value of A → B is 0" (Priest 2001, p. 215).

11.3 Tautologies, Contradictions, and Entailment in Fuzzy Logic

All of the semantic results for $Ł_3$ in Chapter 5 *except* those concerning expressive power (which we'll address in Section 11.6) carry over to $Fuzzy_Ł$ when we consider weak conjunction and disjunction to be the counterparts of the classical conjunction and disjunction operators. In this section we consider tautologousness, contradictoriness, and entailment.

First, the system $Fuzzy_Ł$ is normal, so we have

Result 11.1: Every $Fuzzy_Ł$ tautology is a classical tautology, and every $Fuzzy_Ł$ contradiction is a classical contradiction,

where a ***tautology*** is defined to be a formula that always has the value 1 and a ***contradiction*** is defined to be a formula that always has the value 0. If we define ***entailment*** to hold between a set of formulas Γ and a formula **P** whenever **P** has the value 1 on any fuzzy truth-value assignment on which all the members of Γ have the value 1, normality also guarantees that

Result 11.2: Every $Fuzzy_Ł$ entailment is a classical entailment.

As a consequence of 11.2, every argument that is valid in $Fuzzy_Ł$ is also classically valid.

The converses do not hold:

Result 11.3: Not every classical tautology is a Fuzzy_{L} tautology, and not every classical contradiction is a Fuzzy_{L} contradiction.

Result 11.4: Not every classical entailment is a Fuzzy_{L} entailment.

The classical tautology $P \vee \neg P$ isn't a Fuzzy_{L} tautology, and the classical contradiction $P \wedge \neg P$ isn't a Fuzzy_{L} contradiction. The classically valid argument

$$\frac{\neg (P \leftrightarrow Q)}{(P \leftrightarrow R) \vee (Q \leftrightarrow R)}$$

fails to be valid in Fuzzy_{L}. When P has the value 1, Q has the value 0, and R has the value .5, the premise has the value 1 but the conclusion has the value .5.

Because bold conjunction and disjunction are also normal connectives, Results 11.1 and 11.2 hold as well when the bold connectives are substituted for the classical connectives. Moreover, results 11.3 and 11.4 also hold when we substitute bold connectives for the classical ones. Both $P \rightarrow (P \,\&\, P)$ and $(P \nabla P) \rightarrow P$ fail to be Fuzzy_{L} tautologies although $P \rightarrow (P \wedge P)$ and $(P \vee P) \rightarrow P$ *are* classical tautologies. When P has the value .5, for example, so do both $P \rightarrow (P \,\&\, P)$ and $(P \nabla P) \rightarrow P$. The argument

$$\frac{P \nabla P}{P}$$

is not valid in Fuzzy_{L}—when P has the value .5, the premise of the argument has the value 1 while the conclusion has the value .5—but the argument

$$\frac{P \vee P}{P}$$

is classically valid.

Because every truth-value assignment for Ł3 is also a truth-value assignment for Fuzzy_{L}, we have the following additional result:

Result 11.5: Every Fuzzy_{L} tautology is an Ł3 tautology, and every Fuzzy_{L} contradiction is an Ł3 contradiction.

That is, a formula that has the value 1 on every Fuzzy_{L} assignment must thereby be true on every Ł3 assignment, and similarly for formulas that have the value 0 on every Fuzzy_{L} assignment.[7] The converse, however, does not hold:

Result 11.6: Not all Ł3 tautologies are Fuzzy_{L} tautologies, and not all Ł3 contradictions are Fuzzy_{L} contradictions.

[7] Note that we cannot replace a bold connective of Fuzzy_{L} with a weak connective of Ł3 and expect the result to obtain. For example, $P \nabla \neg P$ is a tautology of Fuzzy_{L} but $P \vee \neg P$ isn't a tautology of Ł3. The result does hold whenever we use exactly the *same* connectives—weak or strong—in both Fuzzy_{L} and Ł3.

An example Ł$_3$ tautology is the formula $(P \rightarrow \neg P) \rightarrow P) \rightarrow P$ (this can be expressed as $(P \rightarrow \neg P) \vee P$ using disjunction: *either P implies ¬P, or P*). When V(P) = .7 in Fuzzy$_Ł$, for example, V($(P \rightarrow \neg P) \rightarrow P) \rightarrow P$) = .7 as well (as the reader can easily verify); so the formula isn't a Fuzzy$_Ł$ tautology. The difference in the formula's behavior in the two systems shouldn't be surprising: the only case where $(P \rightarrow \neg P)$ isn't true in Ł$_3$, that is, where P is truer than $\neg P$, occurs when P has the value **T**, making the second disjunct of $(P \rightarrow \neg P) \vee P$ true. But in Fuzzy$_Ł$, obviously, P will be truer than $\neg P$ whenever P has a value greater than .5 – so the nontruth of $(P \rightarrow \neg P)$ doesn't require P to be true. Comparison of Ł$_3$ and Fuzzy$_Ł$ entailment is left as an exercise.

We found that the set of tautologies of classical propositional logic as well as the sets of tautologies in each of our three-valued propositional systems are all decidable, because in each case we can construct a truth-table to decide the status of a formula. We can't generalize this decision method to *fuzzy* propositional logics because it is impossible to produce a truth-table for all fuzzy truth-value assignments. Consider the simple case of the negated atomic formula $\neg P$: a truth-table for $\neg P$ in a fuzzy system would have to list the infinitely many values from the unit interval that can be assigned to P! So the question arises; Are there *other* decision procedures for the sets of tautologies of fuzzy propositional systems? The answer, interestingly, is *yes* for Fuzzy$_Ł$ (and for each of the systems presented in this chapter)—and that means that the set of theorems of the basic axiomatic system for Fuzzy$_Ł$ will also be decidable.

Several decision procedures for Fuzzy$_Ł$ tautologousness are known; we'll describe a procedure from Aguzzoli and Ciabattoni (2000).[8] First, we define n-valued Łukasiewicz propositional logics for each integer $n \geq 3$: Ł$_n$ is the propositional logic whose truth-values are the n members of the set $\{0, 1 / (n-1), \ldots, (n-2) / (n-1), 1\}$ and whose truth-conditions are the clauses 1–7 for Ł$_3$ (stated at the beginning of Section 11.2). Note that when $n = 3$ the truth-value set is the one we used for Ł$_3$ in Chapter 9: $\{0, {}^1\!/_2, 1\}$.

Second, we define #O(**P**) to be the number of *occurrences* of atomic formulas in **P**. Thus, #O(B) and #O($\neg B$) are both 1, while both #O($B \bigtriangledown C$) and #O($B \bigtriangledown B$) are 2. Aguzzoli and Ciabattoni proved that a formula **P** containing only negation and bold disjunction as connectives is a tautology of Fuzzy$_Ł$ if and only if **P** is a tautology of Ł$_n$ where $n = 2^{\#O(\mathbf{P})} + 1$. Thus, for example, the formula $B \bigtriangledown \neg B$ is a tautology of Fuzzy$_Ł$ if and only if it is a tautology of Ł$_5$, while the formula $\neg B \vee (B \bigtriangledown \neg C)$ is a tautology of Fuzzy$_Ł$ if and only if it is a tautology of Ł$_9$.

Now, the set of tautologies in any finite-valued Łukasiewicz propositional logic is decidable by using truth-tables. Moreover, every Fuzzy$_Ł$ formula can be mechanically converted to an equivalent Fuzzy$_Ł$ formula containing only negation and bold disjunction (proof is left as an exercise), so we can decide whether a formula **P** of

8 Actually, Aguzzoli and Ciabattoni only prove the *if* part; the *only if* part was established in Ackermann (1967, pp. 60–63). Another decision procedure for Fuzzy$_Ł$ is presented in Gottwald (2001, Section 9.1.4).

Fuzzy_{L} is a tautology by converting it to an equivalent formula **Q** containing only negation and bold disjunction and then testing **Q** for tautologousness in L_n, where $n = 2^{\#O(\mathbf{Q})} + 1$.

11.4 *N*-Tautologies, Degree-Entailment, and *N*-Degree-Entailment

We could introduce fuzzy counterparts to our three-valued concepts of quasi-tautology, quasi-contradiction, and quasi-entailment by designating all values that are greater than or equal to .5, that is, all values in [0.5 ... 1], and anti-designating all values that are less than or equal to .5, that is, all values in [0 ... 0.5]. Chapter 5's results comparing these quasi-concepts in L_3 with classical logic also hold for Fuzzy_{L}—for example, every Fuzzy_{L} quasi-tautology is a classical tautology, but not vice versa (and the results also hold when we use the Fuzzy_{L} *bold* connectives in place of the classical ones). However, we will not explore this generalization of the quasi-concepts since it is the custom—and it is also quite illuminating—to introduce a different set of semantic concepts reflecting the greater range of truth-values in fuzzy logical systems.

First we need some definitions. Relative to any set R of real numbers in [0. .1], we define the *greatest lower bound* of R as follows: a real number *n* in [0. .1] is the **greatest lower bound** of R if and only if

a. *n* is less than or equal to each member of R, and
b. there is no real number *m* in [0. .1] such that *m* is also less than or equal to each member of R but *m* is greater than *n*.

We use the expression ***glb(R)*** to denote the greatest lower bound of R.[9] Condition (a) specifies that *n* is a **lower bound** for R, and (b) specifies that *n* is the *greatest* of the lower bounds. Every set R of real numbers in [0. .1] is guaranteed to have a greatest lower bound (see, for example, Stoll [1961], Chapter 3.6). If a set R has a minimum member, then that minimum is glb(R). For example, glb([0. .1]) is 0, and glb({.2, .3, .8}) is .2. But not every set of real numbers has a minimum member, so we need the more general notion of greatest lower bound. For example, consider the set consisting of 1, .5, .25, .125, .0625, and each subsequent real number in this series—each real that is the result of multiplying the previous one by .5. There is no smallest member in this series; for each member, you can get a smaller one by multiplying it by .5. However, there is a greatest lower bound, in this case 0. It is obviously a lower bound. It is the greatest such because for any real number *m* greater than 0, there is a member of the preceding series that is even smaller than *m*.

We can now define the concept of an *n-tautology*, where *n* is any real number in the unit interval: a formula is an ***n*-tautology** in Fuzzy_{L} if (and only if) *n* is the greatest lower bound of the set of truth-values that the formula can have. We must define this

9 The greatest lower bound of R is also called the *infimum* of R, or *inf(R)* for short.

concept in terms of the greatest lower bound of a formula's possible truth-values, rather than the minimum such truth-value, because not all sets of fuzzy truth-values have minimum members. Tautologies per se as we have defined them are then 1-tautologies. By our definition, we can also have 0-tautologies, perhaps somewhat oddly called; these are formulas that can only have the value 0 (or values arbitrarily close to 0).[10]

The formula $P \vee \neg P$ is a .5-tautology of Fuzzy_L. We claimed earlier that this formula has the value .5 when P has the value .5, and that it has no lower value. This is because $V(P \vee \neg P) = \max(V(P), 1 - V(P))$. If $V(P) = .5$, then $\max(V(P), 1 - V(P)) = .5$. If $V(P) \neq .5$, then either $V(P)$ or $1 - V(P)$ is greater than .5, so $\max(V(P), 1 - V(P)) > .5$. In fact, we can use normal forms to prove that every classical tautology involving only the connectives $\neg$, $\wedge$, and $\vee$ is at least a .5-tautology in Fuzzy_L (using the weak connectives for conjunction and disjunction).[11]

In our proof we will also use the concept of an *n-contradiction*, for which we need to define the concept of a *least upper bound*: a number n in [0. .1] is the ***least upper bound*** of a set R of real numbers in [0. .1] (denoted as ***lub(R)***)[12] if and only if (a) n is greater than or equal to each member of R (thus n is an upper bound), and (b) there is no m in [0. .1] such than m is also greater than or equal to each member of R but m is less than n (n is the *least* upper bound). We define a formula of Fuzzy_L to be an ***n-contradiction*** if n is the least upper bound of the set of truth-values that the formula can have.

The following normal form equivalences hold in Fuzzy_L:

$\neg(\mathbf{P} \wedge \mathbf{Q})$	is equivalent to	$\neg\mathbf{P} \vee \neg\mathbf{Q}$	(*DeMorgan's Law*)
$\neg(\mathbf{P} \vee \mathbf{Q})$	is equivalent to	$\neg\mathbf{P} \wedge \neg\mathbf{Q}$	(*DeMorgan's Law*)
$\mathbf{P}$	is equivalent to	$\neg\neg\mathbf{P}$	(*Double Negation*)
$(\mathbf{P} \vee \mathbf{Q}) \wedge \mathbf{R}$	is equivalent to	$(\mathbf{P} \wedge \mathbf{R}) \vee (\mathbf{Q} \wedge \mathbf{R})$	(*Distribution*)
$\mathbf{P} \wedge (\mathbf{Q} \vee \mathbf{R})$	is equivalent to	$(\mathbf{P} \wedge \mathbf{Q}) \vee (\mathbf{P} \wedge \mathbf{R})$	(*Distribution*)
$(\mathbf{P} \wedge \mathbf{Q}) \vee \mathbf{R}$	is equivalent to	$(\mathbf{P} \vee \mathbf{R}) \wedge (\mathbf{Q} \vee \mathbf{R})$	(*Distribution*)
$\mathbf{P} \vee (\mathbf{Q} \wedge \mathbf{R})$	is equivalent to	$(\mathbf{P} \vee \mathbf{Q}) \wedge (\mathbf{P} \vee \mathbf{R})$	(*Distribution*)

(The reader will be asked to verify these in an exercise.) Given these equivalences, it follows that every Fuzzy_L formula containing only negation, weak conjunction, and weak disjunction is equivalent to a formula in conjunctive normal form and to one in disjunctive normal form, where the normal forms are defined as before using the weak connectives:

A literal is a phrase.
If **P** and **Q** are phrases, so is $(\mathbf{P} \wedge \mathbf{Q})$.
Every phrase is in disjunctive normal form.

[10] Novák, Perfilieva, and Močkoř (1999) also acknowledge this oddity, commenting that it is a consequence of the fuzzy "principle of equal importance of [all] truth-values" (p. 102).

[11] This was first proved by Lee and Chang (1971). In particular, Lee and Chang are responsible for Results 11.7–11.11.

[12] The least upper bound of a set is also called the *supremum* of the set, or *sup* for short.

If **P** and **Q** are in disjunctive normal form, so is **(P ∨ Q)**.
A literal is a clause.
If **P** and **Q** are clauses, so is **(P ∨ Q)**.
Every clause is in conjunctive normal form.
If **P** and **Q** are in conjunctive normal form, so is **(P ∧ Q)**.

We now prove

Result 11.7: A clause of Fuzzy_L is a .5-tautology if and only if it contains a complementary pair of literals.

Proof: If a clause **C** is a .5-tautology, then it must contain a complementary pair of literals, for if it didn't then every literal in **C** could be assigned a value less than .5 since the literals would have independent truth-conditions, and so **C** itself would have a value less than .5. Conversely, if **C** contains a complementary pair of literals then on any truth-value assignment at least one of the literals in the pair will have the value .5 or greater, by the truth-condition for negations. Thus the disjunction **C** itself will have at least the value .5. Moreover, if all of the literals in **C** have the value .5 on an assignment, then so will **C**; so .5 is the *greatest* lower bound of the values that **C** can have.

Result 11.8: A phrase of Fuzzy_L is a .5-contradiction if and only if it contains a complementary pair of literals.

Proof: Left as an exercise.

As a consequence of these two results we have:

Result 11.9: A formula **P** of Fuzzy_L that is in conjunctive normal form is a .5-tautology if and only if each clause in **P** contains a complementary pair of literals.

Proof: The value of a (weak) conjunction in Fuzzy_L is the minimum of the values of the conjuncts, so the greatest lower bound of the values the conjunction can have is the greatest lower bound of the values of its conjuncts. Thus **P** is a .5-tautology if and only if all of the conjuncts (clauses) are .5-tautologies (a clause cannot be greater than a .5-tautology because it will have the value .5 whenever all of its literals do), and by Result 11.7 all of these clauses will be .5-tautologies if and only if each of them contains a complementary pair of literals.

Result 11.10: A formula **P** of Fuzzy_L that is in disjunctive normal form is a .5-contradiction if and only if each phrase in **P** contains a complementary pair of literals.

Proof: Left as an exercise.

In Chapter 2 we proved analogues of Results 11.9 and 11.10 for classical propositional logic, except that in the classical case the results state that **P** is, respectively, a tautology or a contradiction. Because every formula in classical logic is equivalent to one in either conjunctive or disjunctive normal form, and every formula of $Fuzzy_L$ that contains only negation, weak conjunction, and/or weak disjunction is equivalent to a formula in either conjunctive or disjunctive normal form, it follows that

> *Result 11.11:* A formula of $Fuzzy_L$ that contains only negation, weak conjunction, and weak disjunction as connectives is a .5-tautology if and only if it is a classical tautology, and is a .5-contradiction if and only if it is a classical contradiction.

Given Result 5.15 of Chapter 5, it also follows that

> *Result 11.12:* A formula of $Fuzzy_L$ that contains only negation, weak conjunction, and weak disjunction as connectives is a .5-tautology if and only if it is a quasi-tautology in L_3 and is a .5-contradiction if and only if it is a quasi-contradiction in L_3.

Classical tautologies that contain the conditional or the biconditional can fail to be at least .5-tautologies in $Fuzzy_L$. One example is the formula $\neg(P \wedge (P \rightarrow \neg P))$.[13] The value of the formula is $1 - \min(V(P), 2 - (2 \cdot V(P)))$, and this is (for example) .4 when $V(P)$ is .6 or .7. It is left to the reader to prove that this formula is specifically a $^1/_3$-tautology. The classically tautologous formulas $\neg(A \rightarrow \neg A) \vee \neg(\neg A \rightarrow A)$ and $\neg(A \leftrightarrow \neg A)$, formulas that can have the value **F** in Ł$_3$ as we showed in Chapter 5, have the value 0 in $Fuzzy_L$ when $V(A) = .5$. So we have

> *Result 11.13:* Not every formula that is a tautology in classical logic is at least a .5-tautology in $Fuzzy_L$; indeed some classical tautologies are 0-tautologies in $Fuzzy_L$.

Analogous counterexamples can be found when we substitute the bold connectives for classical conjunction or disjunction; the reader will be asked to explore these in an exercise.

We revise the three-valued definition of *degree-entailment* so that it applies now to $Fuzzy_L$: We will say that a set of formulas Γ **degree-entails** a formula **P** if on every fuzzy truth-value assignment the value of **P** is greater than or equal to the greatest lower bound of the values of members of Γ on that assignment. (In the case where Γ is a finite set, this can be restated as: on every fuzzy truth-value assignment, **P** is at least as true as the least true member of Γ—because a finite set has a finite number of values on a given assignment, and hence there is a minimum member of that set of values.) An argument is **degree-valid** in $Fuzzy_L$ if the set of its premises degree-entails its conclusion. Obviously, every argument that is degree-valid in $Fuzzy_L$ is

[13] This example is from Kenton F. Machina (1976). Machina attributes the example to Lawrence Eggan.

also valid *simpliciter* in Fuzzy_L; but the converse does not hold. For example, the argument

$$\frac{\text{P}}{\neg(\text{P} \rightarrow \neg\text{P})}$$

is valid *simpliciter* in Fuzzy_L (and therefore also in classical logic), but it is not degree-valid in Fuzzy_L since the conclusion has the value 0 when the single premise has the value .5. More strikingly, the argument

$$\begin{array}{l} \text{P} \\ \underline{\text{P} \rightarrow \text{Q}} \\ \text{Q} \end{array}$$

also fails degree-validity (although it is valid *simpliciter*). If $V(P) = .5$ and $V(Q) = .4$, for example, then $V(P \rightarrow Q) = .9$ and so the least value of the premises is .5 but the conclusion has the value .4.

On the other hand, of course, some valid arguments in Fuzzy_L *are* also degree-valid, for example, the argument

$$\frac{\text{P}}{\text{P} \vee \text{Q}}$$

This is because the value of $P \vee Q$ is the maximum of the values of P and Q and so cannot be less than the value of P. Here then are the relevant results:

Result 11.14: Every Fuzzy_L degree-entailment is a classical entailment.

Result 11.15: Not every classical entailment is a Fuzzy_L degree-entailment.

Degree-validity plays an important role in dealing with arguments with vague premises that may be less than completely true. In such a case (as with the Sorites, which we will discuss in Chapter 14), we may well be interested in whether the conclusion is *at least* as true as the least true premise.

The concept of degree-validity captures whether the conclusion of an argument preserves the smallest degree of truth of its premises. We can go even further and ask, If an argument such as

$$\begin{array}{l} \text{P} \\ \underline{\text{P} \rightarrow \text{Q}} \\ \text{Q} \end{array}$$

is not degree-valid, then *to what extent does it approximate* degree-validity? We shall measure this as a function of how much truth can be lost going from the premises to the conclusion. When the difference between the greatest lower bound of the truth-values of members of a set of formulas Γ on a fuzzy truth-value assignment V and the truth-value of a formula **P** on that assignment is greater than 0, we will call this difference the ***downward distance*** between Γ and **P** on V. If the difference is less than or equal to 0, we will say that the downward distance between Γ and **P** on V is 0. For example, the downward distance between the set of formulas $\{P, P \rightarrow Q\}$ and the formula Q on a fuzzy assignment V such that $V(P) = .9$ and $V(Q) = .2$ is .3 – .2 or .1, since the greatest lower bound of values of formulas in the set is .3, the value of $P \rightarrow Q$. The ***maximum downward distance*** between Γ and **P** is the least upper bound of the downward distances between Γ and **P** on any truth-value assignment. Finally, we will say that a set of formulas Γ ***n-degree-entails*** **P** if $1 - n$ is the maximum downward distance between Γ and **P**, and that an argument is ***n-degree-valid*** if the set of its premises n-degree-entails its conclusion. So, for example, if the maximum downward distance between Γ and **P** is .3, then Γ .7-degree-entails **P**, but if the maximum downward distance is .7, then Γ only .3-degree-entails **P**. We note that 1-degree-validity and 1-degree-entailment coincide with degree-validity and degree-entailment *simpliciter*, since by definition the former cases have a maximum downward distance of 0.

To determine the n-degree-validity of the argument

P
P → Q
———
Q

in Fuzzy$_Ł$, we must find the maximum downward distance between the set of formulas $\{P, P \rightarrow Q\}$ and the formula Q. We first note that when the least truth-value of one of the formulas P and $P \rightarrow Q$ is 1—that is, when both formulas are true, the downward distance to the value of Q is 0 since Q must have the value 1 in this case. More generally, the downward distance must be 0 whenever $V(P) \leq V(Q)$, trivially. What is the downward distance when $V(P) > V(Q)$? Well, in this case $V(P \rightarrow Q)$ is $1 - V(P) + V(Q)$, so $V(Q) = V(P \rightarrow Q) - 1 + V(P)$. Consequently, when $V(P) > V(Q)$ the gap between the value of the least true of the formulas $\{P, P \rightarrow Q\}$ and that of Q is $\min(V(P), V(P \rightarrow Q)) - (V(P \rightarrow Q) + V(P) - 1)$ or $\min(1 - V(P), 1 - V(P \rightarrow Q))$. We now consider three cases:

a. If $V(P) = .5$, then $1 - V(P) = .5$, $V(P \rightarrow Q) \geq 5$, and $1 - V(P \rightarrow Q)) \leq .5$. In this case, then, $\min(1 - V(P), 1 - V(P \rightarrow Q))$ may be as great as .5 (as will happen when $V(Q) = 0$).
b. If $V(P) > .5$ then $1 - V(P) < .5$, and so $\min(1 - V(P), 1 - V(P \rightarrow Q)) < .5$.
c. If $V(P) < .5$ then $V(P \rightarrow Q) > .5$, so $1 - V(P \rightarrow Q) < .5$ and $\min(1 - V(P), 1 - V(P \rightarrow Q)) < .5$.

Thus the greatest downward distance when $V(P) > V(Q)$ is .5. It follows that the maximum downward distance between the set of formulas $\{P, P \rightarrow Q\}$ and the formula Q in any case is .5 and that the argument

P
P → Q
———
Q

is therefore (1 – .5)-degree-valid, or .5-degree-valid.

Interestingly, we have

Result 11.16: Every n-degree-entailment in $\text{Fuzzy}_{\text{Ł}}$ with $n > 0$ is also a classical entailment (whether we use the weak or the bold connectives as the counterparts to the classical ones).

Proof: If a set Γ n-degree-entails a formula **P** with $n > 0$ then by definition the downward distance between the (greater lower bound of) values of members of Γ and **P** on any fuzzy assignment V is less than 1. Because $\text{Fuzzy}_{\text{Ł}}$ is normal, it follows that any classical valuation that makes all of the members of Γ true must make **P** true as well—because if **P** were false the downward distance in this case would not be less than 1.

On the other hand, we have

Result 11.17: Some classical entailments are only n-degree-entailments in $\text{Fuzzy}_{\text{Ł}}$ for very small values of n, including 0.

Proof: An example of a classically valid argument that is 0-degree-valid in $\text{Fuzzy}_{\text{Ł}}$ is

$$\frac{\text{B}}{\neg\,(\text{A} \rightarrow \neg\text{A}) \vee \neg(\neg\text{A} \rightarrow \text{A})}$$

The argument is classically valid because the conclusion is a classical tautology. However, as we saw earlier, the conclusion will have the value in 0 in $\text{Fuzzy}_{\text{Ł}}$ when $\text{V}(A) = .5$. Since $\text{V}(B)$ can be 1 when $\text{V}(A) = .5$, the maximum downward movement from the truth-value of the single premise in $\text{Fuzzy}_{\text{Ł}}$ to the truth-value of the conclusion is 1 and the argument is thus 0-degree-valid.

Another important (for our purposes) example illustrating Result 11.17 is the classically valid argument

$$\begin{array}{l} \text{A} \\ \text{A} \rightarrow \text{B} \\ \text{B} \rightarrow \text{C} \\ \text{C} \rightarrow \text{D} \\ \text{D} \rightarrow \text{E} \\ \text{E} \rightarrow \text{F} \\ \text{F} \rightarrow \text{G} \\ \text{G} \rightarrow \text{H} \\ \text{H} \rightarrow \text{I} \\ \underline{\text{I} \rightarrow \text{J}} \\ \text{J} \end{array}$$

This argument is .1-degree-valid in $\text{Fuzzy}_{\text{Ł}}$. When $\text{V}(A) = .9$, $\text{V}(B) = .8$, $\text{V}(C) = .7, \ldots,$ $\text{V}(I) = .1$ and $\text{V}(J) = 0$, all of the premises have value .9 but the conclusion has the value 0. So this argument is at most .1-degree-valid. It is left as an exercise to prove that it is *exactly* .1-degree-valid. If we add another conditional premise $J \rightarrow K$ and let K be the conclusion of the new argument, we will have an argument with an even smaller n-degree-validity. Assigning $^{10}/_{11}$ to A, $^{9}/_{11}$ to B, $^{8}/_{11}$ to $C, \ldots$, $^{1}/_{11}$ to J, and 0 to K, all of the conditional premises will have the value $^{10}/_{11}$, as will A, and so the downward distance from the premises to the conclusion in this case is $^{10}/_{11}$ giving the new argument at most $^{1}/_{11}$-degree-validity.

More generally, if we increase the number of atomic formulas to m, add corresponding conditional formulas, and set the consequent of the last conditional as the argument's conclusion, the resulting argument will be at most $1/m$-degree-valid. We can assign $(m-1)/m$ to A, $(m-2)/m$ to B, $(m-3)/m$ to $C, \ldots$, and 0 to the conclusion in which case the conditional premises will all have the value $(m-1)/m$. So the downward distance to the conclusion will also be $(m-1)/m$, making the argument at most $1/m$-degree-valid. There is no upper limit on the number of premises an argument can have, so as we add more atomic formulas the n-degree-validity of longer arguments of this form will approach (but never reach) 0. Yet every such argument is classically valid. The astute reader will have noticed the affinity between this pattern of argument and the Sorites arguments—and hence may surmise that when we turn to the first-order version of Fuzzy$_{\text{L}}$, we will find that Sorites arguments have very low n-degree-validity.

11.5 Fuzzy Consequence

It is customary in fuzzy logic to analyze arguments semantically in yet another way, using the concept of *fuzzy consequence*. First, some preliminaries: A fuzzy truth-value assignment defines a fuzzy set of the formulas of Fuzzy$_{\text{L}}$, where each formula's degree of membership in the set is its degree of truth on that truth-value assignment. Let us call every fuzzy truth-value assignment that makes each formula in a fuzzy set Φ at least as true as its degree of membership in Φ a ***consonant truth-value assignment*** for the set Φ. So, for example, for the fuzzy set Φ that contains the formula A to degree 1, the formula $A \wedge B$ to degree .3, and all other formulas to degree 0, the consonant truth-value assignments include all (and only) truth-value assignments that assign 1 to A, any value $\geq$ .3 to B, and any combination of values to the other atomic formulas.[14]

The **fuzzy consequence** of a fuzzy set of formulas Φ is defined to be another fuzzy set, the fuzzy set in which the degree of membership for a formula **P** is the greatest lower bound of the truth-values that **P** can have on any consonant truth-value assignment for Φ. We may summarize this loosely by saying that a formula **P** is a member to degree n in the fuzzy consequence of a fuzzy set of formulas Φ if the

[14] Note that some fuzzy sets have no consonant truth-value assignments. For example, any fuzzy set in which the Fuzzy$_{\text{L}}$ formula $\neg(P \to P)$ has a degree of membership greater than 0 does not have any consonant truth-value assignments since there is no fuzzy truth-value assignment that will give this formula a value greater than 0. Similarly, there are no consonant truth-value assignments for any fuzzy set that contains P with a degree of .5 and $\neg P$ with a degree of .6.

value of **P** is guaranteed to be at least n given the truth-degrees of the members of Φ. We'll use *FC(Φ)* to denote the fuzzy consequence of Φ.

As an example, consider the fuzzy set Φ of formulas in which P is a member to degree 1, $P \rightarrow Q$ is a member to degree .9, and all other formulas of the language are members to degree 0. The consonant truth-value assignments for Φ are those fuzzy assignments that assign P the value 1 and that assign Q a value between .9 and 1 inclusive. FC(Φ) is the fuzzy set that assigns to each formula of Fuzzy$_{Ł}$ the greatest lower bound of its truth-values on the consonant truth-value assignments for Φ. A formula **R** that does not contain P or Q will be a member of FC(Φ) to degree n if it is an n-tautology. This is because the consonant truth-value assignments for Φ are free to assign any combination of values to atomic formulas other than P and Q, and so they will between them assign all the possible values that **R** can have on any truth-value assignment. So, for example, the formula S is a member of FC(Φ) to degree 0, and $S \vee \neg S$ is a member to degree .5. P and $P \rightarrow Q$ are members of FC(Φ) to the degree that they are members of Φ, namely, 1 and .9. Q is a member of FC(Φ) to degree .9 because it has at least that value on any consonant truth-value assignment for Φ. $P \wedge Q$ is also a member of FC(Φ) to degree .9, while $P \wedge \neg Q$ is a member to degree 0. The latter is because some consonant truth-value assignments for Φ assign the value 1 to Q; these assignments give $\neg Q$ (and hence $P \wedge \neg Q$) the value 0.

A special case of fuzzy consequence occurs when the fuzzy set Φ has no consonant truth-value assignments. In this case FC(Φ) is the fuzzy set that contains every formula to degree 1. This is because the set of values assigned to any formula by the consonant truth-value assignments is the empty set $\varnothing$, and 1 is the greatest lower bound (in the unit interval) of all the values in the empty set: it is trivially true that every member of $\varnothing$ is at least as true as 1, since there are no such members.

Another, more interesting, special case of fuzzy consequence arises when a fuzzy set Φ of formulas of Fuzzy$_{Ł}$ is crisp, that is, when every formula of Fuzzy$_{Ł}$ is a member of Φ to either degree 1 or degree 0. The fuzzy consequence of a crisp set is a fuzzy set in which the formulas that are entailed *simpliciter* in Fuzzy$_{Ł}$ by the (degree 1) members of Φ have the value 1, and all other formulas have a degree less than 1. So we may define entailment *simpliciter* in Fuzzy$_{Ł}$ in terms of fuzzy consequence in Fuzzy$_{Ł}$: a set of formulas Γ entails the formula **P** in Fuzzy$_{Ł}$ if **P** is a member to degree 1 of the fuzzy consequence of the crisp set in which the formulas of Γ are members to degree 1 and all other formulas are members to degree 0. More generally, degree-entailment (of which entailment *simpliciter* is a special case) is definable in terms of fuzzy consequence: a set of formulas Γ degree-entails the formula **P** if for every n in [0. .1], **P** is a member to at least degree n of the fuzzy consequence of each fuzzy set in which the greatest lower bound of the values of members of Γ is n and all other formulas of Fuzzy$_{Ł}$ have the value 0. N-degree-entailment is also definable in terms of fuzzy consequence—this is left as an exercise for the reader.

Some theoreticians feature the concept of fuzzy consequence as the centerpiece of fuzzy logic. But others, as we noted in footnote 4 to Section 11.1, merely require an infinite number of truth-values. Our graded axiomatic systems for fuzzy logic will capture the concept of fuzzy consequence, although we will also have ungraded axiomatic systems that simply capture the concept of validity *simpliciter*. In this book we take an inclusive view of what fuzzy logic is, since in our investigations of vagueness it is illuminating to study not only fuzzy consequence, but all of the infinite-valued concepts that we have introduced in this section.

11.6 Fuzzy Generalizations of K^S_3, B^I_3, and B^E_3; the Expressive Power of Fuzzy$_Ł$

In Chapter 5 we saw that the connectives of K^S_3 are definable in $Ł_3$. The definitions of the connectives of K^S_3 for numerical values generalize to an infinite number of truth-values, and all of these generalizations are definable in Fuzzy$_Ł$. Fuzzy$_{KS}$ negation, conjunction, and disjunction coincide with the Fuzzy$_Ł$ (weak) versions of those connectives because the numerical definitions of the connectives coincide in the three-valued case. The clauses for the Fuzzy$_K$ conditional and biconditional are

$$V(\mathbf{P} \rightarrow_K \mathbf{Q}) = \max(1-V(\mathbf{P}), V(\mathbf{Q}))$$
$$V(\mathbf{P} \leftrightarrow_K \mathbf{Q}) = \min(\max(1-V(\mathbf{P}), V(\mathbf{Q})), \max(1-V(\mathbf{Q}), V(\mathbf{P})))$$

and these are the truth-conditions we get when we define these connectives in Fuzzy$_Ł$ just as we did in the three-valued case:

$$\mathbf{P} \rightarrow_K \mathbf{Q} =_{\text{def}} \neg\mathbf{P} \vee \mathbf{Q}$$
$$\mathbf{P} \leftrightarrow_K \mathbf{Q} =_{\text{def}} \neg(\mathbf{P} \wedge \neg\mathbf{Q}) \wedge \neg(\neg\mathbf{P} \wedge \mathbf{Q})$$ [15]

Unlike Fuzzy$_Ł$ conditionals, not all conditionals of the form $\mathbf{P} \rightarrow_K \mathbf{P}$ are 1-tautologies; their value can drop as low as .5 when **P** has the value .5 (but no lower; every conditional of the form $\mathbf{P} \rightarrow_K \mathbf{P}$ is at least a .5-tautology).

Nicholas Rescher (1969) proposed an infinite generalization of B^I_3 in which the truth-clause for negation is identical to Łukasiewicz's infinite-valued negation and the truth-clauses for the binary connectives are based on the rule that a formula will have the value .5 if it contains any components with truth-values other than the classical values 1 and 0. So, for example, the clause for conjunction is

$$V(\mathbf{P} \wedge_{BI} \mathbf{Q}) = \begin{cases} 1/2 & \text{if } 0 < V(\mathbf{P}) < 1 \text{ or } 0 < V(\mathbf{Q}) < 1 \\ \min(V(\mathbf{P}), V(\mathbf{Q})) & \text{otherwise} \end{cases}$$

None of the binary connectives for the resulting system, Fuzzy$_{BI}$, are definable in Fuzzy$_Ł$. This is because the truth-functions for these Fuzzy$_{BI}$ connectives are not *continuous*.

[15] In these definitions as well as everywhere else in this and subsequent chapters we are using the convention, introduced in Chapter 6, that connectives without subscripts are the Łukasiewicz connectives. In this chapter, of course, they are the Łukasiewicz connectives *for fuzzy logic*.

A continuous unary function f (that is, a continuous function of one argument) over the unit interval meets the intuitively stated criterion that as m approaches n, for any value n in the unit interval, f(m) approaches f(n). A precise definition tells us that small changes in m result in small changes in f(m): a unary function f over the unit interval is **continuous** if for any $\varepsilon > 0$ there is a $\delta > 0$ (with ε and δ members of the unit interval) such that whenever $| m_1 - m_2 | < \delta$, $| f(m_1) - f(m_2) | < \varepsilon$.

The negation function defined as $f(m) = 1 - m$ is continuous: as m approaches (gets closer to) 1, for example, f(m) approaches 0, and the same is true for any value n that m may be approaching. But the binary connectives of $Fuzzy_{BI}$ are *not* continuous. A *binary* function f over the unit interval is continuous if for any $\varepsilon > 0$ there is a $\delta > 0$ (with ε and δ members of the unit interval) such that whenever $| m_1 - m_3 | < \delta$ and $| m_2 - m_4 | < \delta$, $| f(m_1, m_2) - f(m_3, m_4) | < \varepsilon$. Intuitively, the binary $Fuzzy_{BI}$ connectives fail to be continuous because as the truth-values of their arguments approach 1 from anywhere above 0, the functions' values stay fixed at .5 and then noncontinuously jump to either 1 or 0 once the arguments have both reached 1. For example, if the truth-values of P and Q are .1 and .2, then $V(P \wedge_{BI} Q) = .5$, and if we increase the values of P and Q the value of the conjunction remains at .5 until the values of P and Q both reach 1, in which case the value of the conjunction noncontinuously jumps to 1. More technically in this case, consider the value .5 for ε. Continuity would require that there is some $\delta > 0$ such that whenever the values of P and of Q change by less than δ, the value of $P \wedge_{BI} Q$ changes by less than .5. But there is a jump in the value of $P \wedge_{BI} Q$ from .5 to 1 when we take *any* values of P and Q that are arbitrarily close to 1 and change them to 1, so there can be no such value.

In 1951 Robert McNaughton proved that a function must at least be continuous to be definable in $Fuzzy_L$, so it follows that the binary $Fuzzy_{BI}$ functions are therefore not definable in $Fuzzy_L$. In fact, McNaughton precisely characterized the functions definable in $Fuzzy_L$:

> *Result 11.18 (McNaughton's Theorem).* For any $n \geq 1$, the n-ary truth-functions definable in $Fuzzy_L$ are exactly the continuous n-ary functions f for which "there are a finite number of distinct polynomials $\lambda_1, \ldots, \lambda_\mu$, each $\lambda_j = b_j + m_{1j}x_1 + \cdots + m_{nj}x_n$, where all the b's and m's are integers, such that for every $(x_1, \ldots, x_n)$, $0 \leq x_i \leq 1$, $1 \leq i \leq n$, there is a j, $1 \leq j \leq \mu$, such that $f(x_1, \ldots, x_n) = \lambda_j(x_1, \ldots, x_n)$." (McNaughton [1951], p. 3)

See McNaughton's article for a discussion of this result along his proof, which is beyond the scope of this text.

In the external system B^E_3 the value .5 is treated in the same way as 0; that suggests that Bochvar's external connectives should treat all values other than 1 and 0 the same way 0 is treated. So, for example, (external) negation in $Fuzzy_{BE}$ maps 1 to 0 and all other values to 1. This function, like the infinite-valued binary functions for the internal connectives, is noncontinuous: as m approaches 1 the negation function remains fixed at 1, but at 1 the function noncontinuously drops to 0. The functions for the other external connectives are similarly noncontinuous. Thus none of the Fuzzy_{BE} connectives are definable in Fuzzy_{L}.

But the infinite-valued generalizations of the Bochvar systems are generally not very interesting for our purposes (except for the Bochvarian external assertion and/or negation, which we'll consider in Section 11.10). They are not very interesting because the various "degrees" of vagueness end up collapsing into one degree, .5, in the internal system as soon as we introduce a binary connective, and they end up disappearing in favor of falsehood for all compound formulas in the external system. To the extent that the infinitely many degrees of fuzzy logic represent useful distinctions for a logic for vagueness, Bochvarian fuzzy systems fall short of the mark. Indeed, if the concept of n-degree-validity is one of the distinctive features of a fuzzy system then the Bochvarian systems don't count as very fuzzy at all, since this concept is indistinguishable from the concepts of validity, quasi-validity, and/or degree-validity that we studied in Chapter 5 if complex formulas cannot have values other than 1, .5, and 0.

In light of noncontinuous truth-functions, Fuzzy_{L} is not functionally complete even though it can represent, for example, all of the Kleene strong truth-functions. Nor would we have expected Fuzzy_{L} to be functionally complete, for at least two other reasons. First, Fuzzy_{L} is a normal system, so nonnormal connectives are not definable. Second, there are *uncountably* many functions with a finite number of arguments[16] that are definable over the unit interval [0. .1]. A set has uncountably many members if it cannot be put in 1–1 correspondence with the positive integers, that is, if you cannot produce a list of the members of the set such that there is a first member, a second member, and so on, so that each member of the set is the nth item in the list for some integer n. There are uncountably many truth-values in the unit interval (see Appendix), and for each truth-value we can define a unary function that maps that truth-value to 1 and all other truth-values to 0, already giving us uncountably many unary functions.[17] Yet the number of formulas in the language Fuzzy_{L} is countable (see the Appendix). So Fuzzy_{L} can't contain formulas expressing each truth-function defined over the unit interval because there aren't enough formulas to go around.

11.7 T-Norms, T-Conorms, and Implication in Fuzzy Logic

In Chapter 5 we listed a set of conditions that define the concepts of *t-norms* and *t-conorms* for three-valued logics. Both weak conjunction and bold conjunction

16 For functional completeness here, as in the classical and three-valued systems, we are only interested in functions with a finite number of arguments because every formula is finitely long, and obviously no finite formula can express a function with an infinite number of arguments.

17 There are uncountably many binary functions, and so on, as well, but the fact that the unary truth-functions over the unit interval are uncountable suffices to show that the set of *all* truth-functions over the unit interval is uncountable.

in Fuzzy_{L} meet infinite-valued t-norm conditions, and both weak disjunction and bold disjunction in Fuzzy_{L} meet infinite-valued t-conorm conditions. Indeed, the t-norm and t-conorm conditions are often taken to be *definitive* of the concepts of conjunction and disjunction generally in fuzzy logics; that is, any fuzzy conjunction is required to be a t-norm and any fuzzy disjunction is required to be a t-conorm.

We restate the "norm" conditions in terms of operations on values in the unit interval. A binary operation **tn** defined over [0. .1] is a **t-norm** if it satisfies the following four conditions:

1. **tn** is *associative*: m **tn** (n **tn** p) = (m **tn** n) **tn** p for all $m, n, p \in [0..1]$
2. **tn** is *commutative*: m **tn** $n = n$ **tn** m for all $m, n \in [0..1]$
3. **tn** is *nondecreasing in both arguments*: if $m \leq n$ then m **tn** $p \leq n$ **tn** p and p **tn** $m \leq p$ **tn** n for all $m, n, p \in [0..1]$ (note that if **tn** is nondecreasing in one argument it will also be nondecreasing in the other by virtue of commutativity)
4. 1 is the identity element for **tn**: 1 **tn** $m = m$ **tn** $1 = m$ for all $m \in [0..1]$ (note that if 1 **tn** $m = m$ then m **tn** 1 must also be m by virtue of commutativity)

A binary operation **tc** defined over [0. .1] is a **t-conorm** if it satisfies the following four conditions:

5. **tc** is *associative*: m **tc** (n **tc** p) = (m **tc** n) **tc** p for all $m, n, p \in [0..1]$
6. **tc** is *commutative*: m **tc** $n = n$ **tc** m for all $m, n \in [0..1]$
7. **tc** is *nondecreasing in both arguments*: if $m \leq n$ then m **tc** $p \leq n$ **tc** p and p **tc** $m \leq p$ **tc** n for all $m, n, p \in [0..1]$
8. 0 is the identity element for **tc**: 0 **tc** $m = m$ **tn** $0 = m$ for all $m \in [0..1]$

Just as in the three-valued case, infinite-valued t-norms and t-conorms have the following properties as a consequence of the previous ones:

9. m **tn** $n = 0$ if $m = 0$ or $n = 0$
10. m **tc** $n = 1$ if $m = 1$ or $n = 1$

As a refresher, the associativity of the operations tells us that the grouping in an iterated conjunction or disjunction doesn't affect the truth-conditions, while commutativity tells us that the order of conjuncts in a conjunction and disjuncts in a disjunction doesn't matter. The requirement that these operations be nondecreasing in the arguments tells us that a conjunction can't be made less true by making one of its conjuncts more true, and similarly for disjunctions. The identity element conditions tell us that the truth-value of a conjunction with a true conjunct coincides with the truth-value of the other conjunct, and similarly for a disjunction with a false disjunct. The last two conditions tell us the falsehood of one conjunct is sufficient to make a conjunction false and the truth of one disjunct is sufficient to make a disjunction true.

In addition to the weak and bold conjunction and disjunction operations in Fuzzy_{L}, another well-studied t-norm and t-conorm over the unit interval are the algebraic product $m \bullet n$ and algebraic sum $m + n - (m \bullet n)$. Subtracting the product in the algebraic sum ensures, among other things, that the value of the algebraic sum for any m, n in the unit interval does not exceed 1 (for example, $.7 + .8$ exceeds 1, but $.7 + .8 - (.7 \bullet .8)$, which is .94, does not). It is left as an exercise to show that the algebraic product meets conditions 1–4 and that the algebraic sum meets conditions 5–8.

Each t-norm, t-conorm pair **tn, tc** that we have considered—weak conjunction and disjunction, bold conjunction and disjunction, and algebraic product and algebraic sum—meets the following *duality condition*:

$$m \ \mathbf{tc}\ n = 1 - ((1 - m)\ \mathbf{tn}\ (1 - n)), \text{ for all } m, n \in [0..1]$$

For example, for weak conjunction and weak disjunction operations min and max we have

$$\min(m, n) = 1 - \max(1 - m, 1 - n)$$

When a t-norm and t-conorm are so related, we say that they are a ***dual t-norm, t-conorm pair***. Note that the operation $1 - m$ is Fuzzy$_L$'s negation operation—and so the duality condition $m\ \mathbf{tc}\ n = 1 - ((1 - m)\ \mathbf{tn}\ (1 - n))$ gives us one of the DeMorgan Laws for both weak and bold connectives: **P** ∨ **Q** is equivalent to ¬(¬ **P** ∧ ¬**Q**) in Fuzzy$_L$, and **P** ∇ **Q** is equivalent to ¬(¬**P** & ¬**Q**). Conjunction and disjunction operations in the major fuzzy systems are *dual* t-norm, t-conorm pairs (although they may not satisfy the DeMorgan Laws for those systems if the negation is defined differently from Fuzzy$_L$).

In Fuzzy$_L$ we have defined *two* dual t-norm, t-conorm pairs: the weak connectives and the bold connectives. An interesting result that points to the desirability of including two such pairs in Fuzzy$_L$ is from George J. Klir and Bo Yuan (Klir and Yuan (1995, pp. 87–88):

Result 11.19: A single dual t-norm, t-conorm pair of operations over the unit interval cannot satisfy both the Law of Excluded Middle $m\ \mathbf{tc}\ (1 - m) = 1$ and the Distribution Law $m\ \mathbf{tn}\ (n\ \mathbf{tc}\ p) = (m\ \mathbf{tn}\ n)\ \mathbf{tc}\ (m\ \mathbf{tn}\ p)$ for all m, n, p in [1..0].

Proof: If both laws hold then in particular,

	$^1/_2\ \mathbf{tc}\ (1 - {}^1/_2) = 1$	(Excluded Middle)
(i)	$^1/_2\ \mathbf{tc}\ {}^1/_2 = 1$	(simplification)

and

	$^1/_2\ \mathbf{tc}\ {}^1/_2 = 1 - ((1 - {}^1/_2)\ \mathbf{tn}\ (1 - {}^1/_2))$	(duality condition)
	$^1/_2\ \mathbf{tc}\ {}^1/_2 = 1 - ({}^1/_2\ \mathbf{tn}\ {}^1/_2)$	(simplification)
	$^1/_2\ \mathbf{tn}\ {}^1/_2 = 1 - ({}^1/_2\ \mathbf{tc}\ {}^1/_2)$	(equivalent)
(ii)	$^1/_2\ \mathbf{tn}\ {}^1/_2 = 1 - 1 = 0$	(from (i))

Then from

$$\tfrac{1}{2}\ \mathbf{tn}\ (\tfrac{1}{2}\ \mathbf{tc}\ \tfrac{1}{2}) = (\tfrac{1}{2}\ \mathbf{tn}\ \tfrac{1}{2})\ \mathbf{tc}\ (\tfrac{1}{2}\ \mathbf{tn}\ \tfrac{1}{2}) \qquad \text{(Distribution)}$$

we get

$$\tfrac{1}{2}\ \mathbf{tn}\ 1 = (\tfrac{1}{2}\ \mathbf{tn}\ \tfrac{1}{2})\ \mathbf{tc}\ (\tfrac{1}{2}\ \mathbf{tn}\ \tfrac{1}{2}) \qquad \text{(from (i))}$$
$$\tfrac{1}{2}\ \mathbf{tn}\ 1 = 0\ \mathbf{tc}\ 0 \qquad \text{(from (ii))}$$
$$\tfrac{1}{2} = 0\ \mathbf{tc}\ 0 \qquad (\mathbf{tn}\ \text{identity})$$
$$\tfrac{1}{2} = 0 \qquad (\mathbf{tc}\ \text{identity})$$

which is a clear contradiction.

We have already shown that the Law of Excluded Middle holds for Fuzzy_L's bold disjunction but not its weak disjunction; we leave it as an exercise to show that Distribution holds for Fuzzy_L's weak disjunction and conjunction but not its bold disjunction and conjunction.

In addition to determining its dual t-conorm, a t-norm can also be used to define the conditional operation for a fuzzy system. We call an operation $\Rightarrow$ on [0. .1] a ***residuation*** operation with respect to the t-norm **tn** (or, more simply, the ***residuum*** for the t-norm) if it meets the **adjointness condition**:

$m\ \mathbf{tn}\ n \leq p$ if and only if $m \leq n \Rightarrow p$, for all $m, n, p \in [0. .1]$

The logical significance of adjointness is this. If we consider the residuation operation to be a conditional operation, then the inequality

$n\ \mathbf{tn}\ (n \Rightarrow p) \leq p$

represents the inference Modus Ponens in the sense that the value p should be at least the value that results from conjoining the values n and $n \Rightarrow p$ (that is, **P** is at least as true as the conjunction $\mathbf{P} \wedge (\mathbf{P} \rightarrow \mathbf{Q})$). Since every t-norm is commutative the Modus Ponens inequality is equivalent to $(n \Rightarrow p)\ \mathbf{tn}\ n \leq p$. Moreover, because every t-norm is nondecreasing in the first (as well as the second) argument, it follows from this form of the Modus Ponens inequality that if $m \leq n \Rightarrow p$ then $m\ \mathbf{tn}\ n \leq p$, which is one-half of the adjointness condition. But we want more than just the earlier the Modus Ponens inequality above. We also want the value $n \Rightarrow p$ to be the *greatest* value q such that $n\ \mathbf{tn}\ q \leq p$ (that is, we want to make $n \Rightarrow p$ as true as it can be while still satisfying the Modus Ponens inequality). Now, if $n \Rightarrow p$ is the greatest value q such that $n\ \mathbf{tn}\ q \leq p$ then it follows from commutativity that $n \Rightarrow p$ is the greatest value q such that $q\ \mathbf{tn}\ n \leq p$. From the last it follows that if $m\ \mathbf{tn}\ n \leq p$, then m must be less than or equal to $n \Rightarrow$ p, and this is the other half of the adjointness condition.

Given the adjointness condition there is a unique residuum, which we will call the ***adjunct residuum***, for any t-norm. The adjunct residuation operations for the t-norms we consider in this text may be defined thus:[18]

$$n \Rightarrow p =_{\text{def}} \max \{m \in [0..1]: m \textbf{ tn } n \leq p\}$$

(i.e., $n \Rightarrow p$ is the maximum value m such that $m \textbf{ tn } n \leq p$). Here are some other properties of residuation operations:

Residium 1: Every residuum is normal; that is, when n and p are both 1 or 0, the residuum of n and p coincides with the classical conditional operation on n and p.

Proof: This is straightforward. For example, let $n = p = 1$. Then $n \Rightarrow p = \max \{m: m \textbf{ tn } 1 \leq 1\}$, which is 1 since $1 \textbf{ tn } 1 \leq 1$. Now let $n = 1$ and $p = 0$. Then $n \Rightarrow p = \max \{m: m \textbf{ tn } 1 \leq 0\}$, which is 0 since 1 is the identity for t-norms (any value of m greater than 0 will have $m \textbf{ tn } 1$, which is m, greater than 0). The other two cases—$n = 0$ and $p = 1$, and $m = 0$ and $p = 0$—are left as an exercise.

Residium 2: For every residuation operation $\Rightarrow$, $n \Rightarrow p = 1$ when $n \leq p$.

Proof: Assume that $n \leq p$. Then $\max \{m: m \textbf{ tn } n \leq p\} = 1$ because $1 \textbf{ tn } n = n$ for any t-norm (1 is the identity element) and therefore $1 \textbf{ tn } n \leq p$.

These additional results underscore the suitability of using the residuum defined by a t-norm to define the conditional operation in a logical system.

That naturally raises the question, What are the residuation operations corresponding to Łukasiewicz weak conjunction, Łukasiewicz bold conjunction, and algebraic product conjunction?

We start with the Łukasiewicz bold conjunction t-norm $\otimes_L$. Because $m \otimes_L n$ is $\max (0, m + n - 1)$, the adjunct residuum for this t-norm is defined as

$$n \Rightarrow_{\otimes L} p =_{\text{def}} \max \{m: \max (0, m + n - 1) \leq p\}.$$

We consider two cases. If $n \leq p$, then 1 is the maximum value that m can take by property Residuum 1. If $n > p$, then the most that m can be and still satisfy the condition $max\ (0,\ m + n - 1) \leq p$ is $1 - n + p$. Because $1 - n + p > 1$ if $n \leq p$, $\max \{m: \max (0, m + n - 1) \leq p\}$ is thus equivalent to $\min (1, 1 - n + p)$ – which is the definition of the conditional operation in Fuzzy_{L}. The role of bold conjunction in this adjunct t-norm/residuum pair explains why bold conjunction is generally introduced as a primitive, rather than defined, conjunction connective in Fuzzy_{L}. (This is not to say it is impossible to define bold conjunction using the other operations—it's left as an exercise to show that bold conjunction can be defined using Fuzzy_{L}'s negation and conditional.)

18 More generally, residuation is defined to produce the least upper bound of $\{m: m \textbf{ tn } n \leq p\}$. But we use the maximum operation because it is well-defined with respect to the t-norms that we consider in this text.

The t-norm that defines weak conjunction in $\text{Fuzzy}_{\text{Ł}}$ is the minimum operation. In this case the adjunct residuation operation is defined as

$$n \Rightarrow_{\wedge} p =_{\text{def}} \max\{m\text{: } \min(m,n) \leq p\}.$$

We consider two cases. If $n \leq p$, then by property Residuum 1 the maximum value that m can take is 1. If $n > p$, then m can be p but no larger. Thus the value of this residuation operation is 1 if $n \leq p$ and is p if $n > p$. Note that this is not the $\text{Fuzzy}_{\text{Ł}}$ conditional; it's the conditional of *Gödel fuzzy logic*, which will be introduced in Section 11.8.

Finally, the algebraic product t-norm's adjunct residuation is defined as:

$$n \Rightarrow_{\cdot} p =_{\text{def}} \max\{m\text{: } m \cdot n \leq p\}.$$

Once again we consider two cases. If $n \leq p$, then m can be as high as 1 by property Residuum 1. If $n > p$, then the maximum value m can have is p/n (notice we are not dividing by 0 in this case, given that $n > p$). So the truth-value for this residuation operation is 1 if $n \leq p$ and is p/n if $n > p$. This is the conditional of *product fuzzy logic*, which will be presented in Section 11.9.

The fuzzy conditional operation that generalizes the K^{S}_3 conditional is called a ***Q-implication*** in fuzzy logic, rather than an ***R-implication***, which is what the three residuation operations are. The *R* stands for *residuum*; the *Q* stands for *quantum*. Q-implications are defined in terms of negation, disjunction, and conjunction thus: $m \Rightarrow_{\text{Q}} n =_{\text{def}}$ *neg m or* (*m and n*). Using the Łukasiewicz bold conjunction and disjunction operations we arrive at the Kleene conditional operation (which we already know is definable using negation and Kleene/Łukasiewicz weak disjunction or disjunction):

$$m \Rightarrow_{K} n =_{\text{def}} (1 - m) \oplus_{L} (m \otimes_{L} n)$$

For $(1 - m) \oplus_{L} (m \otimes_{L} n) = \min(1, (1 - m) + \max(0, m + n - 1)) = \min(1, \max(1 - m, n)) = \max(1 - m, n)$.[19]

11.8 Gödel Fuzzy Propositional Logic

There are three major varieties of fuzzy logic in the literature: Łukasiewicz fuzzy logic, Gödel fuzzy logic, and product fuzzy logic. In this text we focus on Łukasiewicz fuzzy logic—the most widely studied of the three—for several reasons: along with providing a satisfactory solution to the Sorites paradox and other issues arising from vagueness, as we'll see when we turn to fuzzy first-order logic, $\text{Fuzzy}_{\text{Ł}}$ is the only one of the three logics for which it is possible to construct Pavelka-style graded

[19] Klir and Yuan (1995) contains a good introductory discussion of the varieties of fuzzy R- and Q-implications.

derivation systems. But the two other major systems carry some interest as well. We begin with Gödel fuzzy logic,[20] or Fuzzy_G.

The operations for the major connectives of Fuzzy_G are defined as follows:

1. $V(\neg_G \mathbf{P}) = 1$ if $V(\mathbf{P}) = 0$
   0 otherwise
2. $V(\mathbf{P} \&_G \mathbf{Q}) = \min(V(\mathbf{P}), V(\mathbf{Q}))$
3. $V(\mathbf{P} \nabla_G \mathbf{Q}) = \max(V(\mathbf{P}), V(\mathbf{Q}))$
4. $V(\mathbf{P} \rightarrow_G \mathbf{Q}) = 1$ if $V(\mathbf{P}) \leq V(\mathbf{Q})$
   $V(\mathbf{Q})$ otherwise
5. $V(\mathbf{P} \leftrightarrow_G \mathbf{Q}) = 1$ if $V(\mathbf{P}) = V(\mathbf{Q})$
   $\min(V(\mathbf{P}), V(\mathbf{Q}))$ otherwise.

Fuzzy_G bold conjunction and disjunction, $\&_G$ and ∇_G, are identical to Łukasiewicz weak conjunction and disjunction.[21]

The Fuzzy_G conditional operation is the residuum for Fuzzy_G's bold conjunction, as discussed in Section 11.7. The biconditional $\mathbf{P} \leftrightarrow_G \mathbf{Q}$ is defined to be equivalent to $(\mathbf{P} \rightarrow_G \mathbf{Q}) \&_G (\mathbf{Q} \rightarrow_G \mathbf{P})$. The negation clause is new and unfamiliar. It's a general principle in fuzzy logic that the negation for a system should be definable thus: $\neg\mathbf{P} =_{\text{def}} \mathbf{P} \rightarrow \mathbf{0}$, where $\mathbf{0}$ is a special formula that always has the value 0. Note that because the implication in a fuzzy system is generally defined via a t-norm, the negation is also fixed by the t-norm. Given clause 4 for the Fuzzy_G conditional, the formula $\mathbf{P} \rightarrow_G \mathbf{0}$ can only have the value 1 when $\mathbf{P}$ has the value 0, and in all other cases $\mathbf{P} \rightarrow_G \mathbf{0}$ has the value 0. There's an affinity here with the external connectives of Bochvar's three-valued logic. Recall that B^E_3 negation is definable as:

$V(\neg_{BE}\mathbf{P}) = 0$ if $V(\mathbf{P}) = 1$
1 otherwise.

To be sure, in Bochvar's system, $\neg_{BE}\mathbf{P}$ has the value 1 if $\mathbf{P}$ has the value $1/2$, but in Fuzzy_G $\neg_G\mathbf{P}$ has the value 0 when $\mathbf{P}$ has the value .5. The affinity is that negating a formula always results in one of the two values 1 or 0. We note that Gödel fuzzy negation is normal, as are the other Gödel connectives.

It's standard to define weak conjunction in fuzzy logic as

$$\mathbf{P} \wedge \mathbf{Q} =_{\text{def}} \mathbf{P} \& (\mathbf{P} \rightarrow \mathbf{Q}).$$

(We noted in Section 11.2 that $\text{Fuzzy}_Ł$ weak conjunction $\mathbf{P} \wedge \mathbf{Q}$ can be defined as $\mathbf{P} \& (\neg\mathbf{P} \nabla \mathbf{Q})$, and because $\mathbf{P} \rightarrow \mathbf{Q}$ is equivalent to $\neg\mathbf{P} \nabla \mathbf{Q}$ in $\text{Fuzzy}_Ł$ the standard

[20] Gödel fuzzy logic is named for the logician Kurt Gödel, who introduced the major operations used in Fuzzy_G as operations of finite-valued *intuitionistic logics* in Gödel (1932).

[21] Why, you may wonder, are these the *bold* connectives in Fuzzy_G, given that they are *weak* in $\text{Fuzzy}_Ł$? It's because this t-norm, t-conorm pair is used to define the residuum operation giving the Fuzzy_G conditional, and in the fuzzy literature the conjunction and disjunction that play this role in a logical system are considered to be the *bold* connectives.

definition would work there as well.) If we define Fuzzy_G weak conjunction in this way:

$$\mathbf{P} \wedge_G \mathbf{Q} =_{\text{def}} \mathbf{P} \,\&_G\, (\mathbf{P} \rightarrow_G \mathbf{Q})$$

it turns out that weak and bold conjunction are identical in Fuzzy_G—both give us the minimum value of the conjuncts. Fuzzy weak disjunction is standardly defined as

$$\mathbf{P} \vee \mathbf{Q} =_{\text{def}} ((\mathbf{P} \rightarrow \mathbf{Q}) \rightarrow \mathbf{Q}) \wedge ((\mathbf{Q} \rightarrow \mathbf{P}) \rightarrow \mathbf{P}),[22]$$

and using this definition Fuzzy_G weak disjunction is identical to Fuzzy_G bold disjunction. Given these identities, we have not included weak connectives in the truth-condition clauses.

The conditional $P \rightarrow_G P$ is a tautology of Fuzzy_G; it always has the value 1. But some classical tautologies fail to be classical tautologies in Fuzzy_G when we replace the classical conjunction and disjunction with Gödel bold conjunction and disjunction. The Law of Excluded Middle formula $A \nabla_G \neg_G A$ can have any value other than 0. When the value of A is either 1 or 0, $A \nabla_G \neg_G A$ has the value 1. When A has any other value other than 1 or 0, the value of $\neg_G A$ is 0, and so the value of the disjunction, which is the maximum of its disjuncts' values, will be the value of A. (Compare this with the $\text{Fuzzy}_Ł$ bold Law of Excluded Middle, where the smallest possible value is 1, and the $\text{Fuzzy}_Ł$ weak Law of Excluded Middle, where the smallest value is .5.) Fuzzy_G contains some tautologies that are not $\text{Fuzzy}_Ł$ tautologies, for instance, $\neg_G P \vee_G \neg_G \neg_G P$. By the way that negation is defined in Fuzzy_G, one of the two disjuncts will always be true (and the other false). In $\text{Fuzzy}_Ł$ $\neg P \vee \neg\neg P$ has the same truth-conditions as $P \vee \neg P$ – its value can be as small as .5 (when P has the value .5).

The inference Modus Ponens is valid in Fuzzy_G, but the argument

$$\frac{\neg_G \neg_G \mathrm{P}}{\mathrm{P}}$$

isn't valid because the premise will have the value 1 whenever P has any value other than 0; that is, it is possible for the premise to have the value 1 when the conclusion doesn't have this value. The argument is, however, valid when we replace the Gödel negation with the classical connective, and it is also valid when $\text{Fuzzy}_Ł$ negation is substituted. It is thus a significant difference between these systems that the Law of Double Negation fails in Fuzzy_G. Finally, because Fuzzy_G is a normal system, all Fuzzy_G tautologies are classical tautologies and all valid arguments of Fuzzy_G are classically valid.

[22] This definition also works for $\text{Fuzzy}_Ł$. We don't generally use negation and weak conjunction to define weak disjunction in fuzzy systems, because negation is not always defined to be complementation as it is in Łukasiewicz logics and algebras. In Fuzzy_G, for example, if we define $\mathbf{P} \vee_G \mathbf{Q}$ as $\neg_G(\neg_G \mathbf{P} \wedge_G \neg_G \mathbf{Q})$ then $\mathbf{P} \vee_G \mathbf{Q}_G$ can never have a value other than 1 or 0, and that is an odd weak operation for a logic in which formulas can have any value in the unit interval.

Like Fuzzy_L, Fuzzy_G has a decision procedure for tautologies based on truth-tables for finite-valued logics, in this case finite-valued Gödel logics. We define n-valued Gödel propositional logics for each integer $n \geq 3$: G_n is the propositional logic whose truth-values are the n members of the set $\{0, 1 / (n-1), \ldots, (n-2) / (n-1), 1\}$ and whose truth-conditions are the clauses 1–5 for Fuzzy_G. For any propositional formula **P** of Fuzzy_G, let #A(**P**) be the number of distinct atomic formulas in **P**—here we only count *one* occurrence of each atomic subformula. Thus, $\#A(B)$, $\#A(\neg B)$, and $\#A(B \oplus B)$ are all 1, while $\#A(B \oplus C)$ is 2. It turns out that **P** is a tautology of Fuzzy_G if and only if **P** is a tautology of G_n, where $n = \#A(\mathbf{P}) + 2$.[23] This gives us a decision procedure for Fuzzy_G: to determine whether **P** is a tautology of Fuzzy_G just examine the truth-table for **P** in the appropriate finite-valued Gödel system.

11.9 Product Fuzzy Propositional Logic

The truth-conditions for formulas formed with the major connectives of product fuzzy logic,[24] or Fuzzy_P, are

1. $V(\neg_P \mathbf{P}) = 1$ if $V(\mathbf{P}) = 0$
   0 otherwise
2. $V(\mathbf{P} \,\&_P\, \mathbf{Q}) = V(\mathbf{P}) \cdot V(\mathbf{Q})$
3. $V(\mathbf{P} \,\nabla_P\, \mathbf{Q}) = (V(\mathbf{P}) + V(\mathbf{Q})) - (V(\mathbf{P}) \cdot V(\mathbf{Q}))$
4. $V(\mathbf{P} \rightarrow_P \mathbf{Q}) = 1$ if $V(\mathbf{P}) \leq V(\mathbf{Q})$
   $V(\mathbf{Q}) / V(\mathbf{P})$ otherwise
5. $V(\mathbf{P} \leftrightarrow_P \mathbf{Q})$: left as an exercise
6. $V(\mathbf{P} \wedge_P \mathbf{Q}) = \min(V(\mathbf{P}), V(\mathbf{Q}))$
7. $V(\mathbf{P} \vee_P \mathbf{Q}) = \max(V(\mathbf{P}), V(\mathbf{Q}))$

The negation here, defining $\neg_P \mathbf{P}$ as $\mathbf{P} \rightarrow_P \mathbf{0}$, is identical to Fuzzy_G's negation, and so the Law of Double Negation also fails for Fuzzy_P. We've already discussed (in Section 11.7) the rationale behind the algebraic product and algebraic sum that are used to define product bold conjunction and disjunction, and the way the associated residuum operation for the product conditional is derived. The reader will be asked in the exercises to prove that the truth-conditions 6 and 7 for Fuzzy_P weak conjunction and disjunction follow from the definitions

$$\mathbf{P} \wedge_P \mathbf{Q} =_{\text{def}} \mathbf{P} \,\&_P\, (\mathbf{P} \rightarrow_P \mathbf{Q})$$

and

$$\mathbf{P} \vee_P \mathbf{Q} =_{\text{def}} ((\mathbf{P} \rightarrow_P \mathbf{Q}) \rightarrow_P \mathbf{Q}) \wedge_P ((\mathbf{Q} \rightarrow_P \mathbf{P}) \rightarrow_P \mathbf{P}).$$

23 See Hájek (1998b, p. 157), for a proof.

24 Goguen (1968–1969) developed product fuzzy logic and studied its application to Sorites paradoxes. As we noted in Chapter 1, Goguen's article was the beginning of formal fuzzy logic.

The conditional $P \rightarrow_P P$ is a tautology of Fuzzy$_P$. The Law of Excluded Middle formula $A \nabla_P \neg_P A$ can have any value other than 0; indeed, it behaves exactly as it does in Fuzzy$_G$. When the value of A is either 1 or 0, $A \nabla_P \neg_P A$ has the value 1. When A has any other value other than 1 or 0, the value of the disjunction will be the value of A since we are adding 0, the value of $\neg_P A$, to that value and then subtracting the product of 0 and that value, which is 0. Thus some classical tautologies are not Fuzzy$_P$ tautologies (replacing classical conjunction and disjunction with product bold conjunction and disjunction—and since Fuzzy$_P$ weak conjunction and disjunction are the same as the Fuzzy$_G$ weak or bold connectives and the negations are identical, the negative result also holds when Fuzzy$_P$ weak connectives are substituted for the classical ones). As in Fuzzy$_G$, $\neg_P P \vee_P \neg_P \neg_P P$ is a tautology. The Modus Ponens inference is valid in Fuzzy$_P$, but the argument

$$\frac{\neg_P \neg_P \mathrm{P}}{\mathrm{P}}$$

isn't valid in Fuzzy$_P$. So not all classically valid arguments are valid in Fuzzy$_P$. Owing to normality, however, all Fuzzy$_P$ tautologies are classical tautologies and all arguments that are valid in Fuzzy$_P$ are classically valid.

Our examples so far make Fuzzy$_G$ and Fuzzy$_P$ look "logically" identical, but they most certainly are not. The tautologies of the two systems don't coincide—for example, $P \rightarrow_G (P \&_G P)$ is a tautology but $P \rightarrow_P (P \&_P P)$ is not; and $\neg_P \neg_P P \rightarrow_P (((Q \&_P P) \rightarrow_P (R \&_P P)) \rightarrow_P (Q \rightarrow_P R))$ is a tautology but $\neg_G \neg_G P \rightarrow_G (((Q \&_G P) \rightarrow_G (R \&_G P)) \rightarrow_G (Q \rightarrow_G R))$ is not. Another very significant difference between Fuzzy$_G$ and Fuzzy$_P$ is that the Modus Ponens inference is degree-valid in the former but not the latter system. Proof of all three claims is left as an exercise.[25]

The set of tautologies of Fuzzy$_P$ is decidable, although the procedure is more complicated than those we presented for Fuzzy$_Ł$ and Fuzzy$_G$. The interested reader is referred to Baaz, Hájek, Kraníček, and Švejda (1998)—the proof is contained in results leading to their Lemma 9.

11.10 Fuzzy External Assertion and Negation

In Section 11.6 we said we'd have more to say about fuzzy Bochvarian external assertion and external negation, neither of which is definable in the three major fuzzy systems. We are particularly interested in these operations because of Black's Problem of the Fringe, which we'll examine again in Chapter 14: we want to be able to say truly ***not P*** when **P** is vague, and also to say truly ***not not P*** in a sense that is not equivalent to **P**. To be sure, neither $\neg_G \neg_G$**P** nor $\neg_P \neg_P$**P** is equivalent to **P** (thus Gödel/product negation, like Bochvar's external negation, is *non-involutive*, where an **involutive negation** is defined to be one that satisfies the Double Negation

25 See Gottwald (2001) for further comparisons between Gödel and product fuzzy logics.

equivalence), and both double negations (which are identical) are true when **P** has any value other than 0, but Gödel/product negation is not fuzzy Bochvarian external negation. Singly-negated **P** has the value 0 in the Gödel and product systems when **P** has a value strictly between 1 and 0, for example.

The fuzzy versions of Bochvar's external assertion and negation have the following truth-conditions:

$$\begin{aligned} V(\Delta \mathbf{P}) &= 1 \text{ if } V(\mathbf{P}) = 1 \\ & 0 \text{ otherwise} \\ V(\bot \mathbf{P}) &= 1 \text{ if } V(\mathbf{P}) \neq 1 \\ & 0 \text{ otherwise} \end{aligned}$$

Here we have used the symbol **Δ** for external assertion because this is the standard fuzzy symbol, introduced by Matthias Baaz in (1996) (although Baaz did not present the operation as a generalization of Bochvar's three-valued external assertion as we are doing here; he called it *1-projection* and the fuzzy community has subsequently called it *the Baaz delta operation*). We have used the symbol ⊥ for external negation to make it clearly distinguishable from the standard negations for our fuzzy systems.

In Chapter 5 we read Bochvar's external negation formula ***aP*** as: *P is true*. But because we now have degrees of truth, we'll follow Gottwald and Hájek (2005) and read the fuzzy formula **ΔP** as ***P*** *is absolutely true*. Similarly, we'll read the fuzzy formula **⊥P** as ***P*** *is not absolutely true*. Note that these two connectives are interdefinable in Fuzzy_{L} using either $\mathbf{\Delta P} =_{\text{def}} \neg_L \bot \mathbf{P}$ or $\bot \mathbf{P} =_{\text{def}} \neg_L \mathbf{\Delta P}$. Moreover, they are interdefinable in the same way using Fuzzy_{G}/Fuzzy_{P} negation in place of $\neg_L$. So we'll focus on just one of the connectives, fuzzy external assertion.

We have the important result that

Result 11.20: Fuzzy external assertion is not definable in Fuzzy_{L}, Fuzzy_{G}, or Fuzzy_{P}.

Proof: In Section 11.6 we explained that Bochvar's external connectives aren't definable in Fuzzy_{L} because the external operations are not continuous.

However, both Fuzzy_{G} and Fuzzy_{P} include noncontinuous operations, so the same reasoning can't establish the negative result in these latter cases. Rather, we'll use the following facts, each of which can be readily verified, to prove that external assertion isn't definable in either of these two systems:

a. $\neg_G$**P** and $\neg_P$**P** have the value 0 if and only if **P** does not have the value 0.
b. **P** $\&_G$ **Q** and **P** $\&_P$ **Q** have the value 0 if and only if at least one of **P** and **Q** has the value 0.
c. **P** ∇_G **Q** and **P** ∇_P **Q** have the value 0 if and only if both **P** and **Q** have the value 0.
d. **P** $\rightarrow_G$ **Q** and **P** $\rightarrow_P$ **Q** have the value 0 if and only if **P** does not have the value 0 while **Q** does have the value 0.

Because the weak conjunction and disjunction in Fuzzy_G and Fuzzy_P are definable using the other connectives, we have considered in (a)–(d) all of the operations in terms of which other operations of these systems are definable.

Now, for fuzzy external negation to be definable in a system it must be possible to find a formula formed from an atomic formula **P** and zero or more connectives that has the value 1 when **P** has the value 1 and that has the value 0 when **P** has any value other than 1. In particular, focusing on an arbitrary value between 1 and 0, say .5, the formula must have the value 1 when **P** has the value 1 and must have the value 0 when **P** has the value .5. We can show that there is no such formula either Fuzzy_G or Fuzzy_P, because every formula **Q** formed from the single atomic formula **P** and zero or more connectives has the

collapsing property: **Q** has the value 0 when **P** has the value .5 if and only if **Q** has the value 0 when **P** has the value 1.

Every atomic formula **P** obviously has the collapsing property. Moreover, it follows from facts (a)–(d) earlier that any formula **Q** built up from **P** and one of more of the connectives has the collapsing property if its immediate subformulas have that property. For example, if **Q** has the form ¬**R** then from fact (a) we have: ¬**R** has the value 0 when **P** has the value .5 if and only if **R** does not have the value 0 when **P** has the value .5. If **R** has the collapsing property, then **R** does not have the value 0 when **P** has the value .5 if and only if **R** does not have the value 0 when **P** has the value 1. And it follows again from fact (a) that **R** does not have the value 0 when **P** has the value 1 if and only if ¬**R** has the value 0 when **P** has the value 1. Thus in this case **Q** has the collapsing property as well. Similar reasoning using facts (b)–(d) establishes that for any other form that **Q** may have, **Q** will have the collapsing property if each of its immediate components does. It follows, then, that no formula of Fuzzy_G or Fuzzy_L can express fuzzy external assertion.

So if we want fuzzy external assertion (and thus fuzzy external negation) in any of our three fuzzy systems, we explicitly have to add it. Fortunately systems with this additional connective have been developed in recent years, as we shall see when we turn to algebras and derivation systems for fuzzy logic. We do, however, note an interesting positive expressibility result: fuzzy external assertion (and consequently fuzzy external negation) can be defined in terms of $\text{Fuzzy}_G/\text{Fuzzy}_P$ negation $\neg_G$ and Fuzzy_L negation $\neg_L$: $\mathbf{\Delta P} =\text{def}\ {}_{\neg_G \neg_L}\mathbf{P}$. This will be verified in an exercise.

Adding fuzzy external assertion to Fuzzy_L, the formula $\neg\Delta P \wedge \neg\Delta\neg P$ captures the sense of negation needed for a true rendition of Black's formula *P is not true and P is not not true* where *P* is a vague assertion. This formula is true in Fuzzy_L (augmented with the new connective) when *P* has any value other than 0 or 1, and it is false otherwise. Because of the fact that external assertion always produces 1 or 0 as its value, the formula $\neg\Delta P \wedge \neg\Delta\neg P$ is equivalent to $\neg\Delta P \,\&\, \neg\Delta\neg P$ in Fuzzy_L.

We can also express another logically true version of the Law of Excluded Middle: $\Delta P \vee \neg \Delta P$.

11.11 Exercises

SECTION 11.2

1 Given an interpretation on which

$V(P) = 0$
$V(Q) = 0.3$
$V(R) = 0.8$
$V(S) = 0.2$
$V(T) = 0.5$
$V(W) = 1$

what is the fuzzy value assigned to each of the following $Fuzzy_Ł$ formulas?

a. $P \wedge Q$
b. $Q \wedge T$
c. $P \vee S$
d. $P \rightarrow Q$
e. $Q \rightarrow P$
f. $P \vee \neg P$
g. $Q \vee \neg Q$
h. $R \vee \neg R$
i. $R \rightarrow S$
j. $S \rightarrow R$
k. $W \rightarrow P$
l. $W \rightarrow R$
m. $W \rightarrow S$
n. $S \rightarrow S$
o. $P \leftrightarrow Q$
p. $Q \leftrightarrow R$
q. $P \& Q$
r. $Q \& S$
s. $Q \& T$
t. $P \nabla S$
u. $Q \nabla \neg R$
v. $S \rightarrow (R \rightarrow Q)$
w. $S \rightarrow (Q \rightarrow R)$
x. $R \rightarrow (R \& R)$
y. $(R \& R) \rightarrow R$
z. $(R \rightarrow S) \vee (S \rightarrow R)$
z′. $(R \leftrightarrow T) \nabla (R \leftrightarrow \neg T)$

SECTION 11.3

2 Is the argument

$$\frac{P \rightarrow \neg P}{\neg P}$$

valid in $\text{Fuzzy}_{Ł}$? Defend your answer.

3 Is the argument

$$\frac{\neg (P \leftrightarrow Q)}{(P \leftrightarrow R) \triangledown (Q \leftrightarrow R)}$$

valid in $\text{Fuzzy}_{Ł}$? Defend your answer.

4 We noted that $P \rightarrow (P \,\&\, P)$ and $(P \triangledown P) \rightarrow P$ are not $\text{Fuzzy}_{Ł}$ tautologies although $P \rightarrow (P \wedge P)$ and $P \rightarrow (P \vee P)$ are both tautologies in classical logic. Are the latter formulas—using weak connectives—tautologies in $\text{Fuzzy}_{Ł}$? Defend your answer.

5 Are the converse formulas $(P \,\&\, P) \rightarrow P$ and $P \rightarrow (P \triangledown P)$ tautologies in $\text{Fuzzy}_{Ł}$? Defend your answer.

6 Is the argument

$$\frac{P \vee P}{P}$$

valid in $\text{Fuzzy}_{Ł}$? Defend your answer.

7 Compare $Ł_3$ and $\text{Fuzzy}_{Ł}$ entailment: are all $\text{Fuzzy}_{Ł}$ entailments also $Ł_3$ entailments? Does the converse hold? Prove that you are right.

8 Show that the DeMorgan, Double Negation, and Distribution equivalences hold in $\text{Fuzzy}_{Ł}$.

9 Prove that every $\text{Fuzzy}_{Ł}$ formula can be mechanically converted to an equivalent $\text{Fuzzy}_{Ł}$ formula containing only negation and bold disjunction.

10 Use the Aguzzoli-Ciabattoni decision procedure to determine whether the formula $\neg(P \,\&\, \neg P)$ is a tautology of $\text{Fuzzy}_{Ł}$.

SECTION 11.4

11 Prove Result 11.8.

12 Prove Result 11.10.

13 a. Give an example of a formula containing conjunction that is a classical tautology but that is not a 1-tautology of $\text{Fuzzy}_{Ł}$ when bold conjunction is used in place of the classical connective.

b. Give an example of a formula containing disjunction that is a classical tautology but that is not a 1-tautology of $\text{Fuzzy}_{Ł}$ when bold disjunction is used in place of the classical connective.

c. Do the DeMorgan, Double Negation, and Distribution equivalences that are used for converting a formula to normal form hold in $\text{Fuzzy}_{Ł}$ when bold conjunction and disjunction are used in place of weak conjunction and disjunction? For each equivalence that does hold, prove it. For each equivalence that doesn't hold, provide a counterexample.

d. Do Results 11.7 and 11.8 hold when we use bold conjunction and disjunction in place of the weak connectives? If they do, prove it. If they don't, provide counterexamples.

14 Prove that $\neg(P \wedge (P \rightarrow \neg P))$ is a ⅓-tautology in Fuzzy_L.

15 Prove that the argument

$$\begin{array}{l} A \\ A \rightarrow B \\ B \rightarrow C \\ C \rightarrow D \\ D \rightarrow E \\ E \rightarrow F \\ F \rightarrow G \\ G \rightarrow H \\ H \rightarrow I \\ \underline{I \rightarrow J} \\ J \end{array}$$

is exactly .1-degree-valid in Fuzzy_L. We have already shown that it is at most .1-degree-valid, so you need to show that we cannot make the gap between the value of the least true premise and that of the conclusion greater than .9. *Hint*: One way to do this is to consider cases as follows: Case 1: $V(A) \leq .9$; Case 2: $V(A) > .9$ and $V(B) \leq .8$; Case 3: $V(A) > .9$, $V(B) > .8$, and $V(C) \leq .7$; and so on. In each case show that the gap between the value of the least true premise and that of the conclusion cannot be greater than .9.

SECTION 11.5

16 Consider the fuzzy set of Fuzzy_L formulas $\Phi = \{P: .4, \neg P: .4, Q: .6, S \rightarrow Q: .8, R: 1\}$, where **P**: *n* means that formula **P** is a member of the set Φ to degree *n*, and each formula not listed between the set brackets is a member of Φ to degree 0.

a. Describe the consonant fuzzy truth-value assignments for this set by indicating, for each atomic formula of Fuzzy_L, the least value that that formula can have on a truth-value assignment that is consonant for Φ

b. For each of the following formulas, state its degree of membership in the fuzzy consequence FC(Φ):

i. P
ii. ¬P
iii. Q
iv. S
v. R → P
vi. P ∨ ¬P
vii. Q → S
viii. R & Q

ix. $S \rightarrow (Q \rightarrow P)$
x. W
xi. $\neg P \rightarrow W$

17 Show how to define *n*-degree-entailment in terms of fuzzy consequence.

SECTION 11.6

18 The conditional $P \leftrightarrow \neg P$ has the value 1 in Fuzzy_{L} when P has the value .5. What value does $P \leftrightarrow_K \neg P$ have when P has the value .5?

19 Consider the argument

P
$P \rightarrow_K Q$
———
Q

a. Is this argument valid?
b. Is this argument degree-valid?
c. For what value of *n* is the argument *n*-degree-valid?

SECTION 11.7

20 a. Fuzzy_{BI} conjunction and disjunction do not qualify as a t-norm and t-conorm. Which conditions for t-norms and t-conorms are violated by the Fuzzy_{BI} operations?

b. Fuzzy_{BE} conjunction and disjunction do not qualify as a t-norm and t-conorm. Which conditions for t-norms and t-conorms are violated by the Fuzzy_{BE} operations?

21 Show that the algebraic product operation meets conditions 1–4 for t-norms and that the algebraic sum operation meets conditions 5–8 for t-conorms.

22 Show that the condition

$m \ \mathbf{tn} \ n = 1 - ((1-m) \ \mathbf{tc} \ (1-n))$, for all $m, n \in [0..1]$

follows from the duality condition $m \ \mathbf{tc} \ n = 1 - ((1-m) \ \mathbf{tn} \ (1-n))$, for all m, $n \in [0..1]$.

23 Show that

a. bold conjunction and bold disjunction meet the condition for being a dual t-norm, t-conorm pair, and
b. that algebraic product and sum do as well.

24 Show that Distribution holds for Fuzzy_{L}'s weak disjunction and conjunction but not its bold disjunction and conjunction.

25 Complete the proof that every residuation operation is normal, that is, prove the cases where $n = 0$ and $p = 1$, and where $n = 0$ and $p = 0$.

26 Show that given a dual t-norm, t-conorm pair **tn** and **tc**, the residuation operation $\Rightarrow$ adjunct to **tn** satisfies the following equation:

$m \Rightarrow n = (1-m) \ \mathbf{tc} \ n$, for all $m, n \in [0..1]$.

27 Show how to define bold conjunction using Fuzzy_{L}'s negation and conditional.

SECTION 11.8

28 Prove that $Fuzzy_G$ weak conjunction and disjunction are identical, respectively, to $Fuzzy_G$ bold conjunction and disjunction.

29 Using the $Fuzzy_G$ versions of the connectives, what are the truth-values for the formulas in problem 1 when the atomic formulas have the indicated values?

30 a. Prove that $V(\mathbf{P} \vee \mathbf{Q}) = V(((\mathbf{P} \rightarrow \mathbf{Q}) \rightarrow \mathbf{Q}) \wedge ((\mathbf{Q} \rightarrow \mathbf{P}) \rightarrow \mathbf{P}))$ in $Fuzzy_Ł$.

b. Prove that if we define $Fuzzy_G$ weak disjunction as

$$V(\mathbf{P} \vee_G \mathbf{Q}) = V(((\mathbf{P} \rightarrow_G \mathbf{Q}) \rightarrow_G \mathbf{Q}) \wedge_G ((\mathbf{Q} \rightarrow_G \mathbf{P}) \rightarrow_G \mathbf{P}))$$

then weak and bold conjunction are identical in $Fuzzy_G$.

31 Prove that if we were to define $\mathbf{P} \vee_G \mathbf{Q}$ as $\neg_G(\neg_G\mathbf{P} \wedge_G \neg_G\mathbf{Q})$, then the value of any formula $\mathbf{P} \vee_G \mathbf{Q}$ would always be either 1 or 0.

32 Is $\mathbf{P} \rightarrow_G \mathbf{Q}$ equivalent to $\neg\mathbf{P}\ \nabla_G\ \mathbf{Q}$, analogously to $Fuzzy_Ł$'s equivalence?

33 Use the decision procedure presented at the end of Section 11.8 to determine whether $\neg_G(P\ \&_G \neg_G P)$ is a tautology of $Fuzzy_G$.

SECTION 11.9

34 If $\mathbf{P} \leftrightarrow_P \mathbf{Q}$ is defined as $(\mathbf{P} \rightarrow_P \mathbf{Q}) \wedge_P (\mathbf{Q} \rightarrow_P \mathbf{P})$, what are the truth-conditions for $\mathbf{P} \leftrightarrow_P \mathbf{Q}$?

35 Using the $Fuzzy_P$ versions of the connectives, what are the truth-values for the formulas in problem 1 when the atomic formulas have the indicated values?

36 a. Prove that the truth-conditions 6 and 7 for $Fuzzy_P$ weak conjunction and disjunction follow from the definitions

$$\mathbf{P} \wedge_P \mathbf{Q} =_{def} \mathbf{P}\ \&_P (\mathbf{P} \rightarrow_P \mathbf{Q})$$

and

$$\mathbf{P} \vee_P \mathbf{Q} =_{def} ((\mathbf{P} \rightarrow_P \mathbf{Q}) \rightarrow_P \mathbf{Q}) \wedge_P ((\mathbf{Q} \rightarrow_P \mathbf{P}) \rightarrow_P \mathbf{P}).$$

b. What would the truth-conditions for $Fuzzy_P$ weak disjunction be if it were defined as

$$\mathbf{P} \vee_P \mathbf{Q} =_{def} \neg_P(\neg_P\mathbf{P} \wedge_P \neg_P\mathbf{Q})?$$

c. What would the truth-conditions for $Fuzzy_P$ weak conjunction be if it were defined as

$$\mathbf{P} \wedge_P \mathbf{Q} =_{def} \mathbf{P}\ \&_P (\neg_P\mathbf{P}\ \&_P\ \mathbf{Q})?$$

37 a. Prove that $P \rightarrow_G (P\ \&_G\ P)$ is a tautology but $P \rightarrow_P (P\ \&_P\ P)$ is not.

b. Prove that $\neg_P\neg_P P \rightarrow_P (((Q\ \&_P\ P) \rightarrow_P (R\ \&_P\ P)) \rightarrow_P (Q \rightarrow_P R))$ is a tautology but $\neg_G\neg_G P \rightarrow_G (((Q\ \&_G\ P) \rightarrow_G (R\ \&_G\ P)) \rightarrow_G (Q \rightarrow_G R))$ is not.

c. Find another formula that is a tautology in $Fuzzy_G$ but not in $Fuzzy_P$, and defend your answer.

d. Find another formula that is a tautology in $Fuzzy_P$ but not in $Fuzzy_G$, and defend your answer.

e. Prove that the Modus Ponens inference is degree-valid in $Fuzzy_G$, but not in $Fuzzy_P$.

SECTION 11.10

38 a. Can $\mathbf{\Delta P}$ be defined in Fuzzy$_{\mathrm{L}}$ as $\mathbf{\Delta P} =_{\mathrm{def}} \bot\neg_{L}\mathbf{P}$? Defend your answer.

b. Can $\mathbf{\Delta P}$ be defined in Fuzzy$_{\mathrm{G}}$ as $\mathbf{\Delta P} =_{\mathrm{def}} \bot\neg_{G}\mathbf{P}$? Defend your answer.

39 Prove that the definition $\mathbf{\Delta P} =_{\mathrm{def}} \neg_{G}\neg_{L}\mathbf{P}$ correctly defines fuzzy external assertion.

40 a. What is the value of the formula $\neg_{G}\Delta P \wedge_{G} \neg_{G}\Delta\neg_{G}P$ in Fuzzy$_{\mathrm{G}}$ when P has any value other than 1 and 0?

b. What is the value of the formula $\neg_{G}\Delta P \wedge_{G} \neg_{G}\Delta\neg_{G}P$ in Fuzzy$_{\mathrm{G}}$ when P has the value 1? When P has the value 0?

c. What is the value of the formula $\neg_{P}\Delta P \,\&_{P}\, \neg_{P}\Delta\neg_{P}P$ in Fuzzy$_{\mathrm{P}}$ when P has any value other than 1 and 0?

d. What is the value of the formula $\neg_{P}\Delta P \,\&_{P}\, \neg_{P}\Delta\neg_{P}P$ in Fuzzy$_{\mathrm{P}}$ when P has the value 1? When P has the value 0?

12 Fuzzy Algebras

12.1 More on MV-Algebras

In Chapter 9 we introduced *MV-algebras* in connection with Łukasiewicz's three-valued logic. MV-algebras were in fact developed in order to study Łukasiewicz's *infinite*-valued systems, so it will come as no surprise that they capture the algebraic structure of Fuzzy_{L}. Recall that an **MV-algebra** is an algebraic structure $<\text{M}, \oplus, \otimes, ', unit, zero>$ (where *unit* and *zero* are members of M) that meets the following conditions for all x, y, and z in M:

i.	$x \oplus y = y \oplus x$, and $x \otimes y = y \otimes x$	(*commutation*)
ii.	$x \oplus (y \oplus z) = (x \oplus y) \oplus z$, and $x \otimes (y \otimes z) = (x \otimes y) \otimes z$	(*association*)
iii.	$x \oplus zero = x$, and $x \otimes unit = x$	(*identity for join and meet*)
iv.	$x \oplus unit = unit$, and $x \otimes zero = zero$	(*unit* and *zero consumption*)
v.	$x \oplus x' = unit$, and $x \otimes x' = zero$	(*complementation*)
vi.	$(x \oplus y)' = x' \otimes y'$, and $(x \otimes y)' = x' \oplus y'$	(*DeMorgan's Laws*)
vii.	$x = x''$	(*Double Negation*)
viii.	$zero' = unit$	(*duality of zero and unit*)
ix.	$(x' \oplus y)' \oplus y = (y' \oplus x)' \oplus x$	(*lattice meet commutation*)

It is left as an exercise to prove that the algebra $\text{Fuzzy}_{\text{L}}\text{MV} = <[0..1], \oplus_L, \otimes_L, 1-, 1, 0>$, where $\oplus_L$, $\otimes_L$, and $1-$ are Fuzzy_{L}'s bold disjunction, bold conjunction, and negation operations, is an MV-algebra.

The definition

$$x \Rightarrow y =_{\text{def}} x' \oplus y$$

gives us the Fuzzy_{L} conditional operation in the algebra $\text{Fuzzy}_{\text{L}}\text{MV}$. The lattice meet operation defined as

$$x \cap y =_{\text{def}} x \otimes (x' \oplus y)$$

gives us Fuzzy_{L}'s weak conjunction operation in $\text{Fuzzy}_{\text{L}}\text{MV}$. Lattice join, corresponding to Fuzzy_{L}'s weak disjunction, can then be defined as

$$x \cup y =_{\text{def}} (x' \cap y')'$$

which, when spelled out, gives

$$x \cup y = x \oplus (x' \otimes y)$$

(proof is left as an exercise).

We noted in Chapter 11 that the Laws of Excluded Middle and Noncontradiction hold in Fuzzy$_Ł$ when they are expressed using bold disjunction and conjunction but fail when expressed with the weak connectives. The bold version of the Law of Excluded Middle appears in MV-algebras as the first complementation condition $x \oplus x' = \mathit{unit}$, while the bold version of the Law of Noncontradiction appears as the equation $(x \otimes x')' = \mathit{zero}$, which is derivable by complementing both sides of the second complementation condition and then replacing *zero′* with *unit* by virtue of condition viii, the duality of *zero* and *unit.*

In fact, we have a result relating Fuzzy$_Ł$ and MV-algebras that is analogous to Result 4.3 of Chapter 4 relating classical propositional logic and Boolean algebras. First, some definitions. When we interpret formulas of propositional logic in an MV-algebra by assigning a member of the algebra's domain to each atomic formula and using the algebra's bold join, bold meet, and complement operations to define the respective values of bold disjunctions, bold conjunctions, and negations, we call this an ***algebraic interpretation*** based on that MV-algebra. We will say that a formula of propositional logic is a ***tautology of an MV-algebra*** (or an *MV-tautology*) if the formula evaluates to *unit* under every algebraic interpretation based on that algebra. Then, defining the other Fuzzy$_Ł$ connectives in terms of bold disjunction, bold conjunction, and negation (as we've seen we can do), we have

> *Result 12.1:* A formula is a tautology of Fuzzy$_Ł$ if and only if it is an MV-tautology in every MV-algebra.

Proof of 12.1 is beyond the scope of this text; interested readers may consult Gottwald (2001) (also for proofs of Results 12.3 and 12.4 later in this chapter).

It is important to note that this result does *not* say that the set of Fuzzy$_Ł$ tautologies coincides with the set of MV-tautologies for *any* MV-algebra. For example, in Chapter 9 we examined the MV-algebraic structure of Ł$_3$. We know that the set of Ł$_3$ tautologies is distinct from the set of Fuzzy$_Ł$ tautologies, so it follows that the set of MV-tautologies for any three-valued MV-algebra (which, by Result 9.3, coincides with the set of Ł$_3$ tautologies) is different from the set of MV-tautologies for Fuzzy$_Ł$MV. (Moreover, in Section 11.3 of Chapter 11 we introduced Łukasiewicz logics for all finite sets of truth-values taken from the unit interval. Each one of these has an MV-algebraic structure but no two have the same set of tautologies, and each of the tautology sets differs from the set of Fuzzy$_Ł$ tautologies.) A formula must be an MV-tautology of *every* MV-algebra if it is to be a Fuzzy$_Ł$ tautology, and vice versa.

12.2 Residuated Lattices and BL-Algebras

The algebraic structures characterizing $Fuzzy_G$ and $Fuzzy_P$ (based on their respective bold disjunction, bold conjunction, and negation operations) are not MV-algebras.[1] Most notably, neither of those structures satisfies the Double Negation condition (vii) for MV-algebras. This was to be expected, given the nonvalidity of the inference

$$\frac{\neg\neg P}{P}$$

in both systems. It is left as an exercise to determine which other conditions on MV-algebras fail for one or both of these systems.

We'll present algebras for $Fuzzy_G$ and $Fuzzy_P$ that are special cases of *residuated lattices* (MV-algebras are also special cases of residuated lattices). A **residuated lattice** is an algebra $<L, \cup, \cap, \otimes, \Rightarrow, \mathit{unit}, \mathit{zero}>$ that meets the following conditions:[2]

i.	$x \cup y = y \cup x$, and $x \cap y = y \cap x$	(*lattice commutation*)
ii.	$x \cup (y \cup z) = (x \cup y) \cup z$, and $x \cap (y \cap z) = (x \cap y) \cap z$	(*lattice association*)
iii.	$x \cup x = x$, and $x \cap x = x$	(*lattice idempotence*)
iv.	$x \cup (x \cap y) = x$, and $x \cap (x \cup y) = x$	(*lattice absorption*)
v.	$x \cup \mathit{zero} = x$, and $x \cap \mathit{unit} = x$	(*identity for lattice join and meet*)
vi.	$x \otimes y = y \otimes x$	(*bold meet commutation*)
vii.	$x \otimes (y \otimes z) = (x \otimes y) \otimes z$	(*bold meet association*)
viii.	$x \otimes \mathit{unit} = x$	(*identity for bold meet*)
	and, defining $x \le y$ if and only if $x \cap y = x$,	
ix.	if $x \le y$, then $x \otimes z \le y \otimes z$ and $z \otimes x \le z \otimes y$	(*bold meet isotonicity*)
x.	$x \otimes y \le z$ if and only if $x \le y \Rightarrow z$.	(*adjointness*)

Recall that conditions i–v define a lattice with *zero* and *unit* elements.

Conditions vi–viii define the bold meet $\otimes$ as a commutative, associative operation with *unit* as its identity element. Condition ix states that the bold meet operation is isotonic, or nondecreasing in both arguments, and condition x states that $\otimes$ and $\Rightarrow$ form an adjoint pair. The connection with t-norms and their residuation operations should be obvious. Conditions vi–ix are the conditions for t-norm operations, and condition x is the adjointness condition defining the residuum operation for a t-norm.

[1] This section provides just a glimpse of the relations among MV-algebras, residuated lattices, and BL-algebras. An excellent summary of literature exploring the relations among these algebras (along with Boolean algebras) appears in Novák, Perfilieva, and Močkoř (1999, pp. 23–33).

[2] Residuated lattices were first studied in Dilworth and Ward (1939).

Negation is defined in a residuated lattice as

$x' =_{def} x \Rightarrow zero$

This should also look familiar from Chapter 11; it is the way that negation is standardly defined in fuzzy logical systems. Given this operation, a dual operation for $\otimes$ in residuated lattices is definable as

$x \oplus y =_{def} (x' \otimes y')$.

If MV = <M, $\oplus$, $\otimes$, $'$, *unit, zero*> is an MV-algebra, and we define

$$x \cap y =_{def} x \otimes (x' \oplus y)$$
$$x \cup y =_{def} x \oplus (x' \otimes y)$$
$$x \Rightarrow y =_{def} x' \oplus y,$$

as we did for MV-algebras in Section 12.1, then R(MV) = <M, $\cup$, $\cap$, $\otimes$, $\Rightarrow$, *unit, zero*> is a residuated lattice; it is in this sense that we say that MV-algebras are special cases of residuated lattices.[3] For example, we know that $x \leq y$ if and only if $x \cap y = x$ in an MV-algebra, so the definition of inequality required for a residuated lattice holds in R(MV). We can establish the first part of bold meet isotonicity as follows:

First part of Isotonicity of MV-Algebra Meet: If $x \leq y$ then $x \otimes z \leq y \otimes z$.

Proof:
Assume that $x \leq y$ in an MV-algebra. Then:

$x' \oplus y = unit$	(definition of $\leq$)
$(x' \oplus y) \oplus (y' \otimes z') = unit \oplus (y' \otimes z')$	(same operation, both sides)
$(x' \oplus y) \oplus (y' \otimes z') = unit$	(*unit* consumption)
$(x' \oplus (y \oplus (y' \otimes z')) = unit$	(bold meet association)
$(x' \oplus (z' \oplus (z'' \otimes y)) = unit$	(lattice join commutation)
$(x' \oplus z') \oplus (z'' \otimes y) = unit$	(bold meet association)
$(x \otimes z)' \oplus (z'' \otimes y) = unit$	(DeMorgan)
$(x \otimes z)' \oplus (y \otimes z) = unit$	(bold meet commutation, Double Negation)
$x \otimes z \leq y \otimes z$	(definition of $\leq$)

A complete proof that every MV-algebra is a residuated lattice, many pieces of which we have already seen by now, is left as an exercise.

The converse does not generally hold; some residuated lattices are not MV-algebras. One reason is that the conditions defining residuated lattices do not entail Double Negation, so in some residuated lattices it is not true that $x = x''$ for all x in L. This should come as no surprise since we have stated that the algebraic structures for $Fuzzy_G$ and $Fuzzy_P$ are residuated lattices, and we know that Double Negation fails in those logical systems.

[3] Because Boolean algebras are MV-algebras (this was proved in Chapter 9), it follows that Boolean algebras are also special cases of residuated lattices.

The algebraic structure corresponding to Fuzzy_G is a residuated lattice $\text{Fuzzy}_G\text{L} = <[0..1], \cup_G, \cap_G, \otimes_G, \Rightarrow_G, 1, 0>$, where $\otimes_G$ and $\Rightarrow_G$ are Fuzzy_G's bold conjunction and conditional operations (we saw in Section 11.7 of Chapter 11 that $\otimes_G$ and $\Rightarrow_G$ form an adjoint pair), and $\cup_G$ and $\cap_G$ are Fuzzy_G's weak disjunction and conjunction operations (which are identical to Fuzzy_G's bold operations). The algebraic structure $\text{Fuzzy}_P\text{L} = <[0..1], \cup_P, \cap_P, \otimes_P, \Rightarrow_P, 1, 0>$ corresponding to Fuzzy_P is also a residuated lattice. Both of these algebraic structures, along with that of Fuzzy_L, are special types of residuated lattices called *BL-algebras*.[4] A **BL-algebra** is a residuated lattice that satisfies the additional conditions

xi. $x \cap y = x \otimes (x \Rightarrow y)$
xii. $(x \Rightarrow y) \cup (y \Rightarrow x) = \mathit{unit}$

Condition xi should be familiar from the general definition of weak conjunction in fuzzy logics introduced in Chapter 11:

$$\mathbf{P} \wedge \mathbf{Q} =_{\text{def}} \mathbf{P} \mathbin{\&} (\mathbf{P} \rightarrow \mathbf{Q})$$

Condition xii captures the fact that for any formulas **P** and **Q** in a t-norm-based fuzzy logical system, at least one of $\mathbf{P} \rightarrow \mathbf{Q}$ or $\mathbf{Q} \rightarrow \mathbf{P}$ has the value 1 (to be proved in the exercises).

Every MV-algebra is a BL-algebra (also to be proved in the exercises). Conversely, we have

Result 12.2: Every BL-algebra that satisfies Double Negation ($x'' = x$) is an MV-algebra.

Proof: We'll establish that each of the MV-algebra conditions holds in any BL-algebra that satisfies Double Negation. We'll use the following properties, which hold for all BL-algebras with Double Negation (BL-i through BL-iii hold for BL-algebras generally, not just those with Double Negation):

(BL-i) if $\mathit{unit} \leq x$, then $\mathit{unit} = x$

Proof: Assume $\mathit{unit} \leq x$. Then:

$\mathit{unit} \cap x = \mathit{unit}$	(BL-algebra definition of $\leq$)
$x \cap \mathit{unit} = \mathit{unit}$	(lattice meet commutation)
$x = \mathit{unit}$	(identity for lattice meet)

(BL-ii) $x \otimes (x \Rightarrow y) = y$

Proof: Left as an exercise.

(BL-iii) $x \Rightarrow (y \Rightarrow z) = (x \otimes y) \Rightarrow z$

4 BL-algebras were introduced in Hájek (1998a, 1998b) to capture the commonalities of systems of fuzzy logic based on continuous t-norms (thus including Łukasiewicz, Gödel, and product logics). *BL* stands for *basic logic.* The definitions later of Gödel and product algebras as special types of BL-algebras are from Hájek (1998b, pp. 91 and 100).

Proof:

$(x \Rightarrow (y \Rightarrow z)) \cap (x \Rightarrow (y \Rightarrow z)) = x \Rightarrow (y \Rightarrow z)$	(lattice meet idempotence)
$x \Rightarrow (y \Rightarrow z) \leq x \Rightarrow (y \Rightarrow z)$	(definition of $\leq$)
$x \otimes (x \Rightarrow (y \Rightarrow z)) \leq x \otimes (x \Rightarrow (y \Rightarrow z))$	(bold meet isotonicity)
$x \otimes (x \Rightarrow (y \Rightarrow z)) \leq y \Rightarrow z$	(BL-ii)
$y \otimes (x \otimes (x \Rightarrow (y \Rightarrow z))) \leq y \otimes (y \Rightarrow z)$	(bold meet isotonicity)
$y \otimes (x \otimes (x \Rightarrow (y \Rightarrow z))) \leq z$	(BL-ii)
$(x \Rightarrow (y \Rightarrow z)) \otimes (x \otimes y) \leq z$	(bold meet association, commutation)
$x \Rightarrow (y \Rightarrow z) \leq (x \otimes y) \Rightarrow z$	(adjointness)

The rest of the proof, that $(x \otimes y) \Rightarrow z \leq x \Rightarrow (y \Rightarrow z)$, is left as an exercise.

(BL-iv) $x \Rightarrow y = y' \Rightarrow x'$

Proof:

$x \Rightarrow y = x \Rightarrow y''$	(Double Negation)
$x \Rightarrow y = x \Rightarrow ((y \Rightarrow \mathit{zero}) \Rightarrow \mathit{zero})$	(definition of complement)
$x \Rightarrow y = (x \otimes (y \Rightarrow \mathit{zero})) \Rightarrow \mathit{zero}$	(BL-iii)
$x \Rightarrow y = ((y \Rightarrow \mathit{zero}) \otimes x) \Rightarrow \mathit{zero}$	(bold meet commutation)
$x \Rightarrow y = ((y \Rightarrow \mathit{zero}) \Rightarrow (x \Rightarrow \mathit{zero})$	(BL-iii)
$x \Rightarrow y = y' \Rightarrow x'$	(definition of complement)

(BL-v) $x \otimes y = (x \Rightarrow y')'$

Proof:

$x \otimes y = (x \otimes y)''$	(Double Negation)
$x \otimes y = ((x \otimes y) \Rightarrow \mathit{zero})'$	(definition of complement)
$x \otimes y = (x \Rightarrow (y \Rightarrow \mathit{zero}))'$	(BL-iii)
$x \otimes y = (x \Rightarrow y')'$	(definition of complement)

(BL-vi) $x \oplus y = x' \Rightarrow y$

Proof: Left as an exercise.

Now we can establish that all of the conditions defining MV-algebras hold true in any BL-algebra with Double Negation. We'll show this for a few of the MV-algebra conditions; the reader will be asked to establish the remaining conditions in the exercises.

Condition iv:

(i)	$x \oplus \mathit{unit} = x' \Rightarrow \mathit{unit}$	(BL-vi)
	$\mathit{unit} \cap (x' \Rightarrow \mathit{unit}) = x' \Rightarrow \mathit{unit}$	(lattice meet commutation, identity)
	$\mathit{unit} \leq x' \Rightarrow \mathit{unit}$	(definition)
(ii)	$x' \Rightarrow \mathit{unit} = \mathit{unit}$	(BL-i)
	$x \oplus \mathit{unit} = \mathit{unit}$	(by (i) and (ii))

(Proof that $x \otimes \mathit{zero} = \mathit{zero}$ is left as an exercise.)

Condition viii:

$unit \otimes zero = zero$	(bold meet commutation, identity)
$unit \leq zero \Rightarrow zero$	(adjointness)
$unit = zero'$	(definition of complement, BL-i)

Condition ix:

$(x' \oplus y)' \oplus y = (x' \oplus y)'' \Rightarrow y$	(BL-vi)
$(x' \oplus y)' \otimes y = (x'' \Rightarrow y)'' \Rightarrow y$	(BL-vi)
$(x' \oplus y)' \otimes y = (x \Rightarrow y) \Rightarrow y$	(Double Negation)
$(x' \oplus y)' \otimes y = (x \Rightarrow y) \Rightarrow y''$	(Double Negation)
$(x' \oplus y)' \otimes y = (x \Rightarrow y) \Rightarrow ((y \Rightarrow zero) \Rightarrow zero)$	(definition)
$(x' \oplus y)' \otimes y = ((x \Rightarrow y) \otimes (y \Rightarrow zero)) \Rightarrow zero$	(BL-iii)
$(x' \oplus y)' \otimes y = ((x \Rightarrow y) \otimes (y \Rightarrow zero))'$	(definition)
$(x' \oplus y)' \otimes y = ((x \Rightarrow y) \otimes y')'$	(definition)
$(x' \oplus y)' \otimes y = (y' \otimes (x \Rightarrow y))'$	(bold meet commutation)
$(x' \oplus y)' \otimes y = (y' \otimes (y' \Rightarrow x'))'$	(BL-iv)
$(x' \oplus y)' \otimes y = (y' \cap x')'$	(BL-algebra condition (xi))
$(x' \oplus y)' \otimes y = (x' \cap y')'$	(lattice meet commutation)

The rest of this proof reverses the steps leading to $(x' \oplus y)' \otimes y = (y' \cap x')'$.

The lattice Fuzzy$_G$L exemplifies another special case of BL-algebras: the *Gödel-algebras.*[5] A **Gödel algebra** is a BL-algebra that satisfies the additional condition

xiii. (Gödel BL-algebra) $x \otimes x = x$.

This condition says that Gödel bold conjunction (identical to Gödel weak conjunction) is idempotent. Proof that Fuzzy$_G$L is a Gödel algebra is left as an exercise. In addition we have

> *Result 12.3:* A formula is a tautology of Fuzzy$_G$ if and only if it is a G-tautology in every Gödel algebra,

where G-tautologies of Gödel algebras are defined in the by now obvious way.

The algebraic structure Fuzzy$_P$L = <[0. .1], $\cup_P$, $\cap_P$, $\otimes_P$, $\Rightarrow_P$, 1,0>, where $\otimes_P$ and $\Rightarrow_P$ are Fuzzy$_P$'s bold conjunction and conditional operations, and $\cup_P$ and $\cap_P$ are Fuzzy$_P$'s weak disjunction and conjunction operations, is an instance of another special case of BL-algebras, the *product algebras.* A **product algebra** is a BL-algebra that also satisfies the additional conditions

xiii. (Product BL-algebra) $z'' \leq ((x \otimes z) \Rightarrow (y \otimes z)) \Rightarrow (x \Rightarrow y)$
xiv. (Product BL-algebra) $x \cap x' = zero$

where

$$x' =_{def} x \Rightarrow zero.$$

[5] Gödel algebras are identical to so-called *Heyting* algebras that satisfy the "prelinearity" condition: $(x \Rightarrow y) \cup (y \Rightarrow x) = unit$.

Condition xiv gives us the Law of Noncontradiction for weak conjunction: for any formula **P**, at least one of **P** and ¬**P** has the value 0, and so therefore does the weak conjunction of a formula and its negation. It is left as an exercise to prove that Fuzzy_PL is a product algebra, and hence that xiii is characteristic of Fuzzy_P's adjoint bold conjunction and implication operations. Analogously to earlier results, we have

> *Result 12.4:* A formula is a tautology of Fuzzy_P if and only if it is a P-tautology in every product algebra.

12.3 *Zero* and *Unit* Projections in Algebraic Structures

Recall that we introduced fuzzy Bochvarian external negation and assertion, known as 0- and 1-projections in the fuzzy literature, in Section 11.10 of Chapter 11. We will call the corresponding algebraic operations ***zero projection*** and ***unit projection***, and we will use the symbols *!* and α, respectively, to denote these operations. The following algebraic conditions characterize *unit* projection in any BL-algebra:[6]

αi. $\alpha x \cup (\alpha x)' = \mathit{unit}$
αii. $\alpha(x \cup y) \leq \alpha x \cup \alpha y$
αiii. $\alpha x \leq x$
αiv. $\alpha x \leq \alpha\alpha x$
αv. $\alpha x \otimes \alpha(x \Rightarrow y) \leq \alpha y$
αvi. $\alpha\ \mathit{unit} = \mathit{unit}$

and we will call a BL-algebra that meets these additional conditions a ***unit projection algebra***.

Although x and αx are not generally equivalent in a *unit* projection algebra, αx and $\alpha\alpha$x are:

$$\alpha x = \alpha\alpha x$$

Proof:
$\alpha x \leq \alpha\alpha x$, by αiv, and $\alpha\alpha x \leq \alpha x$, by αiii, so $\alpha x = \alpha\alpha x$.

The expressions x and αx are not generally equivalent because the inequality $x \leq \alpha x$ does not hold in all *unit* projection algebras. This is as it should be, if *unit* projection is the algebraic counterpart to fuzzy external assertion: if V(P) = .5, for example, ΔP has the value 0.

We can introduce *zero* projection in a *unit* projection algebra as the algebraic counterpart to fuzzy external negation with the definition

$$!x =_{\text{def}} (\alpha x)'.$$

6 Hájek (1998b) based these algebraic axioms on derivational axioms in Baaz (1996).

so that, for example,

$\alpha x \cup !x = unit$ (by αi)

and

$!unit = zero$

Proof:

$\alpha\ unit = unit$	(αvi)
$(\alpha\ unit)' = unit'$	(same operation, both sides)
(i) $!unit = unit'$	(definition)
and	
$unit \otimes (unit \Rightarrow zero) = zero$	(BL-ii)
$unit \Rightarrow zero = zero$	(identity for bold meet)
(ii) $unit' = zero$	(definition)
so	
$!unit = zero$	(from (i) and (ii))

Adding the external assertion operator Δ to $Fuzzy_L$, the MV-algebraic structure of $Fuzzy_L$ becomes a *unit* projection algebra. To establish this, we need to show that the following hold for all values x and y in [0. .1], where α_Δ stands for the external assertion operation:

α_Δi. $\max(\alpha_\Delta x, 1 - \alpha_\Delta x) = 1$
$\alpha_\Delta$ii. $\alpha_\Delta(\max(x, y)) \le \max(\alpha_\Delta x, \alpha_\Delta y)$
$\alpha_\Delta$iii. $\alpha_\Delta x \le x$
$\alpha_\Delta$iv. $\alpha_\Delta x \le \alpha_\Delta \alpha_\Delta x$
α_Δv. $\max(0, \alpha_\Delta x + \alpha_\Delta(\min(1, 1 - x + y)) - 1) \le \alpha_\Delta y$
$\alpha_\Delta$vi. $\alpha_\Delta 1 = 1$

For (α_Δi), it suffices to point out that $\alpha_\Delta x$ is always either 1 or 0. For ($\alpha_\Delta$iii), we note that if $x \neq 1$ then $\alpha_\Delta x = 0$ and 0 is less than or equal to any value in the unit interval, while if $x = 1$ then $\alpha_\Delta x = 1$ and certainly $1 \le 1$. We leave the remainder, as well as the proof that the Gödel and product algebraic structures for $Fuzzy_G$ and $Fuzzy_P$ augmented with the external assertion operator are *unit* projection algebras, as exercises.

12.4 Exercises

SECTION 12.1

1 Prove that the algebra $Fuzzy_L MV = <[0..1], \oplus_L, \otimes_L, 1-, 1, 0>$, where $\oplus_L$, $\otimes_L$, and $1-$ are $Fuzzy_L$'s bold disjunction, bold conjunction, and negation operations, is an MV-algebra.

2 Prove that it follows from the MV-algebra definition $x \cup y =_{def} (x' \cap y')'$ that $x \cup y = x \oplus (x' \otimes y)$.

SECTION 12.2

3 Which MV-algebra conditions other than Double Negation, if any, fail to hold for the algebra defined by $Fuzzy_G$'s operations?

4 Which MV-algebra conditions other than Double Negation, if any, fail to hold for the algebra defined by $Fuzzy_P$'s operations?

5 Give a complete proof that every MV-algebra is a residuated lattice in the sense described in Section 12.2. You are free to cite previous results along the way.

6 Prove that for any formulas **P** and **Q** in a t-norm-based fuzzy logical system, at least one of $\mathbf{P} \rightarrow \mathbf{Q}$ or $\mathbf{Q} \rightarrow \mathbf{P}$ has the value 1, by showing that this follows from the definition of the residuum operation

$m \,\mathbf{tn}\, n \leq p$ if and only if $m \leq n \Rightarrow p$, for all $m, n, p \in [0..1]$

for any t-norm **tn**. (Hint: consider the case where $m = 1$ and show that in this case either $1 \leq n \Rightarrow p$ or $1 \leq p \Rightarrow n$. You will need to use t-norm properties to establish this.)

7 Prove that every MV-algebra is a BL-algebra.

8 Prove that

(BL-ii) $x \otimes (x \Rightarrow y) = y$

holds in every BL-algebra.

9 Complete the proof of BL-iii:

a. Show that $(x \otimes y) \Rightarrow z \leq x \Rightarrow (y \Rightarrow z)$ in every BL-algebra with Double Negation.

b. Show that if $x \leq y$ and $y \leq x$ in a BL-algebra, then $x = y$—so that the converse inequalities in the proof of BL-3 in fact establish an equality.

10 Prove that

(BL-vi) $x \oplus y = x' \Rightarrow y$

holds in every BL-algebra with Double Negation.

11 Complete the proof of Result 12.2, that every BL-algebra with Double Negation is an MV-algebra, by showing that the following hold true in every BL-algebra with Double Negation:

a. condition i of MV-algebras
b. condition ii of MV-algebras
c. condition iii of MV-algebras
d. the second half of condition iv of MV-algebras: $x \otimes \mathit{zero} = \mathit{zero}$
e. condition v of MV-algebras
f. condition vi of MV-algebras
g. condition vii of MV-algebras.

12 Prove that $Fuzzy_GL$ is a Gödel algebra.

13 Prove that $Fuzzy_PL$ is a product algebra.

14 Why did we not include the condition

xiv. (Gödel BL-algebra) $x \otimes x = x$

in the definition of product algebras?

SECTION 12.3

15 Prove that the following hold in every *unit* projection algebra:

a. α *zero* = *zero*

b. !*zero* = *unit*

c. !x = !!!x, for any x

16 Complete the proof that the MV-algebraic structure of $Fuzzy_L$ augmented with the external assertion operator is a *unit* projection algebra, by showing that $\alpha_\Delta$ii, $\alpha_\Delta$iv, α_Δv, and $\alpha_\Delta$vi all hold.

17 Prove that the Gödel and product algebraic structures $Fuzzy_G$L and $Fuzzy_P$L augmented with the external assertion operator are *unit* projection algebras.

13 Derivation Systems for Fuzzy Propositional Logic

13.1 An Axiomatic System for Tautologies and Validity in Fuzzy$_Ł$

In Chapter 6 we presented an axiomatic system for logical truth and validity in Łukasiewicz's three-valued propositional logic, Ł$_3$A. Łukasiewicz also formulated an axiomatic system for his infinite-valued logic. We shall call this system F$_Ł$A (for *fuzzy Łukasiewicz axiomatic* system). F$_Ł$A includes the first three axiom schemata from Ł$_3$A, plus one more:[1]

F$_Ł$1. $\mathbf{P \rightarrow (Q \rightarrow P)}$
F$_Ł$2. $\mathbf{(P \rightarrow Q) \rightarrow ((Q \rightarrow R) \rightarrow (P \rightarrow R))}$
F$_Ł$3. $\mathbf{(\neg P \rightarrow \neg Q) \rightarrow (Q \rightarrow P)}$
F$_Ł$4. $\mathbf{((P \rightarrow Q) \rightarrow Q) \rightarrow ((Q \rightarrow P) \rightarrow P)}$

The single rule is Modus Ponens:

MP. From **P** and $\mathbf{P \rightarrow Q}$, infer **Q**.

Axiom schema Ł$_3$4 from Ł$_3$A—$\mathbf{((P \rightarrow \neg P) \rightarrow P) \rightarrow P}$ – is *not* included here because, as we saw in Result 11.6 of Chapter 11, it's not a tautologous schema in Fuzzy$_Ł$. On the other hand, simply deleting Ł$_3$4 would leave the system incomplete for Fuzzy$_Ł$, so the axiom F$_Ł$4 is added to give exactly the axiomatic power we need. Given the definition of disjunction

$$\mathbf{P \vee Q =_{def} ((P \rightarrow Q) \rightarrow Q)}$$

F$_Ł$4 can be written as $\mathbf{(P \vee Q) \rightarrow (Q \vee P)}$, which asserts that disjunction is a *commutative* operation. Recall that we derived this formula in Ł$_3$A, as derived axiom schema Ł$_3$D9.

Any derivation in Ł$_3$A that does not use axiom Ł$_3$4 is a legal derivation in F$_Ł$A. So, for example,

Ł$_3$D1. $\mathbf{\neg P \rightarrow (P \rightarrow Q)}$

[1] Łukasiewicz's system included a fifth axiom, which was independently shown to be derivable from the remaining axioms by both Chang (1958a) and Meredith (1928).

is also a derivable axiom schema in $F_ŁA$ because it was justified using only $Ł_3 1$–$Ł_3 3$ and Modus Ponens. We will henceforth refer to this derived axiom as $F_Ł$D1. Similarly, we may conclude that

HS. From **P → Q** and **Q → R**, infer **P → R**

is a derivable rule in $F_ŁA$, since its derivation in $Ł_3A$ used only axiom schema $Ł_3 2$ and Modus Ponens.

On the other hand, although both

¬¬P → P

and

P → ¬¬P

are derivable as axioms in $F_ŁA$, we cannot conclude this from the derivations that we used in $Ł_3A$ since those derivations depend on axiom $Ł_3 4$. We will show how to derive both of these formulas in $F_ŁA$, first introducing for convenience the derived axiom schemata

$F_Ł$D2. ((P → Q) → R) → (¬P → R)
$F_Ł$D3. P → ((P → Q) → Q)
$F_Ł$D4. (((Q → R) → (P → R)) → S) → ((P → Q) → S)
$F_Ł$D5. (P → (Q → R)) → ((S → Q) → (P → (S → R)))
$F_Ł$D6. (P → (Q → R)) → (Q → (P → R))
$F_Ł$D7. P → P

$F_Ł$D4 and $F_Ł$D5 are complicated formulas and perhaps not "interesting" in themselves, but they'll make subsequent derivations more readable. $F_Ł$D2 is justified as follows:

1	**¬P → (P → Q)**	$F_Ł$D1, with **P / P, Q / Q**
2	**(¬P → (P → Q)) → (((P → Q) → R) → (¬P → R))**	$F_Ł$2, with **¬P / P, P → Q / Q, R / R**
3	**((P → Q) → R) → (¬P → R)**	1,2 MP

$F_Ł$D3 is justified as follows:

1	**P → ((Q → P) → P)**	$F_Ł$1, with **P / P, Q → P / Q**
2	**((Q → P) → P) → ((P → Q) → Q)**	$F_Ł$4, with **Q / P, P / Q**
3	**P → ((P → Q) → Q)**	1,2 HS

$F_Ł$D4 is justified as follows:

1	**(P → Q) → ((Q → R) → (P → R))**	$F_Ł$2, with **P / P, Q / Q, R / R**
2	**((P → Q) → ((Q → R) → (P → R))) → ((((Q → R) → (P → R)) → S) → ((P → Q) → S))**	$F_Ł$2, with **P → Q / P, (Q → R) → (P → R) / Q, S / R**
3	**(((Q → R) → (P → R)) → S) → ((P → Q) → S)**	1,2 MP

F$_{\text{Ł}}$D5 is justified as follows:

1	$((((Q \rightarrow R) \rightarrow (S \rightarrow R)) \rightarrow (P \rightarrow (S \rightarrow R))) \rightarrow$	F$_{\text{Ł}}$D4, with $P / P, Q \rightarrow R / Q, S \rightarrow R / R,$
	$((S \rightarrow Q) \rightarrow (P \rightarrow (S \rightarrow R)))) \rightarrow$	$(S \rightarrow Q) \rightarrow (P \rightarrow (S \rightarrow R)) / S$
	$((P \rightarrow (Q \rightarrow R)) \rightarrow ((S \rightarrow Q) \rightarrow (P \rightarrow (S \rightarrow R))))$	
2	$(((Q \rightarrow R) \rightarrow (S \rightarrow R)) \rightarrow (P \rightarrow (S \rightarrow R))) \rightarrow$	F$_{\text{Ł}}$D4, with $S / P, Q / Q, R / R,$
	$((S \rightarrow Q) \rightarrow (P \rightarrow (S \rightarrow R)))$	$P \rightarrow (S \rightarrow R) / S$
3	$(P \rightarrow (Q \rightarrow R)) \rightarrow ((S \rightarrow Q) \rightarrow (P \rightarrow (S \rightarrow R)))$	1,2 MP

F$_{\text{Ł}}$D6 is justified as follows:

1	$Q \rightarrow ((Q \rightarrow P) \rightarrow P)$	F$_{\text{Ł}}$D3, with $Q / P, P / Q$
2	$(Q \rightarrow ((Q \rightarrow R) \rightarrow R)) \rightarrow ((P \rightarrow (Q \rightarrow R)) \rightarrow (Q \rightarrow (P \rightarrow R)))$	F$_{\text{Ł}}$D5, with $Q / P, Q \rightarrow R / Q, R / R, P / S$
3	$(P \rightarrow (Q \rightarrow R)) \rightarrow (Q \rightarrow (P \rightarrow R))$	1,2 MP

F$_{\text{Ł}}$D7 is justified as follows:

1	$(P \rightarrow (Q \rightarrow P)) \rightarrow (Q \rightarrow (P \rightarrow P))$	F$_{\text{Ł}}$D6, with $P / P, Q / Q, P / R$
2	$P \rightarrow (Q \rightarrow P)$	F$_{\text{Ł}}$2, with $P / P, Q / Q$
3	$Q \rightarrow (P \rightarrow P)$	1,2 MP
4	$(P \rightarrow ((Q \rightarrow (P \rightarrow P)) \rightarrow P)) \rightarrow ((Q \rightarrow (P \rightarrow P)) \rightarrow (P \rightarrow P))$	F$_{\text{Ł}}$D6, with $P / P, Q \rightarrow (P \rightarrow P) / Q, P / R$
5	$P \rightarrow ((Q \rightarrow (P \rightarrow P)) \rightarrow P)$	F$_{\text{Ł}}$1, with $P / P, Q \rightarrow (P \rightarrow P) / Q$
6	$(Q \rightarrow (P \rightarrow P)) \rightarrow (P \rightarrow P)$	4,5 MP
7	$P \rightarrow P$	3,6 MP

Now we can derive

F$_{\text{Ł}}$D8. $\neg\neg P \rightarrow P$

and

F$_{\text{Ł}}$D9. $P \rightarrow \neg\neg P$

as follows:

1	$((\neg P \rightarrow \neg(P \rightarrow P)) \rightarrow P) \rightarrow (\neg\neg P \rightarrow P)$	F$_{\text{Ł}}$D2, with $\neg P / P, \neg(P \rightarrow P) / Q, P / R$
2	$(\neg P \rightarrow \neg(P \rightarrow P)) \rightarrow ((P \rightarrow P) \rightarrow P)$	F$_{\text{Ł}}$3, with $P / P, P \rightarrow P / Q$
3	$((\neg P \rightarrow \neg(P \rightarrow P)) \rightarrow ((P \rightarrow P) \rightarrow P)) \rightarrow$	F$_{\text{Ł}}$D6, with $\neg P \rightarrow \neg(P \rightarrow P) / P, P \rightarrow P / Q,$
	$((P \rightarrow P) \rightarrow ((\neg P \rightarrow \neg(P \rightarrow P)) \rightarrow P))$	P / R
4	$(P \rightarrow P) \rightarrow ((\neg P \rightarrow \neg(P \rightarrow P)) \rightarrow P)$	2,3 MP
5	$P \rightarrow P$	F$_{\text{Ł}}$D7, with P / P
6	$(\neg P \rightarrow \neg(P \rightarrow P)) \rightarrow P$	4,5 MP
7	$\neg\neg P \rightarrow P$	1,6 MP

1	$\neg\neg\neg P \rightarrow \neg P$	F$_{\text{Ł}}$D8, with $\neg P / P$
2	$(\neg\neg\neg P \rightarrow \neg P) \rightarrow (P \rightarrow \neg\neg P)$	F$_{\text{Ł}}$3, with $\neg\neg P / P, P / Q$
3	$P \rightarrow \neg\neg P$	1,2 MP

We'll leave it as an exercise to show that other formulas and rules that were derived in the axiomatic system Ł$_3$A in Chapter 6 can also be derived in F$_Ł$A.

Recall that Łukasiewicz bold disjunction and conjunction are definable thus:

$$\mathbf{P} \nabla \mathbf{Q} =_{def} \neg\mathbf{P} \to \mathbf{Q}$$
$$\mathbf{P} \,\&\, \mathbf{Q} =_{def} \neg(\mathbf{P} \to \neg\mathbf{Q})$$

The bold version of the Law of Excluded Middle, $P \nabla \neg P$, is a tautology in Fuzzy$_Ł$ and so we would hope to be able to derive it in F$_Ł$A, as indeed we can. We rewrite the formula as $\neg P \to \neg P$:

1	$\neg P \to \neg P$	F$_Ł$D7, with $\neg P$ / **P**

The bold version of the Law of Noncontradiction, $\neg(P \,\&\, \neg P)$, another tautology in Fuzzy$_Ł$, is also derivable (where we rewrite the formula as $\neg\neg(P \to \neg\neg P)$):

1	$(P \to \neg\neg P) \to \neg\neg(P \to \neg\neg P)$	F$_Ł$D8, with $P \to \neg\neg P$ / **P**
2	$P \to \neg\neg P$	F$_Ł$D8, with P / **P**
3	$\neg\neg(P \to \neg\neg P)$	1,2 MP

The axiomatic system F$_Ł$A is sound for Fuzzy$_Ł$; that is, every theorem (formula derived from no assumptions) is a tautology of Fuzzy$_Ł$; and if a formula **P** can be derived from a set of assumptions Γ, then Γ semantically entails **P**. F$_Ł$A is also *weakly* complete: every tautology of Fuzzy$_Ł$ is a theorem of F$_Ł$A—so it should come as no surprise that we were able to derive the previous tautologies. However, the system is not *strongly* complete for Fuzzy$_Ł$. Strong completeness means that, given any semantic entailment of a formula **P** from a set Γ of formulas, there is a derivation of **P** from those formulas in the axiomatic system. But the best that we can say for F$_Ł$A is that corresponding to any semantical entailment of a formula **P** from a *finite* set of formulas in Fuzzy$_Ł$ there is a derivation in F$_Ł$A of **P** from that set. And it's not just the specific system F$_Ł$A that has a problem here: *no* sound system for Fuzzy$_Ł$ can be strongly complete. For all such systems, strong completeness fails in the case of entailments from infinite sets because for some of these there are no corresponding derivations.

An example of such an entailment in Fuzzy$_Ł$ goes from the set Σ consisting of the formula $\neg P \to Q$ and the infinitely many formulas in the series

$$(\neg P \to P) \to Q$$
$$(\neg P \to (\neg P \to P)) \to Q$$
$$(\neg P \to (\neg P \to (\neg P \to P))) \to Q$$
$$(\neg P \to (\neg P \to (\neg P \to (\neg P \to P)))) \to Q$$
$$\ldots$$

(where the antecedent of each subsequent formula is a conditional whose antecedent is $\neg P$ and whose consequent is the antecedent of the preceding formula) to the formula Q.[2] Q must be true whenever all the formulas in Σ are true.

To see the entailment, assume that all of the formulas in Σ are true and consider the values that P may have. If P has the value 0, then the formula $\neg P \rightarrow Q$ has a true antecedent and so Q must have the value 1 as well. If P has any value greater than or equal to .5, then $\neg P \rightarrow P$ has the value 1, so the formula $(\neg P \rightarrow P) \rightarrow Q$ has a true antecedent and Q must therefore have the value 1. More generally, in Fuzzy$_Ł$ the value of the antecedent of the nth member of the infinite series of formulas $(\neg P \rightarrow P) \rightarrow Q$, $(\neg P \rightarrow (\neg P \rightarrow P)) \rightarrow Q$, $(\neg P \rightarrow (\neg P \rightarrow (\neg P \rightarrow P))) \rightarrow Q$, $(\neg P \rightarrow (\neg P \rightarrow (\neg P \rightarrow (\neg P \rightarrow P)))) \rightarrow Q, \ldots$ is the minimum of 1 and $(n + 1)$ times the value of P (proof is left as an exercise). So if the value of P is any m, $0 < m \leq 1$, then for every n (≥ 1) such that $m \geq {}^1/_{(n+1)}$, the value of the antecedent of the nth formula is 1 and so Q must also have the value 1. Because we can find such an n for any value $m > 0$ that P may have, and because we have also shown that Q must be true when P has the value 0, it follows that Q must be true whenever all of the formulas in Σ are true.

On the other hand, derivations are finite in length and so any single derivation of Q can include only finitely many formulas from the set Σ. But no finite subset Ψ of Σ semantically entails Q in Fuzzy$_Ł$. A finite subset Ψ can contain only finitely many members of the infinite series $(\neg P \rightarrow P) \rightarrow Q$, $(\neg P \rightarrow (\neg P \rightarrow P)) \rightarrow Q$, $(\neg P \rightarrow (\neg P \rightarrow (\neg P \rightarrow P))) \rightarrow Q$, $(\neg P \rightarrow (\neg P \rightarrow (\neg P \rightarrow (\neg P \rightarrow P)))) \rightarrow Q, \ldots$. If Ψ does not contain the formula $\neg P \rightarrow Q$, then Q can be false when all of the formulas in Ψ are true: if P has the value 0 then the antecedents of all the formulas in Ψ are false and thus the formulas are all true, even if Q has the value 0. If Ψ *does* contain the formula $\neg P \rightarrow Q$, then let k be the highest number of any member of the series that appears in Ψ; that is, the kth member of the series occurs in Ψ but no later member does. Whenever the value of P is greater than 0 but less than ${}^1/_{(k+1)}$, it is possible for Q to fail to be true while all of the members of Ψ are true. Q can fail to be true in this case since none of the antecedents of the formulas in Ψ will be true, and in order for the conditionals in Ψ to all be true, Q only needs to be as true as the antecedent that is closest to being true.

F$_Ł$A is a *sound* system and so there can be no derivation when there is no entailment. It therefore follows from the fact that Q is not semantically entailed by any finite subset of Σ that Q cannot be derived from any finite subset within F$_Ł$A. Since the infinite set Σ semantically entails Q but there is no corresponding derivation, F$_Ł$A is *not* strongly complete. Moreover, this reasoning shows that there is *no* sound axiomatic system for Fuzzy$_Ł$ that is strongly complete—because we reasoned generally about derivations in sound systems rather than specifically about derivations in F$_Ł$A.

The example that we have used to demonstrate that no axiomatic system for Fuzzy$_Ł$ can be strongly complete also shows that the semantic entailment relation in

[2] This example is a slight modification of an example in Hájek (1998b, p. 75).

$Fuzzy_L$ is *not compact*, where compactness means that whenever a semantic entailment holds between an infinite set of premises and a conclusion then entailment also holds between at least one finite subset of the premises and the conclusion.[3] To summarize, we have proved

Result 13.1: $Fuzzy_L$ is not compact

and therefore

Result 13.2: No axiomatic system for $Fuzzy_L$ can be strongly complete.

The situation here contrasts with the classical and three-valued logics that we've studied, where both the propositional and the first-order systems are semantically compact and have strongly complete axiomatizations.

We also note that the Deduction Theorem fails for F_LA, just as it did for $Ł_3A$. In fact, the same counterexample will do: whenever $P \wedge ((P \rightarrow Q) \wedge (P \rightarrow (Q \rightarrow R)))$ has the value 1 in $Fuzzy_L$ so does R, but the conditional formula $(P \wedge ((P \rightarrow Q) \wedge (P \rightarrow (Q \rightarrow R)))) \rightarrow R$ is not a $Fuzzy_L$ tautology. Given soundness and completeness for arguments with a *finite* number of premises, this means that R is derivable from $P \wedge ((P \rightarrow Q) \wedge (P \rightarrow (Q \rightarrow R)))$ in F_LA, but $(P \wedge ((P \rightarrow Q) \wedge (P \rightarrow (Q \rightarrow R)))) \rightarrow R$ is not a theorem in this system. However, the following result, yet another **Modified Deduction Theorem**, does hold in F_LA:

Result 13.3 (Modified Deduction Theorem, for F_LA): **Q** is derivable from **P** in F_LA if and only if $\mathbf{P} \rightarrow (\mathbf{P} \rightarrow (\mathbf{P} \rightarrow (\ldots \rightarrow (\mathbf{P} \rightarrow \mathbf{Q}) \ldots)$ is a theorem for some finite number of antecedent **P**'s.[4]

Proof: The *if* part is easy: given the conditional formula we construct a derivation with **P** as an assumption, derive the conditional theorem (for which we do not need any assumptions), and then repeatedly use MP to derive **Q**.

3 Charles G. Morgan and Francis J. Pelletier (1977) pointed out that (Łukasiewicz) fuzzy logic augmented with "J-operators"—operators that state the truth-value of formulas, such as $J_{.5}(P)$, which means: *the truth-value of P is .5*—fails to be compact and therefore cannot be axiomatized by a strongly complete axiomatic system. The example here shows a stronger claim, namely, that Łukasiewicz fuzzy logic *without* J-operators fails to be compact.

We note that compactness is often, equivalently, defined in terms of *satisfiability*. A set of formulas of propositional logic is **satisfiable** if there is at least one truth-value assignment on which all of the members of the set are true. A logical system is satisfaction compact if and only if the following holds for every set Γ of formulas: Γ is satisfiable if and only if every finite subset of Γ is satisfiable. The equivalence of the two definitions follows from the fact, readily confirmable by the reader, that a set of formulas Γ semantically entails the formula **Q** if and only if the set $\Gamma \cup \{\neg \mathbf{Q}\}$ is *not* satisfiable.

4 Note that this is a straightforward generalization of the Modified Deduction Theorem for $Ł_3A$ (Result 6.1 of Chapter 6). The semantic deduction theorem for Łukasiewicz infinite-valued logic was proved in Pogorzelski (1964).

We noted in Chapter 2 that axiomatic derivation systems, rather than natural deduction derivation systems, are the norm in fuzzy logic. We can now say that the lack of a Deduction Theorem means that standard natural deduction systems, in which conditionals are produced by assuming the antecedent, deriving the consequent, and then concluding that the corresponding conditional follows, are not possible in $Fuzzy_L$ or in any other system for which the Deduction Theorem fails as it does here.

To see that the *only if* part holds in F$_Ł$A as well, assume that there is a derivation of **Q** from **P**:

1	**P**	Assumption
2	$\mathbf{R}_1$	
3	$\mathbf{R}_2$	
	...	
k+1	$\mathbf{R}_k$	
k+2	**Q**	

We'll show how to construct a new derivation, without assumptions, in which each of the formulas **S** from the previous derivation appears in a conditional $\mathbf{P} \to (\mathbf{P} \to (\mathbf{P} \to (\ldots \to (\mathbf{P} \to \mathbf{S})\ldots)$ with some finite number of antecedent **P**'s—for then we can conclude that in particular this holds true for the last formula, **Q**. For simplicity, we assume that the given derivation does not use any derived axioms or rules. We begin our *new* derivation (which *may* use derived axioms and rules) with

1	$\mathbf{P} \to \mathbf{P}$	F$_Ł$D7, with **P** / **P**

For each of the formulas $\mathbf{R}_i$ that is an axiom, we include the following lines in the new derivation:

j	$\mathbf{R}_i$	*axiom number*
j+1	$\mathbf{R}_i \to (\mathbf{P} \to \mathbf{R}_i)$	F$_Ł$1, with $\mathbf{R}_i$ / **P**, **P** / **Q**
j+2	$\mathbf{P} \to \mathbf{R}_i$	*j*, *j*+1 MP

If a formula $\mathbf{R}_i$ was derived using Modus Ponens, then there were earlier formulas $\mathbf{R}_e$ and $\mathbf{R}_e \to \mathbf{R}_i$ in the given derivation from which $\mathbf{R}_i$ was derived. In our new derivation we will already have earlier formulas $\mathbf{P} \to (\mathbf{P} \to (\mathbf{P} \to (\ldots \to (\mathbf{P} \to \mathbf{R}_e)\ldots)$ with a finite number *m* of antecedent **P**'s and $\mathbf{P} \to (\mathbf{P} \to (\mathbf{P} \to (\ldots \to (\mathbf{P} \to (\mathbf{R}_e \to \mathbf{R}_i))\ldots)$ with a finite number *n* of antecedent **P**'s. We use TRAN *n* times to derive $\mathbf{R}_e \to (\mathbf{P} \to (\mathbf{P} \to (\ldots \to (\mathbf{P} \to (\mathbf{P} \to \mathbf{R}_i))\ldots)$ from the latter formula, and then we use GHS to derive $\mathbf{P} \to (\mathbf{P} \to (\mathbf{P} \to (\ldots \to (\mathbf{P} \to \mathbf{R}_i)\ldots)$, with *m*+*n* antecedent **P**'s, from the former formula and this new one (both TRAN and GHS are shown to be derivable in F$_Ł$A in Exercise 1).

13.2 A Pavelka-Style Derivation System for Fuzzy$_Ł$

We will now add atomic formulas that stand for truth-values to the language of Fuzzy$_Ł$.[5] When we did this for Ł$_3$, we included a name for each of the three truth-values. However, in the case of Fuzzy$_Ł$ we have an infinite number of truth-values.

5 The resulting system F$_Ł$PA is formulated as in Novák, Perfilieva, and Močkoř (1999). They use simplifications of Pavelka's formulation developed by Hájek (1995a, 1995b).

Recall that our language already has infinitely many atomic formulas, since we can subscript our sentence letters with any integer. So adding another infinite supply of atomic formulas that name truth-values seems unproblematic. However, there is a complication: there are *uncountably many* truth-values in the unit interval of truth-values [0. .1].[6] This means that in order to name every truth-value in the unit interval we would need to add uncountably many atomic formulas to the language. Why do we care about this? Well, the major issue is that many standard and important metatheoretic proof techniques for logical systems assume that the language of those systems is countable.

We will therefore introduce atomic formulas to name only a *countably* infinite subset of values in [0. .1]. The standard subset that is used for this purpose is the set *Rat* of *rational* numbers in the unit interval, numbers that can be expressed as fractions with integral numerators and denominators such as $^1/_2$, $^2/_5$, and $^{599}/_{6432}$. 0 and 1 are included here, since 0 is $0/_n$ for any integer *n*, and 1 is $^1/_1$ (or $^n/_n$ for any positive, nonzero *n*).[7] We will call the system that results from adding to the language of Fuzzy$_Ł$ an atomic formula that denotes each value in *Rat*, *RFuzzy*$_Ł$. Our atomic formulas will be boldface fractional expressions, such as, $\mathbf{^1/_2}$ and $\mathbf{^{599}/_{6432}}$, except that we will denote 0 and 1 with the boldface numerals **0** and **1**. Note that there are infinitely many fractions that generate each rational number, for example, $^1/_2 = {}^2/_4 = {}^3/_6 = {}^4/_8 = {}^5/_{10} = \ldots$. We will always use the expression with the smallest numerator, so for example $\mathbf{^1/_2}$ will be the unique atomic formula that we use to denote $^1/_2$.

The graded axiomatic system F$_Ł$PA contains Łukasiewicz's axioms F$_Ł$1–F$_Ł$4, all with the grade 1 and renamed to reflect the fact that we are now working in a Pavelka-style system:

F$_Ł$P1. [**P** → (**Q** → **P**), 1]
F$_Ł$P2. [(**P** → **Q**) → ((**Q** → **R**) → (**P** → **R**)), 1]
F$_Ł$P3. [(¬**P** → ¬**Q**) → (**Q** → **P**), 1]
F$_Ł$P4. [((**P** → **Q**) → **Q**) → ((**Q** → **P**) → **P**), 1]

In addition, there are infinitely many (*countably* infinitely many) axioms, graded with 1, relating the truth-values under the conditional and negation operations:

F$_Ł$P5.1. *Includes every graded formula* [(**m** → **n**) → **p**, 1] *where* **m**, **n**, *and* **p** *are atomic formulas denoting rational truth-values m, n, and p in the unit interval such that p = min (1, 1−m + n)*
F$_Ł$P5.2. *Includes every graded formula* [**p** → (**m** → **n**), 1] *where* **m**, **n**, *and* **p** *are as in F$_Ł$P5.1*

6 See the Appendix for a definition of *uncountability* and a proof that there are uncountably many truth-values in the unit interval.
7 See the Appendix for a proof that the set Rat is countable.

F$_Ł$P6.1. *Includes every graded formula* [¬**m** → **p**, 1] *where* **m** *and* **p** *are atomic formulas denoting rational truth-values m and p such that p = 1 − m*
F$_Ł$P6.2. *Includes every graded formula* [**p** → ¬**m**, 1] *where* **m** *and* **p** *are as in F$_Ł$P6.1.*

Some of these new axioms are

[(**1** → **1**) → **1**, 1]
[**1** → (**1** → **1**), 1]
[(**$^1/_2$** → **$^3/_4$**) → **1**, 1]
[**1** → (**$^1/_2$** → **$^3/_4$**), 1]
[(**$^1/_2$** → **$^1/_3$**) → **$^5/_6$**, 1]
[**$^5/_6$** → (**$^1/_2$** → **$^1/_3$**), 1]
[¬**1** → **0**, 1]
[**0** → ¬**1**, 1]
[¬**$^1/_2$** → **$^1/_2$**, 1]
[**$^1/_2$** → ¬**$^1/_2$**, 1]
[**$^3/_5$** → ¬**$^2/_5$**, 1]
[¬**$^2/_5$** → **$^3/_5$**, 1]

We also add the special axiom schema

F$_Ł$P7. *Includes* [**m**, *m*] *for any rational value m in the unit interval,*
where **m** *is the atomic formula that denotes the value m*

and the truth-value constant introduction rule:

TCI. From [**P**, *m*] infer [**m** → **P**, 1],
where **m** is the atomic formula that denotes the value *m*.

The graded Modus Ponens rule for F$_Ł$PA is

MP. From [**P**, *m*] and [**P** → **Q**, *n*], infer [**Q**, *p*]
where *p* is the result of applying the Łukasiewicz bold conjunction operation to *m* and *n*, i.e., $p = \max(0, m + n - 1)$

To understand the graded value *p* inferred for **Q**, recall the Adjointness Condition from Chapter 11 for a t-norm **tn** and its corresponding residuation operation ⇒:

m **tn** $n \le p$ if and only if $m \le n \Rightarrow p$, for all $m, n, p \in [0..1]$

This condition defines the value produced by the residuation operation. The inequalities in the Adjointness Condition can be restated thus:

the value *p* is at least *m* **tn** *n* if and only if the value $n \Rightarrow p$ is at least *m*

or:

the value *p* is at least the value *m* **tn** *n*, where $n \Rightarrow p$ is at least *m*

When the t-norm and residuation operations are Łukasiewicz's bold conjunction and conditional, this condition entails:

> the value *p* of **Q** is at least the value that results from applying the bold conjunction operation to the least value *m* that **P** → **Q** might have and the least value *n* that **P** might have

or:

> if the value of **P** is at least *n* and the value of **P** → **Q** is at least *m*, then the value *p* of **Q** is at least max (0, $m + n - 1$).

And this is the formula used in the graded Modus Ponens rule.

Before proceeding to derivations, we pause to address an important issue raised by our specification of the axioms F_LP5.1, F_LP5.2, F_LP6.1, and F_LP6.2 relating truth-values under the operations. We said in F_LP5.1, for example: *where* ***m****,* ***n****, and* ***p*** *are atomic formulas denoting rational truth-values m, n, and p in the unit interval such that p* = *min (1, 1*−*m* + *n).* The issue is, Given any pair of rational truth-values *m* and *n* from the unit interval, will the value *p* defined as (1, $1-m+n$) also be a *rational* value from the unit interval? The answer is *yes*, and proof is left as exercise. Similarly, we would like to be certain that if *m* is a rational value in the unit interval then so is $1-m$. Proof of this latter claim, which is also true, is left as an exercise.

The following derivation produces a graded formula asserting that the Law of Excluded Middle using bold disjunction has at least the value 1 in RFuzzy_L. We reexpress *A* ∇¬*A* as ¬*A* → ¬*A*, since we define **P** ∇ **Q** as ¬**P** → **Q**.

1	[¬A → ¬A, 1]	F_LPD7, with ¬A / **P**

Note that we have used an axiom shown to be derivable in F_LA and given it the graded value 1. All axioms derivable in F_LA may be used as derived axioms in F_LPA with graded value 1, since they are derivable in F_LPA from axioms, all of which have graded value 1—and the rules of F_LPA always preserve the value 1. For a similar reason, all rules derivable in F_LA may be used in F_LPA *with all of the grades set to* 1 (we may want to tune the grades more finely, but in that case we need to produce a derivation justifying the fine-tuning, as we will do shortly).

Recall from Chapter 6 that a formula of the form **n** → **P**, where **n** is an atomic formula denoting the truth-value *n*, means *the truth-value of* ***P*** *is at least n*, and a formula of the form **P** → **n** means: *the truth-value of* ***P*** *is at most n*. The following derivation, with all formulas graded with 1, establishes the theoremhood (in F_LPA) of a formula claiming that if *Q* has at least the value $^9/_{10}$ then *P* → *Q* also has at least the value $^9/_{10}$:

1	[($^9/_{10}$ → Q) → (P → ($^9/_{10}$ → Q)), 1]	F_LP1, with $^9/_{10}$ → Q / **P**, P / **Q**
2	[(P → ($^9/_{10}$ → Q)) → ($^9/_{10}$ → (P → Q)), 1]	F_LPD6, with P / **P**, $^9/_{10}$ / **Q**, Q / **R**
3	[($^9/_{10}$ → Q) → ($^9/_{10}$ → (P → Q)), 1]	1,2 HS

We can sandwich this into a longer derivation that derives $[P \rightarrow Q, .9]$ from $[Q, .9]$:

1	$[Q, .9]$	Assumption
2	$[{}^{9}/_{10} \rightarrow Q, 1]$	1, TCI
3	$[({}^{9}/_{10} \rightarrow Q) \rightarrow (P \rightarrow ({}^{9}/_{10} \rightarrow Q)), 1]$	F$_{Ł}$P1, with ${}^{9}/_{10} \rightarrow Q$ / **P**, P / **Q**
4	$[(P \rightarrow ({}^{9}/_{10} \rightarrow Q)) \rightarrow ({}^{9}/_{10} \rightarrow (P \rightarrow Q)), 1]$	F$_{Ł}$PD6, with P / **P**, ${}^{9}/_{10}$ / **Q**, Q / **R**
5	$[({}^{9}/_{10} \rightarrow Q) \rightarrow ({}^{9}/_{10} \rightarrow (P \rightarrow Q)), 1]$	3,4 HS
6	$[{}^{9}/_{10} \rightarrow (P \rightarrow Q), 1]$	2,5 MP
7	$[{}^{9}/_{10}, .9]$	F$_{Ł}$P7
8	$[P \rightarrow Q, .9]$	6,7 MP

The following derivation establishes that if P has at most the value ${}^{1}/_{3}$ then $P \rightarrow Q$ has at least the value ${}^{2}/_{3}$:

1	$[(P \rightarrow {}^{1}/_{3}, 1]$	Assumption
2	$[\neg{}^{1}/_{3} \rightarrow \neg P, 1]$	1, GCON
3	$[{}^{2}/_{3} \rightarrow \neg{}^{1}/_{3}, 1]$	F$_{Ł}$P6.2
4	$[{}^{2}/_{3} \rightarrow \neg P, 1]$	2,3 HS
5	$[({}^{2}/_{3} \rightarrow \neg P) \rightarrow ((\neg P \rightarrow (P \rightarrow Q)) \rightarrow$ $({}^{2}/_{3} \rightarrow (P \rightarrow Q)))]$	F$_{Ł}$P1, with ${}^{2}/_{3}$ / **P**, $\neg$P / **Q**, P $\rightarrow$ Q / **R**
6	$[(\neg P \rightarrow (P \rightarrow Q)) \rightarrow ({}^{2}/_{3} \rightarrow (P \rightarrow Q)), 1]$	4,5 MP
7	$[\neg P \rightarrow (P \rightarrow Q), 1]$	F$_{Ł}$PD1, with P / **P**, Q / **Q**
8	$[{}^{2}/_{3} \rightarrow (P \rightarrow Q), 1]$	7,8 MP
9	$[{}^{2}/_{3}, {}^{2}/_{3}]$	F$_{Ł}$P7
10	$[P \rightarrow Q, {}^{2}/_{3}]$	8,9 MP

(When possible, we list the values grading formulas in derivations as decimal reals to distinguish clearly between the values and their names that occur within the formulas; when it isn't possible, as in the case of ${}^{2}/_{3}$ in the preceding derivation, we use a fraction.)

If we are analyzing arguments with bold conjunctions, as we will be in Chapter 15, a useful derived rule is

BCI (Bold Conjunction Introduction). From [**P**, *m*] and [**Q**, *n*] infer [**P & Q**, *p*] where $p = \max(0, m + n - 1)$.

In our justification we'll use the derived rule

HS. From [**P** $\rightarrow$ **Q**, *m*] and [**Q** $\rightarrow$ **R**, *n*] infer [**P** $\rightarrow$ **R**, *p*] where $p = \max(0, m + n - 1)$

whose derivation is assigned as an exercise. We've rewritten **P & Q** as $\neg(P \rightarrow \neg Q)$:

1	$[P, m]$	Assumption
2	$[Q, n]$	Assumption
3	$[P \rightarrow ((P \rightarrow \neg Q) \rightarrow \neg Q), 1]$	$F_Ł$PD3, with $P / P, \neg Q / Q$
4	$[(P \rightarrow \neg Q) \rightarrow \neg Q, m]$	1,3 MP
	{note that max (0, $m + 1 - 1) = m$}	
5	$[\neg\neg(P \rightarrow \neg Q) \rightarrow (P \rightarrow \neg Q), 1]$	$F_Ł$PD8, with $P \rightarrow \neg Q / P$
6	$[\neg\neg(P \rightarrow \neg Q) \rightarrow \neg Q, m]$	4,5 HS
7	$[(\neg\neg(P \rightarrow \neg Q) \rightarrow \neg Q) \rightarrow (Q \rightarrow \neg(P \rightarrow \neg Q)), 1]$	$F_Ł$P3, with $\neg(P \rightarrow \neg Q) / P, Q / Q$
8	$[(Q \rightarrow \neg(P \rightarrow \neg Q)), m]$	6,7 MP
9	$[\neg(P \rightarrow \neg Q), \max(0, m + n - 1)]$	2,8 MP

A related derived rule for weak conjunction is

WCI (Weak Conjunction Introduction). From $[P, m]$ and $[Q, n]$ infer $[P \wedge Q, p]$ where $p = \min(m, n)$

To justify this, we first derive a helping axiom:

$F_Ł$PD18. $[(P \rightarrow Q) \rightarrow (((P \rightarrow R) \rightarrow S) \rightarrow ((Q \rightarrow R) \rightarrow S)), 1]$

Justification:

1	$[(((Q \rightarrow R) \rightarrow (P \rightarrow R)) \rightarrow (((P \rightarrow R) \rightarrow S) \rightarrow ((Q \rightarrow R) \rightarrow S))) \rightarrow ((P \rightarrow Q) \rightarrow (((P \rightarrow R) \rightarrow S) \rightarrow ((Q \rightarrow R) \rightarrow S))), 1]$	$F_Ł$PD4, with $P / P, Q / Q, R / R, ((P \rightarrow R) \rightarrow S) \rightarrow ((Q \rightarrow R) \rightarrow S) / S$
2	$[((Q \rightarrow R) \rightarrow (P \rightarrow R)) \rightarrow (((P \rightarrow R) \rightarrow S) \rightarrow ((Q \rightarrow R) \rightarrow S)), 1]$	$F_Ł$P2, with $Q \rightarrow R / P, P \rightarrow R / Q, S / R$
3	$[(P \rightarrow Q) \rightarrow (((P \rightarrow R) \rightarrow S) \rightarrow ((Q \rightarrow R) \rightarrow S)), 1]$	1,2 MP

and the rule

DC. From $[P \rightarrow R, 1]$ and $[Q \rightarrow R, 1]$ infer $[(P \vee Q) \rightarrow R, 1]$

(This rule, Disjunctive Consequence, was derived in L_3A in Chapter 6, but the derivation there used L_3A's fourth axiom, which is not an axiom of $F_Ł$PA. Do note that axiom schema $F_Ł$P4 makes the following derivation of DC simpler here than it was in L_3A, where we *used* DC *to derive* the F$_{Ł}$P4 formula $((\mathbf{P} \rightarrow \mathbf{Q}) \rightarrow \mathbf{Q}) \rightarrow ((\mathbf{Q} \rightarrow \mathbf{P}) \rightarrow \mathbf{P})$ as a derived axiom schema.) Recall that $\mathbf{P} \vee \mathbf{Q}$ is defined as $(\mathbf{P} \rightarrow \mathbf{Q}) \rightarrow \mathbf{Q}$:

1	$[\mathbf{P} \rightarrow \mathbf{R}, 1]$	Assumption
2	$[\mathbf{Q} \rightarrow \mathbf{R}, 1]$	Assumption
3	$[(\mathbf{P} \rightarrow \mathbf{R}) \rightarrow (((\mathbf{P} \rightarrow \mathbf{Q}) \rightarrow \mathbf{Q}) \rightarrow ((\mathbf{R} \rightarrow \mathbf{Q}) \rightarrow \mathbf{Q})), 1]$	F$_{Ł}$PD18, with **P / P, R / Q, Q / R, Q / S**
4	$[((\mathbf{R} \rightarrow \mathbf{Q}) \rightarrow \mathbf{Q})) \rightarrow ((\mathbf{Q} \rightarrow \mathbf{R}) \rightarrow \mathbf{R})), 1]$	F$_{Ł}$P4, with **R / P, Q / Q**
5	$[(\mathbf{P} \rightarrow \mathbf{R}) \rightarrow (((\mathbf{P} \rightarrow \mathbf{Q}) \rightarrow \mathbf{Q}) \rightarrow ((\mathbf{Q} \rightarrow \mathbf{R}) \rightarrow \mathbf{R})), 1]$	3,4 GHS
6	$[((\mathbf{P} \rightarrow \mathbf{Q}) \rightarrow \mathbf{Q}) \rightarrow ((\mathbf{Q} \rightarrow \mathbf{R}) \rightarrow \mathbf{R}), 1]$	1,5 MP
7	$[(\mathbf{Q} \rightarrow \mathbf{R}) \rightarrow (((\mathbf{P} \rightarrow \mathbf{Q}) \rightarrow \mathbf{Q}) \rightarrow \mathbf{R}), 1]$	6, TRAN
8	$[((\mathbf{P} \rightarrow \mathbf{Q}) \rightarrow \mathbf{Q}) \rightarrow \mathbf{R}, 1]$	2,7 MP

To justify WCI, in which we rewrite $\mathbf{P} \wedge \mathbf{Q}$ as $\neg((\neg\mathbf{P} \rightarrow \neg\mathbf{Q}) \rightarrow \neg\mathbf{Q})$, we consider two cases: either $m \leq n$ or $n < m$. In the former case, we begin the derivation with

1	$[\mathbf{P}, m]$	Assumption
2	$[\mathbf{Q}, n]$	Assumption
3	$[\mathbf{m} \rightarrow \mathbf{P}, 1]$	1, TCI
4	$[\mathbf{n} \rightarrow \mathbf{Q}, 1]$	2, TCI
5	$[\mathbf{1} \rightarrow (\mathbf{m} \rightarrow \mathbf{n}), 1]$	F$_{Ł}$P5.1
	{This is an instance of F$_{Ł}$P5.1 because of the assumption in this case that $m \leq n$}	
6	$[\mathbf{1}, 1]$	F$_{Ł}$P7
7	$[\mathbf{m} \rightarrow \mathbf{n}, 1]$	5,6 MP
8	$[\mathbf{m} \rightarrow \mathbf{Q}, 1]$	4,7 HS

where **m** stands for the atomic formula that denotes the value m and **n** stands for the atomic formula that denotes the value n. Because $m \leq n$ in this case, $m = \min(m, n)$ and so we can rewrite the preceding as

1	$[\mathbf{P}, m]$	Assumption
2	$[\mathbf{Q}, n]$	Assumption
3	$[\mathbf{min}\ (\mathbf{m}, \mathbf{n}) \rightarrow \mathbf{P}, 1]$	1, TCI
4	$[\mathbf{n} \rightarrow \mathbf{Q}, 1]$	2, TCI
5	$[\mathbf{1} \rightarrow (\mathbf{min}\ (\mathbf{m}, \mathbf{n}) \rightarrow \mathbf{n}), 1]$	F$_{Ł}$P5.1
6	$[\mathbf{1}, 1]$	F$_{Ł}$P7
7	$[\mathbf{min}\ (\mathbf{m}, \mathbf{n}) \rightarrow \mathbf{n}, 1]$	5,6 MP
8	$[\mathbf{min}\ (\mathbf{m}, \mathbf{n}) \rightarrow \mathbf{Q}, 1]$	4,7 HS

(where **min (m, n)** is standing in for the atomic formula whose value is $\min(m, n)$) In the case where $n < m$, we begin the derivation with

1	$[\mathbf{P}, m]$	Assumption
2	$[\mathbf{Q}, n]$	Assumption
3	$[\mathbf{min\ (m, n)} \rightarrow \mathbf{Q}, 1]$	2, TCI
4	$[\mathbf{m} \rightarrow \mathbf{P}, 1]$	1, TCI
5	$[\mathbf{1} \rightarrow (\mathbf{min\ (m, n)} \rightarrow \mathbf{m}), 1]$	$F_{Ł}P5.1$
6	$[\mathbf{1}, 1]$	$F_{Ł}P7$
7	$[\mathbf{min\ (m, n)} \rightarrow \mathbf{m}, 1]$	5,6 MP
8	$[\mathbf{min\ (m, n)} \rightarrow \mathbf{P}, 1]$	4,7 HS

In either case the derivation continues (with $\{x, y\}$ meaning line x if we began the first way, and line y if we began the second way):

9	$[\neg\mathbf{P} \rightarrow \neg\mathbf{min\ (m, n)}, 1]$	$\{3, 8\}$ GCON
10	$[\neg\mathbf{Q} \rightarrow \neg\mathbf{min\ (m, n)}, 1]$	$\{8, 3\}$ GCON
11	$[((\neg\mathbf{P} \rightarrow \neg\mathbf{Q}) \rightarrow \neg\mathbf{Q}) \rightarrow \neg\mathbf{min\ (m, n)}, 1]$	9,10 DC
12	$[\neg\neg((\neg\mathbf{P} \rightarrow \neg\mathbf{Q}) \rightarrow \neg\mathbf{Q}) \rightarrow \neg\mathbf{min\ (m, n)}, 1]$	11, DN
13	$[\mathbf{min\ (m, n)} \rightarrow \neg((\neg\mathbf{P} \rightarrow \neg\mathbf{Q}) \rightarrow \neg\mathbf{Q}), 1]$	12, GCON
14	$[\mathbf{min\ (m, n)}, \min(m, n)]$	$F_{Ł}P7$
15	$[\neg((\neg\mathbf{P} \rightarrow \neg\mathbf{Q}) \rightarrow \neg\mathbf{Q}), \min(m, n)]$	13,14 MP

In Chapter 11 we noted that the conclusion *J* of the argument

A
A → B
B → C
C → D
D → E
E → F
F → G
G → H
H → I
I → J
———
J

has the value 0 when *A* has the value .9 and each of the conditional premises has the value .9. We can capture this in $F_{Ł}$PA. First, we'll establish that when *A* has at least the value .9 and each of the conditional premises has at least the value .9, then *J* has at least the value 0. We do note that the claim that *J* has at least the value 0 follows immediately from the derivable axiom

$F_{Ł}$PD19: [**P**, 0] for any formula **P**

Justification:

1	[**1**, 1]	$F_ŁP7$
2	[**1** → (¬**P** → **1**), 1]	$F_ŁP1$, with **1** / **P**, ¬**P** / **Q**
3	[¬**P** → **1**, 1]	1,2 MP
4	[**1** → ¬¬**1**, 1]	$F_ŁPD3$, with **1** / **P**
5	[¬**P** → ¬¬**1**, 1]	3,4 HS
6	[¬**1** → **P**, 1]	5, GCON
7	[**0** → ¬**1**, 1]	$F_ŁP6.1$
8	[**0** → **P**, 1]	6,7 HS
9	[**0**, 0]	$F_ŁP7$
10	[**P**, 0]	8,9 MP

but we produce a longer derivation here in order to draw some more general conclusions:

1	[A, .9]	Assumption
2	[A → B, .9]	Assumption
3	[B → C, .9]	Assumption
4	[C → D, .9]	Assumption
5	[D → E, .9]	Assumption
6	[E → F, .9]	Assumption
7	[F → G, .9]	Assumption
8	[G → H, .9]	Assumption
9	[H → I, .9]	Assumption
10	[I → J, .9]	Assumption
11	[B, .8]	1,2 MP
	{max (0, .9 + .9 − 1) = .8; similar comment for remaining steps}	
12	[C, .7]	3,11 MP
13	[D, .6]	4,12 MP
14	[E, .5]	5,13 MP
15	[F, .4]	6,14 MP
16	[G, .3]	7,15 MP
17	[H, .2]	8,16 MP
18	[I, .1]	9,17 MP
19	[J, 0]	10,18 MP

Now, this derivation shows only that *J* has at least the value 0. The conclusion that *J* has at least the value 0 is in fact consistent with a stronger claim: that *J* has at least the value .1, and even a much stronger claim: that *J* has at least the value 1! Now, because of the soundness of $F_ŁPA$ (more on this later), we know that there is no derivation from the graded premises to these stronger conclusions since we know from Chapter 11 that the argument is .1-degree-valid. But keep in mind the general point, which we may restate as: the claim that *J* has at least the value 0 does not entail that it has at most the value 0. If we want to support the latter claim in F_LPA, we will need a derivation with that claim as its conclusion. We will produce such a derivation in a moment.

But first, note that if we had ended the derivation at line 18 we would have a nontrivial conclusion about *I* (in contrast with the trivial conclusion about *J* on line 19: *trivial* because *every* formula has at least the value 0). The derivation ending at line 18 shows that if each of $A, A \rightarrow B, B \rightarrow C, \ldots, H \rightarrow I$ has at least the value .9, then *I* must have at least the value .1. So even though our longer derivation establishes a trivial conclusion, the "subconclusions" reached along the way and used in subsequent reasoning in the derivation are not themselves trivial. This, by the way, meshes with what we noted about Modus Ponens in Chapter 11: that longer chains of Modus Ponens reasoning have lower *n*-degree-validities than do shorter ones.

We will now produce a derivation that shows that when *A* and each of the conditional premises have *at most* the value .9, the conclusion *J* has (*at most*) the value 0. We introduce two derived rules to capture repeated inference patterns in our derivation. The first is

> **CV (Consequent Value).** From $[\mathbf{P} \rightarrow \mathbf{m}, 1]$, $[\mathbf{Q} \rightarrow \mathbf{m}, 1]$ and $[(\mathbf{P} \rightarrow \mathbf{Q}) \rightarrow \mathbf{n}, 1]$, infer $[\mathbf{Q} \rightarrow \neg(\mathbf{n} \rightarrow \neg\mathbf{m}), 1]$

The derived rule does not require that *m* and *n* be atomic formulas denoting rational values—in fact we can use any formula in their place—but we shall only be using this rule when atomic formulas denoting rational values *are* used in place of *m* and *n*, hence the notation and the name. Given, for example, $[A \rightarrow {}^9/_{10}, 1]$, $[B \rightarrow {}^9/_{10}, 1]$, and $[(A \rightarrow B) \rightarrow {}^9/_{10}, 1]$, the rule allows us to derive $[B \rightarrow \neg({}^9/_{10} \rightarrow \neg{}^9/_{10}), 1]$. Once we do this, we shall use a second derived rule to simplify the conditional containing the names of truth-values to a single truth-value so that from $[B \rightarrow \neg({}^9/_{10} \rightarrow \neg{}^9/_{10}), 1]$ we will be able to derive $[B \rightarrow {}^9/_{10}, 1]$. Here's the derivation justifying CV:

1	$[\mathbf{P} \rightarrow \mathbf{m}, 1]$	Assumption
2	$[\mathbf{Q} \rightarrow \mathbf{m}, 1]$	Assumption
3	$[(\mathbf{P} \rightarrow \mathbf{Q}) \rightarrow \mathbf{n}, 1]$	Assumption
4	$[(\mathbf{P} \rightarrow \mathbf{m}) \rightarrow ((\mathbf{m} \rightarrow \mathbf{Q}) \rightarrow (\mathbf{P} \rightarrow \mathbf{Q})), 1]$	F_LP2, with **P** / **P**, **m** / **Q**, **Q** / **R**
5	$[(\mathbf{m} \rightarrow \mathbf{Q}) \rightarrow (\mathbf{P} \rightarrow \mathbf{Q}), 1]$	1,4 MP
6	$[(\mathbf{m} \rightarrow \mathbf{Q}) \rightarrow \mathbf{n}, 1]$	3,5 HS
7	$[(\neg\mathbf{Q} \rightarrow \neg\mathbf{m}) \rightarrow \mathbf{n}, 1]$	6, GCON
8	$[\mathbf{n} \rightarrow ((\mathbf{n} \rightarrow \neg\mathbf{m}) \rightarrow \neg\mathbf{m}), 1]$	F_LPD3, with **n** / **P**, ¬**m** / **Q**
9	$[(\neg\mathbf{Q} \rightarrow \neg\mathbf{m}) \rightarrow ((\mathbf{n} \rightarrow \neg\mathbf{m}) \rightarrow \neg\mathbf{m}), 1]$	7,8 HS
10	$[(\mathbf{n} \rightarrow \neg\mathbf{m}) \rightarrow ((\neg\mathbf{Q} \rightarrow \neg\mathbf{m}) \rightarrow \neg\mathbf{m}), 1]$	9, TRAN
11	$[((\neg\mathbf{Q} \rightarrow \neg\mathbf{m}) \rightarrow \neg\mathbf{m}) \rightarrow ((\neg\mathbf{m} \rightarrow \neg\mathbf{Q}) \rightarrow \neg\mathbf{Q}), 1]$	F_LP4, with ¬**Q** / **P**, ¬**m** / **Q**
12	$[(\mathbf{n} \rightarrow \neg\mathbf{m}) \rightarrow ((\neg\mathbf{m} \rightarrow \neg\mathbf{Q}) \rightarrow \neg\mathbf{Q}), 1]$	10,11 HS
13	$[\neg\mathbf{m} \rightarrow \neg\mathbf{Q}, 1]$	2, GCON
14	$[(\mathbf{n} \rightarrow \neg\mathbf{m}) \rightarrow \neg\mathbf{Q}, 1]$	12,13 GMP
15	$[\neg\neg(\mathbf{n} \rightarrow \neg\mathbf{m}) \rightarrow \neg\mathbf{Q}, 1]$	14, DN
16	$[\mathbf{Q} \rightarrow \neg(\mathbf{n} \rightarrow \neg\mathbf{m}), 1]$	15, GCON

Note that we used several derived rules in this derivation, rules that have been shown in Section 13.1 and Exercise 1 to be derivable in F$_Ł$A. In each case the derived rule has been applied to formulas with grade 1 and the derived formula also has grade 1, which we pointed out earlier is always allowed since the axioms used in the rule derivations in F$_Ł$A have grade 1 in F$_Ł$PA and the rule MP preserves grade 1 in F$_Ł$PA.

The second derived rule, for deriving a simplified value, is

VS (Value Summary). From $[\mathbf{m} \rightarrow \neg\mathbf{n}, 1]$, $[\mathbf{p} \rightarrow (\mathbf{q} \rightarrow \mathbf{m}), 1]$ and $[\neg\mathbf{r} \rightarrow \mathbf{p}, 1]$, infer $[\neg(\mathbf{q} \rightarrow \neg\mathbf{n}) \rightarrow \mathbf{r}, 1]$

This will allow, for example, the move from $[B \rightarrow \neg({}^{9}/_{10} \rightarrow \neg{}^{9}/_{10}), 1]$ to $[B \rightarrow {}^{9}/_{10}, 1]$. It is left as an exercise to show that that this rule is derivable in F$_Ł$PA. As in the previous case, any atomic formulas can be used in place of ***m***, ***n***, ***p***, ***q***, and ***r***—but since we'll be using this rule when these are all formulas denoting rational values, we choose the name and notation appropriate for this intended use.

Here, then, is the derivation showing that when A and each of the conditional premises in the earlier argument have at most the value .9, the conclusion J has at most the value 0.[8] The basic strategy is that once we have placed an upper bound n on the antecedent of one of the conditional premises, then we then derive an upper bound of $n - {}^{1}/_{10}$ on the consequent of that premise, and hence on the antecedent of the next conditional premise in the chain—we can derive this upper bound because the greatest value that m can be when the formula $(\mathbf{n} \rightarrow \mathbf{m}) \rightarrow {}^{9}/_{10}$ is true is ${}^{1}/_{10}$ less than n. (The reader will be asked to explain this in an exercise.) Having an upper bound of ${}^{9}/_{10}$ for A on line 1, we can then derive an upper bound for B that is ${}^{1}/_{10}$ less by first deriving an upper bound of ${}^{9}/_{10}$ for B (lines 11–12), and then an upper bound of ${}^{8}/_{10}$ (lines 13–18), and so on....

1	$[A \rightarrow {}^{9}/_{10}, 1]$	Assumption
2	$[(A \rightarrow B) \rightarrow {}^{9}/_{10}, 1]$	Assumption
3	$[(B \rightarrow C) \rightarrow {}^{9}/_{10}, 1]$	Assumption
4	$[(C \rightarrow D) \rightarrow {}^{9}/_{10}, 1]$	Assumption
5	$[(D \rightarrow E) \rightarrow {}^{9}/_{10}, 1]$	Assumption
6	$[(E \rightarrow F) \rightarrow {}^{9}/_{10}, 1]$	Assumption
7	$[(F \rightarrow G) \rightarrow {}^{9}/_{10}, 1]$	Assumption
8	$[(G \rightarrow H) \rightarrow {}^{9}/_{10}, 1]$	Assumption
9	$[(H \rightarrow I) \rightarrow {}^{9}/_{10}, 1]$	Assumption
10	$[(I \rightarrow J) \rightarrow {}^{9}/_{10}, 1]$	Assumption
11	$[B \rightarrow (A \rightarrow B), 1]$	F$_Ł$P1, with B / **P**, A / **Q**
12	$[B \rightarrow {}^{9}/_{10}, 1]$	2,11 HS
13	$[B \rightarrow \neg({}^{9}/_{10} \rightarrow \neg{}^{9}/_{10}), 1]$	1,2,12 CV

8 I am grateful to Petr Hájek (personal correspondence dated March 1, 2005) for helping me to produce this derivation. It was he who suggested the derivation steps involved in the CV rule. Any inelegance is mine, not his, since he showed me these steps using the bold connectives of Fuzzy$_Ł$ rather than just the two connectives—negation and implication—that I have chosen to include in F$_Ł$PA.

14	$[{}^{1}/_{10} \rightarrow \neg{}^{9}/_{10}, 1]$	F_LP6.2
15	$[{}^{2}/_{10} \rightarrow ({}^{9}/_{10} \rightarrow {}^{1}/_{10}), 1]$	F_LP5.2
16	$[\neg{}^{8}/_{10} \rightarrow {}^{2}/_{10}, 1]$	F_LP6.1
17	$[\neg({}^{9}/_{10} \rightarrow \neg{}^{9}/_{10}) \rightarrow {}^{8}/_{10}, 1]$	14,15,16 VS
18	$B \rightarrow {}^{8}/_{10}, 1]$	13,17 HS
19	$[C \rightarrow (B \rightarrow C), 1]$	F_LP1, with C / **P**, B / **Q**
20	$[C \rightarrow {}^{9}/_{10}, 1]$	3,19 HS
21	$[C \rightarrow \neg({}^{9}/_{10} \rightarrow \neg{}^{9}/_{10}), 1]$	3,12,20 CV
22	$[C \rightarrow {}^{8}/_{10}, 1]$	17,21 HS
23	$[C \rightarrow \neg({}^{9}/_{10} \rightarrow \neg{}^{8}/_{10}), 1]$	3,18,22 CV
24	$[{}^{2}/_{10} \rightarrow \neg{}^{8}/_{10}, 1]$	F_LP6.2
25	$[{}^{3}/_{10} \rightarrow ({}^{9}/_{10} \rightarrow {}^{2}/_{10}), 1]$	F_LP5.2
26	$[\neg{}^{7}/_{10} \rightarrow {}^{3}/_{10}, 1]$	F_LP6.1
27	$[\neg({}^{9}/_{10} \rightarrow \neg{}^{8}/_{10}) \rightarrow {}^{7}/_{10}, 1]$	24,25,26 VS
28	$[C \rightarrow {}^{7}/_{10}, 1]$	23,27 HS
29	$[D \rightarrow (C \rightarrow D), 1]$	F_LP1, with D / **P**, C / **Q**
30	$[D \rightarrow {}^{9}/_{10}, 1]$	4,29 HS
31	$[D \rightarrow \neg({}^{9}/_{10} \rightarrow \neg{}^{9}/_{10}), 1]$	4,20,30 CV
32	$[D \rightarrow {}^{8}/_{10}, 1]$	17,31 HS
33	$[D \rightarrow \neg({}^{9}/_{10} \rightarrow \neg{}^{8}/_{10}), 1]$	4,22,32 CV
34	$[D \rightarrow {}^{7}/_{10}, 1]$	27,33 HS
35	$[D \rightarrow \neg({}^{9}/_{10} \rightarrow \neg{}^{7}/_{10}), 1]$	4,28,34 CV
36	$[{}^{3}/_{10} \rightarrow \neg{}^{7}/_{10}, 1]$	F_LP6.2
37	$[{}^{4}/_{10} \rightarrow ({}^{9}/_{10} \rightarrow {}^{3}/_{10}), 1]$	F_LP5.2
38	$[\neg{}^{6}/_{10} \rightarrow {}^{4}/_{10}, 1]$	F_LP6.1
39	$[\neg({}^{9}/_{10} \rightarrow \neg{}^{7}/_{10}) \rightarrow {}^{6}/_{10}, 1]$	36,37,38 VS
40	$[D \rightarrow {}^{6}/_{10}, 1]$	35,39 HS
41	$[E \rightarrow (D \rightarrow E), 1]$	F_LP1, with E / **P**, D / **Q**
42	$[E \rightarrow {}^{9}/_{10}, 1]$	5,41 HS
43	$[E \rightarrow \neg({}^{9}/_{10} \rightarrow \neg{}^{9}/_{10}), 1]$	5,30,42 CV
44	$[E \rightarrow {}^{8}/_{10}, 1]$	17,43 HS
45	$[E \rightarrow \neg({}^{9}/_{10} \rightarrow \neg{}^{8}/_{10}), 1]$	5,32,44 CV
46	$[E \rightarrow {}^{7}/_{10}, 1]$	27,45 HS
47	$[E \rightarrow \neg({}^{9}/_{10} \rightarrow \neg{}^{7}/_{10}), 1]$	5,34,46 CV
48	$[E \rightarrow {}^{6}/_{10}, 1]$	39,47 HS
49	$[E \rightarrow \neg({}^{9}/_{10} \rightarrow \neg{}^{6}/_{10}), 1]$	5,40,48 CV
50	$[{}^{4}/_{10} \rightarrow \neg{}^{6}/_{10}, 1]$	F_LP6.2
51	$[{}^{5}/_{10} \rightarrow ({}^{9}/_{10} \rightarrow {}^{4}/_{10}), 1]$	F_LP5.2
52	$[\neg{}^{5}/_{10} \rightarrow {}^{5}/_{10}, 1]$	F_LP6.1
53	$[\neg({}^{9}/_{10} \rightarrow \neg{}^{6}/_{10}) \rightarrow {}^{5}/_{10}, 1]$	50,51,52 VS
54	$[E \rightarrow {}^{5}/_{10}, 1]$	49,53 HS
...	...	
154	$[J \rightarrow \mathbf{0}, 1]$	149,153 HS

where we have continued after line 54 by deriving, for each of the atomic formulas *F* through *J*, the formulas placing respective upper bounds of $^4/_{10}$, $^3/_{10}$, $^2/_{10}$, $^1/_{10}$, and 0 on their values. For each atomic formula, we first derive the upper bound we did for the previous atomic formula, using the earlier formulas with the obvious substitutions; then we add instances of F$_Ł$P6.2, F$_Ł$P5.2, and F$_Ł$P6.1 that will give an upper bound that is $^1/_{10}$ less than the final upper bound for the previous letter. Again, it should be obvious which instances of F$_Ł$P6.2, F$_Ł$P5.2, and F$_Ł$P6.1 are used in each case. This derivation, along with the derivation establishing that *J* has at least the value 0 when *A* and each of the conditional premises all have at least the value $^9/_{10}$, allows us to conclude that *J* has exactly the value 0 when *A* and the conditional premises all have exactly the value $^9/_{10}$.

The definitions of theoremhood and derivability in the graded Pavelka system F$_Ł$PA are more complicated than in F$_Ł$A, for reasons similar to those for the Pavelka system for L$_3$ in Chapter 6. We say that a formula **P** is a ***theorem to degree n*** in F$_Ł$PA if *n* is the least upper bound of the values *m* for which there is a derivation (without assumptions) of the graded formula [**P**, *m*]. As in the three-valued system, there may be two different values *j* and *k* such that there are derivations of the graded formula [**P**, *j*] and of the same formula graded as [**P**, *k*]. In this case we want to say that **P** is a theorem to at least the greater of the two values (particularly when we recall that for any formula **P**, including Fuzzy$_Ł$ tautologies, there is always a derivation of [**P**, 0]!). But unlike in the three-valued case, in F$_Ł$PA we may have a single formula for which there are derivations with an infinite number of different grades, and there may be no value *m* that is the largest. For this reason, we define the degree *n* to which **P** is a theorem as the least upper bound of all of those grades.

In a similar vein, we say that a formula **P** is ***derivable to degree n*** from a set Γ of graded formulas if *n* is the least upper bound of the degrees *m* such that [**P**, *m*] is derivable from the graded formulas in Γ. It follows from these definitions that in general we cannot directly establish the degree of theoremhood or derivability of a formula with a single derivation, since the single derivation does not imply that there are no other derivations with higher graded values for that formula. Again, this is most obvious when we recall that *any* formula can be derived with the graded value 0: it doesn't follow that 0 is the highest graded value derivable for that formula. All that a single derivation can show is that a formula is a theorem, or derivable from a graded set of formulas, to *at least* the degree established by the derivation. There is one exception: if we can derive a formula with graded value 1, then we may conclude that it *is* a theorem, or derivable, to degree 1—since no higher degree is possible.

Special case aside, the *fuzzy soundness* of F$_Ł$PA may sometimes allow us to reason more generally about degrees of theoremhood and derivability. F$_Ł$PA is **fuzzy sound** in the following sense: every formula that is a theorem to degree *n* in F$_Ł$PA is a tautology to degree *n* in RFuzzy$_Ł$, and if a formula **P** is derivable to degree *n* in F$_Ł$PA from a graded set Γ of formulas then *n* is the greatest lower bound of the values that **P** can have in $RFuzzy_L$ given the graded values of formulas in Γ.[9] In the latter case, **P** is a member to degree *n* of the fuzzy consequence of the fuzzy set in which each formula of Γ is a member to the degree indicated by its graded value and all other formulas are members to degree 0. (We note that the truth-value assignments for $RFuzzy_L$ are like those for $Fuzzy_L$ except that, in addition, each of the special formulas added to denote rational truth-values is always assigned the truth-value that it denotes. This isn't a surprise, but we need to state it explicitly!)

So, for example, we have produced a derivation showing that $A \vee \neg A$ is derivable in F_LPA (without assumptions) to at least degree .5. Given fuzzy soundness we can conclude something stronger: because this formula is a .5-tautology fuzzy soundness tells us that there is *no* derivation of this formula with a higher graded value; $A \vee \neg A$ is a theorem to degree .5 in F_LPA. Similarly, we have produced a derivation showing that *J* is derivable to at least the degree 0 from the graded formulas [*A*, .9], [$A \rightarrow B$, .9], [$B \rightarrow C$, .9], [$C \rightarrow D$, .9], [$D \rightarrow E$, .9], [$E \rightarrow F$, .9], [$F \rightarrow G$, .9], [$G \rightarrow H$, .9], [$H \rightarrow I$, .9], and [$I \rightarrow J$, .9]. From the fuzzy soundness of the system we can draw this stronger conclusion: Because the set $\{A, A \rightarrow B, B \rightarrow C, C \rightarrow D, D \rightarrow E, E \rightarrow F, F \rightarrow G, G \rightarrow H, H \rightarrow I, I \rightarrow J\}$ .1-degree-entails *J*, there is *no* derivation of *J* from these graded formulas in which *J* has a graded value higher than 0.

The converse of fuzzy soundness holds as well: F_LPA is *fuzzy complete* for $RFuzzy_L$. F_LPA is **fuzzy complete** in the sense that every formula that is a tautology to degree *n* in $RFuzzy_L$ is a theorem to degree *n* in F_LPA, and a formula **P** is derivable to degree *n* from a graded set Γ of formulas if *n* is the greatest lower bound of the values that **P** can have in $RFuzzy_L$ given the graded values of the formulas in Γ.

F_LPA is also **sound** in the traditional sense: if there is a derivation (without assumptions) of the graded formula [**P**, 1] then **P** is a tautology of $RFuzzy_L$, and if a formula **P** is derivable in F_LPA with graded value 1 from a set Γ of formulas all of which have the graded value 1, then the (ungraded) set of formulas in Γ entails the formula **P** in $RFuzzy_L$. And the system is **weakly complete** in the traditional sense: for every tautology **P** of $RFuzzy_L$ there is a derivation of **P** (without assumptions) with graded value 1 in F_LPA. Strong completeness fails for the same reason that it failed for our non-Pavelka axiomatic system F_LA for $Fuzzy_L$: semantic entailment is not compact in $Fuzzy_L$. On the other hand, traditional completeness does hold in F_LPA for entailments from finite sets: if a finite set of formulas Γ entails the formula **P** in $RFuzzy_L$, then there is a derivation in F_LPA of [**P**,1], from the set consisting of each of the formulas in Γ with graded value 1.

There is one possibly confusing relationship between fuzzy completeness and traditional completeness for F_LPA. In the case of fuzzy completeness, we have *strong* fuzzy completeness: completeness holds for entailments from infinite sets as well as entailments from finite sets. Why doesn't it follow that we have strong

9 See Hájek (1998b, pp. 80–83), for a proof of these claims and of the claims about fuzzy completeness, traditional soundness, and traditional weak completeness later.

traditional completeness as well? Consider our earlier counterexample to compactness: the set Σ consisting of the infinite sequence of formulas $(\neg P \rightarrow P) \rightarrow Q$, $(\neg P \rightarrow (\neg P \rightarrow P)) \rightarrow Q$, $(\neg P \rightarrow (\neg P \rightarrow (\neg P \rightarrow P))) \rightarrow Q$, $(\neg P \rightarrow (\neg P \rightarrow (\neg P \rightarrow (\neg P \rightarrow P)))) \rightarrow Q, \ldots$ semantically entails the formula Q, but Q is not entailed by any finite subset of Σ. Strong fuzzy completeness tells us that the least upper bound of the graded values with which Q is derivable is 1 when the formulas in the set Σ are themselves graded with the value 1. Then how can this semantic entailment be a counterexample to strong *traditional* completeness for F$_Ł$PA, as we just claimed?

The answer is that there is no derivation of Q in F$_Ł$PA with *exactly* the graded value 1 from a finite subset of formulas from Σ with the graded value 1, but there are derivations with values that approach as close to 1 as you can without actually getting there, just as there are n-entailments of Q from finite subsets of Σ in RFuzzy$_Ł$ with n as close to 1 as you can be without actually getting there (this will be examined in the exercises). The *least upper bound* of a set of values—which is featured in the definition of fuzzy completeness—need not be a member of that set, and in this case it is not. So F$_Ł$PA can be strongly fuzzy complete but only weakly traditionally complete, since strong traditional completeness would require a derivation of Q with *exactly* the value 1.

As a consequence of the strong fuzzy completeness of F$_Ł$PA we have fuzzy compactness. Recall that a fuzzy set of the formulas of a language is a fuzzy set to which every formula of the language belongs to some degree (including possibly degree 0). We will say that a fuzzy set of formulas Σ is a ***finite base*** of a fuzzy set of formulas Γ if a *finite subset* of the formulas that are members of Γ to a degree greater than 0 are members of Σ to the same degree (as in Γ), and all other formulas of the language are members of Σ to degree 0. Here is the result:

> *Result 13.4:* Fuzzy$_Ł$ and RFuzzy$_Ł$ are ***fuzzy compact***: for every formula **P** and fuzzy set of formulas Γ, **P** is a member to degree n of the fuzzy consequence of Γ if and only if **P** is a member to degree n of the fuzzy consequence of at least one finite base of Γ.

Degree-entailment and n-degree-entailment are not standardly studied in connection with fuzzy Pavelka derivation systems, where fuzzy consequence is the predominant semantic relation. Unlike entailment *simpliciter* and fuzzy consequence, these semantic concepts talk about all possible combinations of values that the members of a set Γ of formulas might have. In the case of entailment *simpliciter*, we are only interested in what happens when all of the members of Γ have the value 1. For fuzzy consequence, we are only interested in one combination of values for the formulas in Γ. But for degree-validity and n-degree-validity we are interested in the relationship between the values that the members of Γ might have, *no matter what those values might be*, and the value that the entailed formula has.

However, just as we can define degree-entailment in terms of fuzzy consequence—a set of formulas Γ degree-entails the formula **P** if for every n in [0. .1], **P** is a member to at least degree n of the fuzzy consequence of each fuzzy set in which the greatest lower bound of the values of members of Γ is *n* and all other formulas of Fuzzy_L have the value 0—we *can* define a syntactic counterpart to degree-entailment as: **P** is *degree-derivable* from a set of formulas Γ if for every *n* in [0. .1], [**P**, *n*] is derivable from each set Γ* in which the formulas of Γ have all been graded with values whose greatest lower bound is *n*. In the context of three-valued logic, we were able to produce derivations establishing degree-validity because we only had a finite number of combinations of values to consider. But in general, no finite number of derivations can establish a claim about all possible combinations of values for some set of formulas in RFuzzy_L (an exception to this general rule is the case where we can establish that **P** is a theorem to degree 1, for here it doesn't matter what graded values may be associated with members of Γ).

Before closing this section we would like to address the question, In RFuzzy_L and the axiomatic system F_LPA we've included constants to name only the rational truth-values in the unit interval so that our supply of atomic formulas will be countable, but do we *lose* anything important by failing to include constants denoting the *non*rational members of [0. .1]? Well, we won't be able to prove things like *if* **P** *has some specific irrational value i and* **Q** *has a larger irrational value j, then* **P** → **Q** *has the value 1.* Is this a significant loss? Is it important to be able to state explicitly relationships between irrational degrees of membership in fuzzy sets? As a first answer, we note that in our analyses of vagueness all of our examples have used *rational* truth-values, and it is not clear that issues of vagueness require even finer distinctions involving irrational truth-values (in fact, some have claimed that only finitely many truth-values are required for issues of vagueness).[10]

Second, Petr Hájek, Jeff Paris, and John Shepherdson (2000) have proved an important relevant theorem. Let RealFuzzy_L be the logical system Fuzzy_L with the addition of atomic formulas naming each value, irrational and rational alike, in the unit interval [0. .1]. So RealFuzzy_L is an extension of Rfuzzy_L in the sense that the formulas of the latter are a subset of those of the former. Let RealF_LPA be the Pavelka-style axiomatic system that is like F_LPA except that it also includes axioms analogous to $F_LP5.1$, $F_LP5.2$, $F_LP6.1$, $F_LP6.2$, and F_LP for *all* of the real values in [0. .1], not just the rational ones. RealF_LPA is thus an extension of F_LPA in the sense that every derivation in F_LPA is also a derivation in RealF_LPA. Hájek and colleagues' theorem states that RealF_LPA is a ***conservative extension*** of F_LPA:

> *Conservative Extension Theorem for F_LPA:* If a formula **P** of Rfuzzy_L is a theorem to degree *n* of RealF_LPA then it is also a theorem to degree *n* of F_LPA, and if a formula **P** of Rfuzzy_L is derivable to degree *n* from a graded set of Rfuzzy_L formulas in RealF_LPA then **P** is also derivable to degree *n* from that graded set of formulas in F_LPA.

10 E.g., Morgan and Pelletier (1977). They suggest that this is a reason for sticking to finitely many-valued logics. We won't go that far, since fuzzy logics are interesting in their own right.

That is, if we restrict our attention to rational truth-values then everything we can demonstrate in RealF$_Ł$PA can already be demonstrated in F$_Ł$PA; we do not need the additional power of the full real-valued system. So if it is correct that we lose little of interest on the score of vagueness by restricting our attention to rational truth-degrees, then it is also correct that we lose little by using the system F$_Ł$PA rather than RealF$_Ł$PA.

13.3 An Alternative Axiomatic System for Tautologies and Validity in Fuzzy$_Ł$ Based on BL-Algebras

We will briefly present axiomatic systems for product and Gödel fuzzy propositional logics in Sections 13.4 and 13.5. The systems, developed in Hájek (1998b), are extensions of his "basic" fuzzy logic axiomatic system. In this section we present the basic system, which we'll call *BLA*, and explain how BLA can be expanded to an axiomatic system for Łukasiewicz fuzzy logic.

BLA axiomatizes the basic properties of continuous t-norms and their related residuum operations as captured in BL-algebras. In addition to t-norm and residuum operators & and → we also introduce a special constant formula **0**, which always has the value 0. The axiom schemata for BLA are:

BL1. (P → Q) → ((Q → R) → (P → R))
BL2. (P & Q) → P
BL3. (P & Q) → (Q & P)
BL4. (P & (P → Q)) → (Q & (Q → P))
BL5. (P → (Q → R)) → ((P & Q) → R)
BL6. ((P & Q) → R) → (P → (Q → R))
BL7. ((P → Q) → R) → (((Q → P) → R) → R)
BL8. 0 → P

and the single inference rule is, once again, Modus Ponens.

BL1 should come as no surprise, since we have repeatedly seen it in our axiomatic systems. BL2 tells us that we can infer the first conjunct from a t-norm conjunction, and BL3, which formally captures the commutativity of the t-norm operation, will also allow us to infer the second conjunct—we'll see this in the derivation later. Turning to BL4, we recall that weak conjunction **P ∧ Q** in a fuzzy system is standardly defined as **P & (P → Q)**, so BL4 formally captures the commutativity of weak conjunction. BL5 and BL6 capture the equivalence of the formulas **P → (Q → R)** and **(P & Q) → R**. For BL7, note that if the conditional → denotes a residuum operation then for any pair of formulas **P** and **Q**, either **P → Q** or **Q → P** will have the value 1. So if each of these implies some formula **R**, then we may conclude that **R** has the value 1 as well. BL8 says that **0** is the least truth-value.

It may be surprising that BLA has no axioms with a negation operator. But recall that negation is standardly defined in fuzzy logic as **¬P** $=_{\text{def}}$ **P → 0**, so that is the definition that we use here. Here is an example of a derivation of the negated formula $\neg\mathbf{0}$, which is defined to be $\mathbf{0} \rightarrow \mathbf{0}$:

1	$\mathbf{0} \rightarrow \mathbf{0}$	BL8, with **0** / **P**

and here is a derivation of the formula $P \rightarrow \neg\neg P$, which is defined to be $P \rightarrow ((P \rightarrow \mathbf{0}) \rightarrow \mathbf{0})$:

1	$((\mathbf{0} \rightarrow P) \,\&\, (P \rightarrow \mathbf{0})) \rightarrow ((P \rightarrow \mathbf{0}) \,\&\, (\mathbf{0} \rightarrow P))$	BL3, with $\mathbf{0} \rightarrow P$ / **P**, $P \rightarrow \mathbf{0}$ / **Q**
2	$((P \rightarrow \mathbf{0}) \,\&\, (\mathbf{0} \rightarrow P)) \rightarrow (P \rightarrow \mathbf{0})$	BL2, with $P \rightarrow \mathbf{0}$ / **P**, $\mathbf{0} \rightarrow P$ / **Q**
3	$((\mathbf{0} \rightarrow P) \,\&\, (P \rightarrow \mathbf{0})) \rightarrow (P \rightarrow \mathbf{0})$	1,2 HS
4	$(((\mathbf{0} \rightarrow P) \,\&\, (P \rightarrow \mathbf{0})) \rightarrow (P \rightarrow \mathbf{0})) \rightarrow$ $((\mathbf{0} \rightarrow P) \rightarrow ((P \rightarrow \mathbf{0}) \rightarrow (P \rightarrow \mathbf{0})))$	BL6, with $\mathbf{0} \rightarrow P$ / **P**, $P \rightarrow \mathbf{0}$ / **Q**, $P \rightarrow \mathbf{0}$ / **R**
5	$(\mathbf{0} \rightarrow P) \rightarrow ((P \rightarrow \mathbf{0}) \rightarrow (P \rightarrow \mathbf{0}))$	3,4 MP
6	$\mathbf{0} \rightarrow P$	BL8, with P / **P**
7	$(P \rightarrow \mathbf{0}) \rightarrow (P \rightarrow \mathbf{0})$	5,6 MP
8	$((P \rightarrow \mathbf{0}) \rightarrow (P \rightarrow \mathbf{0})) \rightarrow (((P \rightarrow \mathbf{0}) \,\&\, P) \rightarrow \mathbf{0})$	BL5, with with $P \rightarrow \mathbf{0}$ / **P**, P / **Q**, **0** / **R**
9	$((P \rightarrow \mathbf{0}) \,\&\, P) \rightarrow \mathbf{0}$	7,8 MP
10	$(P \,\&\, (P \rightarrow \mathbf{0})) \rightarrow ((P \rightarrow \mathbf{0}) \,\&\, P)$	BL3, with P / **P**, $P \rightarrow \mathbf{0}$ / **Q**
11	$(P \,\&\, (P \rightarrow \mathbf{0})) \rightarrow \mathbf{0}$	9,10 HS
12	$((P \,\&\, (P \rightarrow \mathbf{0})) \rightarrow \mathbf{0}) \rightarrow (P \rightarrow ((P \rightarrow \mathbf{0}) \rightarrow \mathbf{0}))$	BL6, with P / **P**, $P \rightarrow \mathbf{0}$ / **Q**, **0** / **R**
13	$P \rightarrow ((P \rightarrow \mathbf{0}) \rightarrow \mathbf{0})$	11,12 MP

We used Hypothetical Syllogism in this proof; it is derived in BLA exactly as it is derived in F_LA. By virtue of this proof, we have the derived axiom schema

BLD1. $\mathbf{P} \rightarrow \neg\neg\mathbf{P}$, or $\mathbf{P} \rightarrow ((\mathbf{P} \rightarrow \mathbf{0}) \rightarrow \mathbf{0})$

The converse formula, $\neg\neg P \rightarrow P$ or $((P \rightarrow \mathbf{0}) \rightarrow \mathbf{0}) \rightarrow \mathbf{0}$, is not derivable in BLA. This is a good thing, since the converse formula is not a tautology in every variety of fuzzy logic. In particular, it isn't a tautology in either Fuzzy_G or Fuzzy_L.

We said earlier that the system BLA axiomatizes the basic properties of continuous t-norms and their related residuum operations as captured in BL-algebras. We (and the reader, in some exercises) will show that the algebraic counterparts of the axioms of BLA all evaluate to *unit* in every BL-algebra—they are all BL-tautologies—and that, corresponding to MP, the algebraic rule

If $unit \leq x$ and $unit \leq x \Rightarrow y$, then $unit \leq y$

is true in every BL-algebra, from which it follows that the BL-algebraic expressions corresponding to theorems of BLA are all BL-tautologies.[11]

Recall that a BL-algebra is an algebra $\{L, \cup, \cap, \otimes, \Rightarrow, unit, zero\}$ that meets the following conditions:

[11] The proofs later owe largely to Hájek (1998b). The converse also holds: every expression that evaluates to *unit* in every BL-algebra is the algebraic counterpart of a theorem of the axiomatic system BLA (Hájek 1998b, pp. 49–54).

i.	$x \cup y = y \cup x$, and $x \cap y = y \cap x$	(*lattice commutation*)
ii.	$x \cup (y \cup z) = (x \cup y) \cup z$, and $x \cap (y \cap z) = (x \cap y) \cap z$	(*lattice association*)
iii.	$x \cup x = x$, and $x \cap x = x$	(*lattice idempotence*)
iv.	$x \cup (x \cap y) = x$, and $x \cap (x \cup y) = x$	(*lattice distribution*)
v.	$x \cup \mathit{zero} = x$, and $x \cap \mathit{unit} = x$	(*identity for lattice join and meet*)
vi.	$x \otimes y = y \otimes x$	(*bold meet commutation*)
vii.	$x \otimes (y \otimes z) = (x \otimes y) \otimes z$	(*bold meet association*)
viii.	$x \otimes \mathit{unit} = x$	(*identity for bold meet*)
	and, defining $x \leq y$ if and only if $x \cap y = x$,	
ix.	if $x \leq y$, then $x \otimes z \leq y \otimes z$ and $z \otimes x \leq y \otimes x$	(*bold meet isotonicity*)
x.	$x \otimes y \leq z$ if and only if $x \leq y \Rightarrow z$	(*adjointness*)
xi.	$x \cap y = x \otimes (x \Rightarrow y)$	
xii.	$(x \Rightarrow y) \cup (y \Rightarrow x) = \mathit{unit}$	

For axiom schema BL1, we will show that for any elements x, y, and z in a BL-algebra, $\mathit{unit} \leq (x \Rightarrow y) \Rightarrow ((y \Rightarrow z) \Rightarrow (x \Rightarrow z))$. (BL-i in Section 12.2 of Chapter 12 proved that if $\mathit{unit} \leq x$, then $\mathit{unit} = x$, so it will follow that $(x \Rightarrow y) \Rightarrow ((y \Rightarrow z) \Rightarrow (x \Rightarrow z)) = \mathit{unit}$. We'll refer to results from Chapter 12 by the numbering used there.) We'll help ourselves to the fact that if $x = y$ in a BL-algebra, then $x \leq y$, because lattice orderings are reflexive ($x \leq x$).

	BL1:	
	$x \otimes (x \Rightarrow y) \leq y$	(BL-ii: Chapter 12)
(i)	$(x \otimes (x \Rightarrow y)) \otimes (y \Rightarrow z) \leq y \otimes (y \Rightarrow z)$	(bold meet isotonicity)
	and	
(ii)	$y \otimes (y \Rightarrow z) \leq z$	(BL-ii: Chapter 12)
	so	
	$(x \otimes (x \Rightarrow y)) \otimes (y \Rightarrow z) \leq z$	(from (i) and (ii) by $\leq$ transitivity)
	$((x \Rightarrow y) \otimes (y \Rightarrow z)) \otimes x \leq z$	(bold meet association, commutation)
	$((\mathit{unit} \otimes (x \Rightarrow y)) \otimes (y \Rightarrow z)) \otimes x \leq z$	(bold meet identity, association)
	$(\mathit{unit} \otimes (x \rightarrow y)) \otimes (y \Rightarrow z) \leq (x \Rightarrow z)$	(adjointness)
	$\mathit{unit} \otimes (x \Rightarrow y) \leq (y \Rightarrow z) \Rightarrow (x \Rightarrow z)$	(adjointness)
	$\mathit{unit} \leq (x \Rightarrow y) \Rightarrow ((y \Rightarrow z) \Rightarrow (x \Rightarrow z))$	(adjointness)

For axiom schema BL2, we show that $\mathit{unit} \leq (x \otimes y) \Rightarrow x$:

BL2:	
$y \cap \mathit{unit} \leq y$	(identity for lattice meet)
$y \leq \mathit{unit}$	(definition of $\leq$)
$x \otimes y \leq x \otimes \mathit{unit}$	(bold meet isotonicity)
$x \otimes y \leq x$	(bold meet identity)
$\mathit{unit} \otimes (x \otimes y) \leq x$	(bold meet identity, commutation)
$\mathit{unit} \leq (x \otimes y) \Rightarrow x$	(adjointness)

For BL5, we show that $unit \leq (x \Rightarrow (y \Rightarrow z)) \Rightarrow ((x \otimes y) \Rightarrow z)$:

BL5:

$x \Rightarrow (y \Rightarrow z) \leq (x \otimes y) \Rightarrow z$	(BL-iii: Chapter 12)
$unit \otimes (x \Rightarrow (y \Rightarrow z)) \leq (x \otimes y) \Rightarrow z$	(bold meet identity, commutation)
$unit \leq (x \Rightarrow (y \Rightarrow z)) \Rightarrow ((x \otimes y) \Rightarrow z)$	(adjointness)

For axiom schema BL7, we show that $unit \leq ((x \Rightarrow y) \Rightarrow z) \Rightarrow (((y \Rightarrow x) \Rightarrow z) \Rightarrow z)$:

BL7:

$(((x \Rightarrow y) \Rightarrow z) \otimes (x \Rightarrow y)) \otimes ((y \Rightarrow x) \Rightarrow z) \leq ((x \Rightarrow y) \Rightarrow z) \otimes (x \Rightarrow y)$
(since $x \otimes y \leq x$; proof is left as an exercise)

so (i) $(((x \Rightarrow y) \Rightarrow z) \otimes ((y \Rightarrow x) \Rightarrow z)) \otimes (x \Rightarrow y) \leq ((x \Rightarrow y) \Rightarrow z) \otimes (x \Rightarrow y)$
(bold meet commutation, association)

and (ii) $(((x \Rightarrow y) \Rightarrow z) \otimes ((y \Rightarrow x) \Rightarrow z)) \otimes (y \Rightarrow x) \leq ((y \Rightarrow x) \Rightarrow z) \otimes (y \Rightarrow x)$
(same reason)

thus (iii) $((((x \Rightarrow y) \Rightarrow z) \otimes ((y \Rightarrow x) \Rightarrow z)) \otimes (x \Rightarrow y)) \cup ((((x \Rightarrow y) \Rightarrow z) \otimes ((y \Rightarrow x) \Rightarrow z)) \otimes (y \Rightarrow x)) \leq (((x \Rightarrow y) \Rightarrow z) \otimes (x \Rightarrow y)) \cup (((y \Rightarrow x) \Rightarrow z) \otimes (y \Rightarrow x))$
(by (i) and (ii) since if $x \leq z$ and $y \leq w$, then $x \cup y \leq z \cup w$—proof is left as an exercise)

In addition,

$((x \Rightarrow y) \cap z) \cap z = (x \Rightarrow y) \cap z$ (lattice meet idempotence, association)

so $(x \Rightarrow y) \cap z \leq z$ (by definition)

and similarly $(y \Rightarrow x) \cap z \leq z$

thus $((x \Rightarrow y) \cap z) \cup ((y \Rightarrow x) \cap z) \leq z$ (since $x \leq z$ and $y \leq z$ if and only if $x \cup y \leq z$—proof is left as an exercise)

and (iv) $(((x \Rightarrow y) \Rightarrow z) \otimes (x \Rightarrow y)) \cup (((y \Rightarrow x) \Rightarrow z) \otimes (y \Rightarrow x)) \leq z$
(bold meet commutation, BL-algebra condition (xi))

Therefore

$((((x \Rightarrow y) \Rightarrow z) \otimes ((y \Rightarrow x) \Rightarrow z)) \otimes (x \Rightarrow y)) \cup ((((x \Rightarrow y) \Rightarrow z) \otimes ((y \Rightarrow x) \Rightarrow z)) \otimes (y \Rightarrow x)) \leq z$
(from (iii) and (iv), by $\leq$ transitivity)

$(((x \Rightarrow y) \Rightarrow z) \otimes ((y \Rightarrow x) \Rightarrow z)) \otimes ((x \Rightarrow y) \cup (y \Rightarrow x)) \leq z$
(since $x \otimes (y \cup z) = (x \otimes y) \cup (x \otimes z)$—proof is left as an exercise)

$(((x \Rightarrow y) \Rightarrow z) \otimes ((y \Rightarrow x) \Rightarrow z)) \otimes unit \leq z$
(BL-algebra condition (xii))

$((x \Rightarrow y) \Rightarrow z) \otimes ((y \Rightarrow x) \Rightarrow z) \leq z$ (bold meet identity)

$(x \Rightarrow y) \Rightarrow z \leq ((y \Rightarrow x) \Rightarrow z) \Rightarrow z$ (adjointness)

$unit \otimes (x \Rightarrow y) \Rightarrow z) \leq ((y \Rightarrow x) \Rightarrow z) \Rightarrow z$
(bold meet identity, commutation)

$unit \leq ((x \Rightarrow y) \Rightarrow z) \Rightarrow (((y \Rightarrow x) \Rightarrow z) \Rightarrow z)$
(adjointness)

We leave it as an exercise to show that the algebraic counterparts of axiom schemata BL3, BL4, BL6, and BL8 are all BL-tautologies, and that the rule MP, interpreted algebraically as

If $unit \leq x$ and $unit \leq x \Rightarrow y$, then $unit \leq y$,

is true in every BL-algebra.

We can extend the axiomatic system BLA with one axiom schema to obtain a system $BL_ŁA$ that is equivalent (in a sense to be explained later) to the axiomatic system $F_ŁA$ for $Fuzzy_Ł$:

$BL_Ł$9. $\neg\neg P \rightarrow P$

Note that this explicitly parallels a result presented in Chapter 12: that every BL-algebra with Double Negation is an MV-algebra (recall that the algebraic structures corresponding to $Fuzzy_Ł$ are MV-algebras). We have already shown that the converse formula to $BL_Ł$9 is derivable in the basic system BLA; this additional axiom schema provides full Double Negation.

We can use the definition of bold conjunction:

$$P \,\&\, Q =_{def} \neg(P \rightarrow \neg Q)$$

to introduce bold conjunction into $F_ŁA$. The systems $F_ŁA$ and $BL_ŁA$ are then equivalent in the sense that every theorem of $F_ŁA$ is a theorem of $BL_ŁA$, and vice versa (see Hájek 1998b, pp. 65–70). Rather than produce examples of derivations in $BL_ŁA$, we'll illustrate the use of BLA in the following axiomatic systems for Gödel and product logic.

13.4 An Axiomatic System for Tautologies and Validity in $Fuzzy_G$

We obtain an axiomatic system BL_GA for $Fuzzy_G$ by adding the following axiom schema

BL_G9. $P \rightarrow_G (P \,\&_G\, P)$

to BLA (we *exclude* the Łukasiewicz axiom $BL_Ł$9), with all of the connectives now subscripted with *G* to reflect the fact that we are working in $Fuzzy_G$. It was left as an exercise in Chapter 12 to show that this axiom, interpreted algebraically, is true in every Gödel algebra. (So are the axioms BL1–BL8, since every Gödel algebra is a BL-algebra.) BL_GA is both sound and *strongly* complete for $Fuzzy_G$ (Hajek 1998b, pp. 101–102). The latter means that all entailments in $Fuzzy_G$ from infinite, as well as finite, sets have corresponding derivations in BL_GA.

The *strong* completeness owes to the fact that Gödel fuzzy logic, unlike Łukasiewicz fuzzy logic, is compact (Baaz and Zach 1998). So, for example, consider our earlier example of a set Σ consisting of the formula $\neg P \rightarrow Q$ and the infinitely many formulas in the series

$(\neg P \rightarrow P) \rightarrow Q$
$(\neg P \rightarrow (\neg P \rightarrow P)) \rightarrow Q$
$(\neg P \rightarrow (\neg P \rightarrow (\neg P \rightarrow P))) \rightarrow Q$
$(\neg P \rightarrow (\neg P \rightarrow (\neg P \rightarrow (\neg P \rightarrow P)))) \rightarrow Q$
...

We noted that in Fuzzy$_{Ł}$ this set, but none of its finite subsets, entails the formula Q. In Fuzzy$_G$ (using the Fuzzy$_G$ negation and conditional) Σ also entails Q, but so does at least one finite subset. In fact, the subset consisting of just $\neg_G P \rightarrow_G Q$ and $(\neg_G P \rightarrow_G P) \rightarrow_G Q$ entails Q, as does every subset that contains $\neg_G P \rightarrow_G Q$ and at least one formula from the infinite series (proof is left as an exercise).

We'll show that Q is derivable in BL$_G$A from the formulas $\neg_G P \rightarrow_G Q$ and $(\neg_G P \rightarrow_G (\neg_G P \rightarrow_G P)) \rightarrow Q_G$. (Note that this is a different example from the entailment mentioned in the previous paragraph—a derivation corresponding to that entailment is assigned as one of the exercises.) First we derive some useful axiom schemata.

BL$_G$D2. $(\mathbf{P} \rightarrow_G \neg_G \mathbf{P}) \rightarrow_G \neg_G \mathbf{P}$

Justification (where the formula is reexpressed as $(\mathbf{P} \rightarrow_G (\mathbf{P} \rightarrow_G \mathbf{0})) \rightarrow_G (\mathbf{P} \rightarrow_G \mathbf{0})$):

1	$(\mathbf{P} \rightarrow_G (\mathbf{P} \rightarrow_G \mathbf{0})) \rightarrow_G ((\mathbf{P} \,\&_G\, \mathbf{P}) \rightarrow_G \mathbf{0})$	BL$_G$5, with **P / P, P / Q, 0 / R**
2	$\mathbf{P} \rightarrow_G (\mathbf{P} \,\&_G\, \mathbf{P})$	BL$_G$9, with **P / P**
3	$(\mathbf{P} \rightarrow_G (\mathbf{P} \,\&_G\, \mathbf{P})) \rightarrow_G (((\mathbf{P} \,\&_G\, \mathbf{P}) \rightarrow_G \mathbf{0}) \rightarrow_G (\mathbf{P} \rightarrow_G \mathbf{0}))$	BL$_G$1, with **P / P, P &$_G$ P / Q, 0 / R**
4	$((\mathbf{P} \,\&_G\, \mathbf{P}) \rightarrow_G \mathbf{0}) \rightarrow_G (\mathbf{P} \rightarrow_G \mathbf{0})$	2,3 MP
5	$(\mathbf{P} \rightarrow_G (\mathbf{P} \rightarrow_G \mathbf{0})) \rightarrow_G (\mathbf{P} \rightarrow_G \mathbf{0})$	1,4 HS

BL$_G$D3. $(\mathbf{P} \rightarrow_G \mathbf{Q}) \rightarrow_G (\neg_G \mathbf{Q} \rightarrow_G \neg_G \mathbf{P})$

Justification (where the formula is reexpressed as $(\mathbf{P} \rightarrow_G \mathbf{Q}) \rightarrow_G ((\mathbf{Q} \rightarrow_G \mathbf{0}) \rightarrow_G (\mathbf{P} \rightarrow_G \mathbf{0}))$):

1	$(\mathbf{P} \rightarrow_G \mathbf{Q}) \rightarrow_G ((\mathbf{Q} \rightarrow_G \mathbf{0}) \rightarrow_G (\mathbf{P} \rightarrow_G \mathbf{0}))$	BL$_G$1, with **P / P, Q / Q, 0 / R**

BL$_G$D4. $\mathbf{P} \rightarrow_G (\mathbf{Q} \rightarrow_G \mathbf{P})$

Justification:

1	$(\mathbf{P} \,\&_G\, \mathbf{Q}) \rightarrow_G \mathbf{P}$	BL$_G$2, with **P / P, Q / Q**
2	$((\mathbf{P} \,\&_G\, \mathbf{Q}) \rightarrow_G \mathbf{P}) \rightarrow_G (\mathbf{P} \rightarrow_G (\mathbf{Q} \rightarrow_G \mathbf{P}))$	BL$_G$6, with **P / P, Q / Q, P / R**
3	$\mathbf{P} \rightarrow_G (\mathbf{Q} \rightarrow_G \mathbf{P})$	1,2 MP

TRANS: From $\mathbf{P} \rightarrow_G (\mathbf{Q} \rightarrow_G \mathbf{R})$, infer $\mathbf{Q} \rightarrow_G (\mathbf{P} \rightarrow_G \mathbf{R})$.

Justification:

1	$\mathbf{P} \rightarrow_G (\mathbf{Q} \rightarrow_G \mathbf{R})$	*given*
2	$(\mathbf{P} \rightarrow_G (\mathbf{Q} \rightarrow_G \mathbf{R})) \rightarrow_G ((\mathbf{P} \,\&_G\, \mathbf{Q}) \rightarrow_G \mathbf{R})$	BL$_G$5, with $\mathbf{P}$ / $\mathbf{P}$, $\mathbf{Q}$ / $\mathbf{Q}$, $\mathbf{R}$ / $\mathbf{R}$
3	$(\mathbf{P} \,\&_G\, \mathbf{Q}) \rightarrow_G \mathbf{R}$	1,2 MP
4	$(\mathbf{Q} \,\&_G\, \mathbf{P}) \rightarrow_G (\mathbf{P} \,\&_G\, \mathbf{Q})$	BL$_G$3, with $\mathbf{Q}$ / $\mathbf{P}$, $\mathbf{P}$ / $\mathbf{Q}$
5	$(\mathbf{Q} \,\&_G\, \mathbf{P}) \rightarrow_G \mathbf{R}$	3,4 MP
6	$((\mathbf{Q} \,\&_G\, \mathbf{P}) \rightarrow_G \mathbf{R}) \rightarrow_G (\mathbf{Q} \rightarrow_G (\mathbf{P} \rightarrow_G \mathbf{R}))$	BL$_G$6, with $\mathbf{Q}$ / $\mathbf{P}$, $\mathbf{P}$ / $\mathbf{Q}$, $\mathbf{R}$ / $\mathbf{R}$
7	$\mathbf{Q} \rightarrow_G (\mathbf{P} \rightarrow_G \mathbf{R})$	5,6 MP

And here's the derivation—we omit the subscripted G for readability:

1	$\neg P \rightarrow Q$	Assumption
2	$(\neg P \rightarrow (\neg P \rightarrow P)) \rightarrow Q$	Assumption
3	$((\neg P \rightarrow P) \rightarrow \neg(\neg P \rightarrow P)) \rightarrow \neg(\neg P \rightarrow P)$	BL$_G$D2, with $\neg P \rightarrow P$ / $\mathbf{P}$
4	$P \rightarrow (\neg P \rightarrow P)$	BL$_G$D4, with P / $\mathbf{P}$, $\neg P$ / $\mathbf{Q}$
5	$(P \rightarrow (\neg P \rightarrow P)) \rightarrow (\neg(\neg P \rightarrow P) \rightarrow \neg P)$	BL$_G$D3, with P / $\mathbf{P}$, $\neg P \rightarrow P$ / $\mathbf{Q}$
6	$\neg(\neg P \rightarrow P) \rightarrow \neg P$	4,5 MP
7	$((\neg P \rightarrow P) \rightarrow \neg(\neg P \rightarrow P)) \rightarrow \neg P$	3,6 HS
8	$(\neg P \rightarrow P) \rightarrow ((P \rightarrow \neg\neg P) \rightarrow (\neg P \rightarrow \neg\neg P))$	BL$_G$1, with $\neg P$ / $\mathbf{P}$, P / $\mathbf{Q}$, $\neg_G\neg P$ / $\mathbf{R}$
9	$(P \rightarrow \neg\neg P) \rightarrow ((\neg P \rightarrow P) \rightarrow (\neg P \rightarrow \neg\neg P))$	8, TRANS
10	$P \rightarrow \neg\neg P$	BL$_G$D1, with $\mathbf{P}$ / $\mathbf{P}$
11	$(\neg P \rightarrow P) \rightarrow (\neg P \rightarrow \neg\neg P)$	9,10 MP
12	$(\neg P \rightarrow \neg\neg P) \rightarrow \neg\neg P$	BL$_G$D2, with $\neg P$ / $\mathbf{P}$
13	$(\neg P \rightarrow P) \rightarrow \neg\neg P$	11,12 HS
14	$((\neg P \rightarrow P) \rightarrow \neg\neg P) \rightarrow (\neg\neg\neg P \rightarrow \neg(\neg P \rightarrow P))$	BL$_G$D3, with $\neg P \rightarrow P$ / $\mathbf{P}$, $\neg\neg P$ / $\mathbf{Q}$
15	$\neg\neg\neg P \rightarrow \neg(\neg P \rightarrow P)$	13,14 MP
16	$\neg P \rightarrow \neg\neg\neg P$	BL$_G$D1, with $\neg P$ / $\mathbf{P}$
17	$\neg P \rightarrow \neg(\neg P \rightarrow P)$	15,16 HS
18	$((\neg P \rightarrow P) \rightarrow \neg P) \rightarrow$ $((\neg P \rightarrow \neg(\neg P \rightarrow P)) \rightarrow ((\neg P \rightarrow P) \rightarrow \neg(\neg P \rightarrow P)))$	BL$_G$1, with $\neg P \rightarrow P$ / $\mathbf{P}$, $\neg P$ / $\mathbf{Q}$, $\neg(\neg P \rightarrow P)$ / $\mathbf{R}$
19	$(\neg P \rightarrow \neg(\neg P \rightarrow P)) \rightarrow$ $(((\neg P \rightarrow P) \rightarrow \neg P) \rightarrow ((\neg P \rightarrow P) \rightarrow \neg(\neg P \rightarrow P)))$	18, TRANS
20	$((\neg P \rightarrow P) \rightarrow \neg P) \rightarrow ((\neg P \rightarrow P) \rightarrow \neg(\neg P \rightarrow P))$	17,19 MP
21	$((\neg P \rightarrow P) \rightarrow \neg P) \rightarrow \neg P$	7,20 HS
22	$((\neg P \rightarrow P) \rightarrow \neg P) \rightarrow Q$	1,21 HS
23	$((\neg P \rightarrow (\neg P \rightarrow P)) \rightarrow Q) \rightarrow$ $((((\neg P \rightarrow P) \rightarrow \neg P) \rightarrow Q) \rightarrow Q)$	BL$_G$7, with $\neg P$ / $\mathbf{P}$, $\neg P \rightarrow P$ / $\mathbf{Q}$, Q / $\mathbf{R}$
24	$(((\neg P \rightarrow P) \rightarrow \neg P) \rightarrow Q) \rightarrow Q$	2,23 MP
25	Q	22,24 MP

In contrast with F$_Ł$PA (and BL$_Ł$A), we have a full Deduction Theorem for BL$_G$A:

Result 11.4 (Deduction Theorem for BL$_G$A): $\mathbf{Q}$ is derivable from $\mathbf{P}$ in BL$_G$A if and only if $\mathbf{P} \rightarrow \mathbf{Q}$ is a theorem.

Proof: Left as an exercise.

Although BL_GA is strongly complete, no *Pavelka-style* axiomatic system can be strongly fuzzy complete for $Fuzzy_G$. This negative result is a consequence of the fact that Gödel implication is noncontinuous: Pavelka proved that a logical system whose algebraic structure over the unit interval of real numbers is a residuated lattice (as are the algebraic structure of $Fuzzy_G$ and those of $Fuzzy_Ł$ and $Fuzzy_P$) is Pavelka-axiomatizable only if its operations are continuous.[12] The reader will understand that this is an unfortunate limitation, since Pavelka-style systems help us to see formally the seductiveness of Modus Ponens chains of reasoning with vague predicates (not too bad when the chain of reasoning is short) while showing how they can falter (pretty bad as the chain of reasoning gets much longer).

13.5 An Axiomatic System for Tautologies and Validity in $Fuzzy_P$

The system BLA can be extended to a sound and weakly complete system BL_PA for $Fuzzy_P$ by adding the following two axiom schemata (bold conjunction and implication operators in the eight BLA axiom schemata will now be the product operators subscripted with *P*):

$BL_P9.$ $\neg_P\neg_P\mathbf{P} \rightarrow_P (((\mathbf{Q} \,\&_P\, \mathbf{P}) \rightarrow_P (\mathbf{R} \,\&_P\, \mathbf{P})) \rightarrow_P (\mathbf{Q} \rightarrow_P \mathbf{R}))$
$BL_P10.$ $(\mathbf{P} \,\&_P\, \neg_P\mathbf{P}) \rightarrow_P \mathbf{0}$

It is left as an exercise to prove that the axiom schemata BL_P9 and BL_P10 of BL_PA, interpreted algebraically, are true in every product algebra. And since every product algebra is a BL-algebra, the axioms BL1–BL8 are also true in every product algebra when interpreted algebraically.

As an example of a derivation in BL_PA we'll derive the formula *Q* from the formulas *P* and $\neg_P P$:

1	P	
2	$\neg_P$P	
3	(P $\&_P$ $\neg_P$P) $\rightarrow_P$ ($\neg_P$P $\&_P$ P)	BL_P3, with P / **P**, $\neg_P$P / **Q**
4	($\neg_P$P $\&_P$ P) $\rightarrow_P$ (P $\&_P$ $\neg_P$P)	BL_P3, with ¬P / **P**, P / **Q**
5	(P $\&_P$ $\neg_P$P) $\rightarrow_P$ (P_P $\&_P$ $\neg_P$P)	3,4 HS
6	((P $\&_P$ $\neg_P$P) $\rightarrow_P$ (P $\&_P$ $\neg_P$P)) $\rightarrow_P$ (P $\rightarrow_P$ ($\neg_P$P $\rightarrow_P$ (P $\&_P$ $\neg_P$P)))	BL_P6, with P / **P**, $\neg_P$P / **Q**, P $\&_P$ $\neg_P$P / **R**
7	P $\rightarrow_P$ ($\neg_P$P $\rightarrow_P$ (P $\&_P$ $\neg_P$P))	5,6 HS
8	$\neg_P$P $\rightarrow_P$ (P $\&_P$ $\neg_P$P)	1,7 MP
9	P $\&_P$ $\neg_P$P	2,8 MP
10	(P $\&_P$ $\neg_P$P) $\rightarrow_P$ **0**	BL_P10, with P / **P**
11	**0**	9,10 MP
12	**0** $\rightarrow_P$ Q	BL_P8, with Q / **P**
13	Q	11,12 MP

[12] Pavelka (1979, Part III).

As was the case for F_LA, the standard deduction theorem does not hold for BL_PA. But we do have the same Modified Deduction Theorem:

Result 13.5 (Modified Deduction Theorem for BL_PA): **Q** is derivable from **P** in BL_PA if and only if $\mathbf{P} \rightarrow (\mathbf{P} \rightarrow (\mathbf{P} \rightarrow (\ldots \rightarrow (\mathbf{P} \rightarrow \mathbf{Q}) \ldots)$ is a theorem for some finite number of antecedent **P**'s.

Proof: Left as an exercise.

Also like Fuzzy$_L$, no axiomatic system can be *strongly* complete for Fuzzy$_P$. An example establishing this negative fact can be constructed from the example we used for Fuzzy$_L$: Let Σ^* consist of the formulas $\neg_P\neg_P R$, $(P \vee_P R) \rightarrow_P R) \rightarrow_P (Q \vee_P R)$, and the infinitely many formulas in the series

$(((P \vee_P R) \rightarrow_P R) \rightarrow_P (P \vee_P R)) \rightarrow_P (Q \vee_P R)$
$(((P \vee_P R) \rightarrow_P R) \rightarrow_P (((P \vee_P R) \rightarrow_P R) \rightarrow_P (P \vee_P R))) \rightarrow_P (Q \vee_P R)$
$(((P \vee_P R) \rightarrow_P R) \rightarrow_P (((P \vee_P R) \rightarrow_P R) \rightarrow_P (((P \vee_P R) \rightarrow_P R) \rightarrow_P (P \vee_P R)))) \rightarrow_P$
$(Q \vee_P R)$
$(((P \vee_P R) \rightarrow_P R) \rightarrow_P (((P \vee_P R) \rightarrow_P R) \rightarrow_P$
$(((P \vee_P R) \rightarrow_P R) \rightarrow_P (((P \vee_P R) \rightarrow_P R) \rightarrow_P (P \vee_P R))))) \rightarrow_P (Q \vee_P R)$
...

Σ^* thus consists of $\neg\neg R$ and each formula in Σ modified as follows: first $\neg P$ is replaced with $P \rightarrow_P R$, and then P and Q, wherever they occur, are, respectively, replaced with $P \vee_P R$ and $Q \vee_P R$.[13] The infinite set Σ^*, but none of its finite subsets, semantically entails the formula $Q \vee_P R$ in Fuzzy$_P$, so Fuzzy$_P$ is not compact (proof is left as an exercise). It follows, just as it did for Fuzzy$_L$, that no axiomatic derivation system can be strongly complete for Fuzzy$_P$:

Result 13.6: Fuzzy$_P$ is not compact

and therefore

Result 13.7: No axiomatic system for Fuzzy$_P$ can be strongly complete.

Neither is there a fuzzy complete Pavelka-style axiomatic system for Fuzzy$_P$.[14] The reason is the same as for Fuzzy$_G$, namely, Fuzzy$_P$'s conditional is not continuous.

13 The transformation of the formulas in Σ is from Hájek (1998b, p. 94). Hájek uses the transformation to show that various semantic results in Fuzzy$_L$ carry over to Fuzzy$_P$.

14 However, Pavelka-style systems have been developed for extensions of Gödel and product fuzzy logics that include an involutive negation operator such as Łukasiewicz's; see Esteva, Godo, Hájek, and Navara (2000). Recall that an *involutive* negation is one that satisfies Double Negation.

13.6 Summary: Comparison of $\text{Fuzzy}_{Ł}$, Fuzzy_{G}, and Fuzzy_{P} and Their Derivation Systems

In this section we place various results about the three major fuzzy propositional logics in a table, for overall comparison:

	Semantically compact	*Deduction theorem*	*Weakly complete axiomatization*	*Tautologies/ theorems decidable*	*Strongly complete axiomatization*	*Fuzzily complete axiomatization*
$Fuzzy_L$	No	Modified	Yes	Yes	No	Yes
$Fuzzy_G$	Yes	Yes	Yes	Yes	Yes	No
$Fuzzy_P$	No	Modified	Yes	Yes	No	No

(The set of theorems for each of $F_ŁPA$, $BL_ŁA$, BL_GA, and BL_PA is decidable, as a consequence of the decidability of the tautologies of $\text{Fuzzy}_Ł$, Fuzzy_G, and Fuzzy_P as noted in Chapter 11).

Recalling that the strong completeness of an axiomatic system is tied to semantic compactness, we can characterize the major differences as follows: Fuzzy_G is the only one of the three systems that is semantically compact and in which the standard Deduction Theorem holds, while $\text{Fuzzy}_Ł$ is the only one of the three for which an adequate Pavelka-style axiomatization is possible. As a consequence of this latter fact, and the strong fuzzy completeness of $F_ŁPA$, we also have fuzzy compactness for $\text{Fuzzy}_Ł$.

13.7 External Assertion Axioms

Because external assertion is not definable in any of our three fuzzy logics, none of the the axiom systems we've examined is sufficient for derivations involving this operation. Matthias Baaz (1996) formulated the following axioms, which may be added either to $F_ŁA$ (and $F_ŁPA$)[15] or to BLA (and any system on which BLA is based), for external assertion:

Δ1. $\Delta P \vee \neg\Delta P$
Δ2. $\Delta(P \vee Q) \rightarrow (\Delta P \vee \Delta Q)$
Δ3. $\Delta P \rightarrow P$
Δ4. $\Delta P \rightarrow \Delta\Delta P$
Δ5. $\Delta(P \rightarrow Q) \rightarrow (\Delta P \rightarrow \Delta Q)$

[15] But with an important caveat in this case: although $F_ŁPA$ with these axioms and rule will be fuzzy sound for $\text{Fuzzy}_{FŁ}$ augmented with the external assertion operation, it will no longer be (strongly) fuzzy complete. Again, the problem is that external assertion is not a continuous operation.

It may strike the reader as unusual that we will actually produce derivations in the incomplete $F_Ł\Delta PA$—knowing that it is not a fully adequate system. We do so not only to illustrate the external assertion axioms and rules but also to demonstrate the potential usefulness of an axiomatic system that isn't complete. The usefulness stems from the fact that there is no decision procedure for n-tautologousness or n-degree-entailment in $\text{Fuzzy}_Ł$ (the former is proved in Hájek (1995a) but sometimes a derivation might be quite simple to produce.

We also add the rule

EA (External Assertion). From $\mathbf{P}$ infer $\mathbf{\Delta P}$.

We will denote derivation systems augmented with these axiom schemata and the rule EA by adding the symbol Δ to the names of those systems, for instance, $F_L\Delta A$.

In Section 11.9 of Chapter 11 we noted that axiom schema $\Delta 1$ is a tautology in Fuzzy$_L$ augmented with the external assertion operator. By including axiom schema $\Delta 1$ it becomes a theorem of $F_L\Delta A$. The converse of axiom schema $\Delta 3$ is not included as an axiom because it is not a tautology in any of our fuzzy systems. If $\mathbf{P}$ has the value .5, for example, then $\mathbf{P} \rightarrow \mathbf{\Delta P}$ has the value .5 in Fuzzy$_L$ (augmented with Δ) and the value 0 in Fuzzy$_G$ and Fuzzy$_P$.

We also noted in Section 11.9 that if a formula $\mathbf{P}$ has any value other than 1 or 0, then the formula $\neg\mathbf{\Delta P} \wedge \neg\mathbf{\Delta}\neg\mathbf{P}$ is true in Fuzzy$_L$. Here's a derivation in $F_L\Delta PA$ that shows that when the formula P has the value .5, $\neg\Delta P \wedge \neg\Delta\neg P$ has the value 1:

1	$[{}^1\!/_2 \rightarrow P, 1]$	Assumption
2	$[P \rightarrow {}^1\!/_2, 1]$	Assumption
3	$[\Delta P \rightarrow P, 1]$	$\Delta 3$, with P / $\mathbf{P}$
4	$[\Delta P \rightarrow {}^1\!/_2, 1]$	2,3 HS
5	$[\neg{}^1\!/_2 \rightarrow \neg\Delta P, 1]$	4, GCON
6	$[{}^1\!/_2 \rightarrow \neg{}^1\!/_2, 1]$	F_LP6.2
7	$[{}^1\!/_2 \rightarrow \neg\Delta P, 1]$	5,6 HS
8	$[\Delta P \rightarrow \neg\Delta P, 1]$	4,7 HS
9	$[(\Delta P \rightarrow \neg\Delta P) \rightarrow \neg\Delta P, 1]$	$\Delta 1$, with P / $\mathbf{P}$
10	$[\neg\Delta P, 1]$	8,9 MP
11	$[\neg P \rightarrow \neg{}^1\!/_2, 1]$	1, GCON
12	$[\neg{}^1\!/_2 \rightarrow {}^1\!/_2, 1]$	F_LP6.1
13	$[\neg P \rightarrow {}^1\!/_2, 1]$	11,12 HS
14	$[\Delta\neg P \rightarrow \neg P, 1]$	$\Delta 3$, with $\neg$P / $\mathbf{P}$
15	$[\Delta\neg P \rightarrow {}^1\!/_2, 1]$	13,14 HS
16	$[\neg{}^1\!/_2 \rightarrow \neg\Delta\neg P, 1]$	15, GCON
17	$[{}^1\!/_2 \rightarrow \neg\Delta\neg P, 1]$	6,16 HS
18	$[\Delta\neg P \rightarrow \neg\Delta\neg P, 1]$	15,17 HS
19	$[(\Delta\neg P \rightarrow \neg\Delta\neg P) \rightarrow \neg\Delta\neg P, 1]$	$\Delta 1$, with $\neg$P / $\mathbf{P}$
20	$[\neg\Delta\neg P, 1]$	18,19 MP
21	$[\neg\Delta P \wedge \neg\Delta\neg P, 1]$	10,20 WCI

As one more example in $F_L\Delta PA$, we'll show that if the value of P is at most .5, then $\neg\Delta P$ is true. The reasoning is that if P has at most the value .5, then ΔP has at most the value .5 (line 3), and it follows from the disjunctive axiom schema $\Delta 1$ that $\Delta\neg P$ must be true. The latter is established by first deriving the formula on line 9, then showing that it follows from this formula that $\Delta\neg P$ holds. In the derivation we switch freely between the equivalent forms $\mathbf{P} \vee \mathbf{Q}$ and $(\mathbf{P} \rightarrow \mathbf{Q}) \rightarrow \mathbf{Q}$, depending on whether we are viewing the formula as a disjunction or as a conditional:

1	$P \rightarrow {}^1\!/_2$	Assumption
2	$\Delta P \rightarrow P$	$\Delta 3$, with P / $\mathbf{P}$
3	$\Delta P \rightarrow {}^1\!/_2$	1,2 HS
4	${}^1\!/_2 \rightarrow (\neg\Delta P \vee {}^1\!/_2)$	$\mathrm{F_L}$P1, with ${}^1\!/_2$ / $\mathbf{P}$, $\neg\Delta P \rightarrow {}^1\!/_2$ / $\mathbf{Q}$
	{Note: $\neg\Delta P \vee {}^1\!/_2$ is defined to be $(\neg\Delta P \rightarrow {}^1\!/_2) \rightarrow {}^1\!/_2$}	
5	$\Delta P \rightarrow (\neg\Delta P \vee {}^1\!/_2)$	3,4 HS
6	$\neg\Delta P \rightarrow (\neg\Delta P \vee {}^1\!/_2)$	$\mathrm{F_L}$PD3, with $\neg\Delta P$ / $\mathbf{P}$, ${}^1\!/_2$ / $\mathbf{Q}$
7	$(\Delta P \vee \neg\Delta P) \rightarrow (\neg\Delta P \vee {}^1\!/_2)$	5,6 DC
8	$\Delta P \vee \neg\Delta P$	$\Delta 1$, with P / $\mathbf{P}$
9	$(\neg\Delta P \rightarrow {}^1\!/_2) \rightarrow {}^1\!/_2$	7,8 MP
10	$((\neg\Delta P \rightarrow {}^1\!/_2) \rightarrow {}^1\!/_2) \rightarrow (({}^1\!/_2 \rightarrow \neg\Delta P) \rightarrow \neg\Delta P))$	$\mathrm{F_L}$P4, with $\neg\Delta P$ / $\mathbf{P}$, ${}^1\!/_2$ / $\mathbf{Q}$
11	$(({}^1\!/_2 \rightarrow \neg\Delta P) \rightarrow \neg\Delta P))$	9,10 MP
12	$\neg{}^1\!/_2 \rightarrow \neg\Delta P$	3, GCON
13	${}^1\!/_2 \rightarrow \neg{}^1\!/_2$	$\mathrm{F_L}$P6.2
14	${}^1\!/_2 \rightarrow \neg\Delta P$	12,13 HS
15	$\neg\Delta P$	11,14 MP

13.8 Exercises

SECTION 13.1

1 Show that the following are derivable as rules in the axiomatic system $\mathrm{F_L}$A for $\mathrm{Fuzzy_L}$ (you will probably find it useful to refer to proofs in Chapter 6 as a guide, but recall that $\mathrm{L_3}$4 is not an axiom schema in $\mathrm{F_L}$A):

a. **CON**. From $\neg\mathbf{P} \rightarrow \neg\mathbf{Q}$ infer $\mathbf{Q} \rightarrow \mathbf{P}$.

b. **MT**. From $\neg\mathbf{P}$ and $\mathbf{Q} \rightarrow \mathbf{P}$ derive $\neg\mathbf{Q}$.

c. **LSIMP**. From $\mathbf{P} \wedge \mathbf{Q}$ infer $\mathbf{P}$.

d. **RSIMP**. From $\mathbf{P} \wedge \mathbf{Q}$ infer $\mathbf{Q}$.

e. **SUB**. From $\mathbf{P} \rightarrow \mathbf{Q}$, $\mathbf{Q} \rightarrow \mathbf{P}$ and a formula **R** that contains **P** as a subformula, infer any formula **R*** that is the result of replacing one or more occurrences of **P** in **R** with **Q**.

f. **DN**. From any formula **R** that contains **P** as a subformula, infer any formula **R*** that is the result of replacing one or more occurrences of **P** in **R** with $\neg\neg\mathbf{P}$, and vice versa.

g. **TRAN**. From any formula **R** that contains $\mathbf{P} \rightarrow (\mathbf{Q} \rightarrow \mathbf{S})$ as a subformula, infer any formula **R*** that is the result of replacing one or more occurrences of $\mathbf{P} \rightarrow (\mathbf{Q} \rightarrow \mathbf{S})$ in **R** with $\mathbf{Q} \rightarrow (\mathbf{P} \rightarrow \mathbf{S})$.

h. **GCON**. From any formula **R** that contains $\mathbf{P} \rightarrow \mathbf{Q}$ as a subformula, infer any formula **R*** that is the result of replacing one or more occurrences of $\mathbf{P} \rightarrow \mathbf{Q}$ in **R** with $\neg\mathbf{Q} \rightarrow \neg\mathbf{P}$, and vice versa.

i. **GHS**. From a conditional $(\mathbf{P}_1 \rightarrow (\mathbf{P}_2 \rightarrow (\mathbf{P}_3 \rightarrow \ldots (\mathbf{P}_{n-1} \rightarrow \mathbf{P}_n) \ldots)$ and $\mathbf{P}_n \rightarrow \mathbf{Q}$, infer $(\mathbf{P}_1 \rightarrow (\mathbf{P}_2 \rightarrow (\mathbf{P}_3 \rightarrow \ldots (\mathbf{P}_{n-1} \rightarrow \mathbf{Q}) \ldots)$.

j. **GMP**. From a conditional $(\mathbf{P}_1 \rightarrow (\mathbf{P}_2 \rightarrow (\mathbf{P}_3 \rightarrow \ldots (\mathbf{P}_{n-1} \rightarrow \mathbf{P}_n) \ldots)$ and one of the antecedents $\mathbf{P}_i$, $1 \leq i \leq n-1$, infer the conditional that results from deleting $\mathbf{P}_i$, the conditional arrow following $\mathbf{P}_i$, and associated parentheses

k. **DS**. From $\mathbf{P} \vee \mathbf{Q}$, $\mathbf{P} \rightarrow \mathbf{R}$ and $\mathbf{Q} \rightarrow \mathbf{R}$, infer $\mathbf{R}$.

2 Explain why we do *not* want the following as a derived rule in F$_{Ł}$A (although it is derivable in Ł$_3$A):

From $\mathbf{P} \rightarrow \mathbf{Q}$ and $(\mathbf{P} \rightarrow \neg\mathbf{P}) \rightarrow \mathbf{Q}$, infer $\mathbf{Q}$.

3 Show that the following are derivable as axiom schemata in F$_{Ł}$A (similar hint to that in Exercise 1):

a. **F$_{Ł}$D10.** $((\mathbf{P} \rightarrow \mathbf{P}) \rightarrow \mathbf{Q}) \rightarrow \mathbf{Q}$

b. **F$_{Ł}$D11.** $\neg(\mathbf{P} \rightarrow \mathbf{Q}) \rightarrow \mathbf{P}$

c. **F$_{Ł}$D12.** $\neg(\mathbf{P} \rightarrow \mathbf{Q}) \rightarrow \neg\mathbf{Q}$

d. **F$_{Ł}$D13.** $(\mathbf{P} \rightarrow \mathbf{Q}) \vee (\mathbf{Q} \rightarrow \mathbf{P})$

4 Show that the following are derivable as rules in F$_{Ł}$A:

a. **L&SIMP**. From **P & Q**, infer **P**.

b. **R&SIMP**. From **P & Q**, infer **Q**.

c. **BCF (Bold Conjunction Formation)**. From **P** and **Q**, infer **P & Q**.

5 Show that the following are derivable axiom schemata in F$_{Ł}$A:

a. **F$_{Ł}$D14.** $\mathbf{P} \rightarrow (\mathbf{P} \triangledown \mathbf{Q})$

b. **F$_{Ł}$D15.** $\mathbf{Q} \rightarrow (\mathbf{P} \triangledown \mathbf{Q})$

c. **F$_{Ł}$D16.** $(\mathbf{P} \vee \mathbf{Q}) \rightarrow (\mathbf{P} \triangledown \mathbf{Q})$

d. **F$_{Ł}$D17.** $(\mathbf{P}\ \&\ \mathbf{Q}) \rightarrow (\mathbf{P} \wedge \mathbf{Q})$

6 Prove that the following claim is true in Fuzzy$_{Ł}$: the value of the antecedent of the *n*th member of the infinite series of formulas $(\neg P \rightarrow P) \rightarrow Q$, $(\neg P \rightarrow (\neg P \rightarrow P)) \rightarrow Q$, $(\neg P \rightarrow (\neg P \rightarrow (\neg P \rightarrow P))) \rightarrow Q$, $(\neg P \rightarrow (\neg P \rightarrow (\neg P \rightarrow (\neg P \rightarrow P)))) \rightarrow Q, \ldots$, is the minimum of 1 and $(n+1)$ times the value of *P*. *Hint*: Show that the value of the antecedent of each formula after the first one is a particular function of the value of the antecedent of the previous formula.

7 a. Show that *R* is semantically entailed by $P \wedge ((P \rightarrow Q) \wedge (P \rightarrow (Q \rightarrow R)))$ in Fuzzy$_{Ł}$.

b. Show that R is derivable from $P \wedge ((P \rightarrow Q) \wedge (P \rightarrow (Q \rightarrow R)))$ in the axiomatic system F$_{Ł}$A.

c. Show that $(P \wedge ((P \rightarrow Q) \wedge (P \rightarrow (Q \rightarrow R)))) \rightarrow R$ is not a tautology of Fuzzy$_{Ł}$.

SECTION 13.2

8 Prove that

a. if *m* and *n* are rational values in the unit interval so is max $(1, 1-m+n)$.

b. if *m* is a rational value in the unit interval then so is $1-m$.

9 a. Show in $F_{Ł}PA$ that the instance [**1**, 1] of $F_{Ł}P7$ is derivable from the other axioms.

b. Show that all instances of $F_{Ł}P7$ are derivable from the other axioms in $F_{Ł}PA$.

10 Derive $[A \vee \neg A, 1/2]$ in $F_{Ł}PA$. (*Hint*: Use the corresponding derivation in Chapter 6 as a guide. Before using any derived rules or axioms from Chapter 6, be sure to establish that they are also derivable in $F_{Ł}PA$.)

11 Show that the rule

HS. From [**P** → **Q**, *m*] and [**Q** → **R**, *n*] infer [**P** → **R**, *p*]
where $p = \max(0, m + n - 1)$

is derivable in $F_{Ł}PA$.

12 Derive graded versions of the following rules for $F_{Ł}PA$ (where the graded value of the inferred formula should be the least value that it can have given the least values of the formulas from which it is derived):

a. TRAN

b. GHS

c. CON

d. MT

13 Show that the following are derivable as rules in $F_{Ł}PA$:

a. **BDF (Bold Disjunction Formation)**. From [**P**, *m*] and [**Q**, *n*] infer [**P** ∇ **Q**, *p*]
where $p = \min(1, m + n)$

b. **WCF (Weak Conjunction Formation)**. From [**P**, *m*] and [**Q**, *n*] infer [**P** ∧ **Q**, *p*] where $p = \min(m, n)$

c. **WCI (Weak Conjunction Inference)**. From [**P** ∧ **Q**, *n*] infer either of [**P**, *n*] or [**Q**, *n*]

d. **DS (Disjunctive Syllogism)**. From [**P** ∨ **Q**, *m*] and [¬**P**, *n*] infer [**Q**, *p*] and from [**P** ∨ **Q**, *m*] and [¬**Q**, *n*] infer [**P**, *p*]
where $p = \max(0, m + n - 1)$

14 Show that the rule

VS (Value Summary). From [**m** → ¬**n**, **1**], [**p** → (**q** →**m**), **1**] and [¬**r** →**p**, **1**], infer [¬(**q** → ¬**n**) →**r**, **1**]

is derivable in $F_{Ł}PA$.

15 Consider the derived rule

DC. From [**P** → **R**, 1] and [**Q** → **R**, 1] infer [(**P** ∨ **Q**) → **R**, 1].

a. Modify the derivation of this rule in Section 13.2 so that it begins with graded formulas [**P** → **R**, *m*] and [**Q** → **R**, *n*], showing the correct graded values for each subsequent formula in the derivation.

b. The graded value obtained for (**P** ∨ **Q**) → **R** in the derivation in part (a) is too weak. Prove this by showing (semantically) that the least graded value that (**P** ∨ **Q**) → **R** can have is min (*m*, *n*) and then giving an example of specific values *m* and *n* such that the graded value for (**P** ∨ **Q**) → **R** in your derivation in part (a) is *less* than min (*m*, *n*).

c. Produce a justification for the fully graded rule

FDC. From [**P** → **R**, *m*] and [**Q** → **R**, *n*] infer [(**P** ∨ **Q**) → **R**, min (*m*, *n*)].

Hint: Look at the derivation of MCD for L_3PA in Section 6.2 of Chapter 6 for an idea about how to modify your derivation in part (a).

16 Produce a derivation of *[D, .7]* from *[A →* **.5**, *1]*, *[B →* **.5**, *1]*, *[(A → B) →* **.9**, *1]*, *[C → B, 1]*, *[D →* **.4**, *1]*, and *[(C → ¬D) →* **.9**, *1]* in $F_Ł$PA.

17 Explain why the greatest value that *m* can be when the formula (**n** → **m**) → $^9/_{10}$ is true is $^1/_{10}$ less than *n* (the rational value of the formula **n**).

18 In Section 13.3 we claimed that corresponding to the entailment in $RFuzzy_Ł$ of *Q* from the set consisting of the formula *¬P→ Q* and the infinitely many formulas in the series *(¬P→ P)→ Q*, *(¬P→ (¬P→ P))→ Q*, *(¬P→ (¬P→ (¬P→ P)))→ Q*, *(¬P → (¬P → (¬P → (¬P → P)))) → Q*, ... there are *n*-entailments of *Q* from finite subsets of Γ in $RFuzzy_Ł$ with *n* as close to 1 as you can be without actually getting there. Produce a series of entailments that support this claim.

SECTION 13.3

19 a. Prove that for any x and y in a BL-algebra, x ⊗ y ≤ x.

b. Prove that for any x, y and z in a BL-algebra, x ≤ z and y ≤ z if and only if x ∪ y ≤ z.

c. Prove that for any x, y, z, and w in a BL-algebra, if x ≤ z and y ≤ w, then x ∪ y ≤ z ∪ w.

d. Prove that for any x, y and z in a BL-algebra, x ⊗ (y ∪ z) = (x ⊗ y) ∪ (x ⊗ z).

20 Show that each of the following axioms, when interpreted algebraically, evaluates to *unit* in every BL-algebra:

a. BL3

b. BL4

c. BL6

d. BL8.

21 Show that the algebraic interpretation of Modus Ponens,

If *unit* ≤ x and *unit* ≤ x ⇒ y, then *unit* ≤ y,

is true in every BL-algebra.

22 Show that the algebraic interpretation of the axiom schema

$BL_Ł9$. ¬¬P → P

holds true in every MV-algebra.

SECTION 13.4

23 Show that the algebraic interpretation of the axiom schema

BL_G9. P → (P & P)

is true in every Gödel algebra.

24 Prove that every set that contains the formula $\neg_G P \rightarrow_G Q$ and at least one formula from the infinite series

$(\neg_G P \rightarrow_G P) \rightarrow_G Q$

$(\neg_G P \rightarrow_G (\neg_G P \rightarrow_G P)) \rightarrow_G Q$

$(\neg_G P \rightarrow_G (\neg_G P \rightarrow_G (\neg_G P \rightarrow_G P))) \rightarrow_G Q$

$(\neg_G P \rightarrow_G (\neg_G P \rightarrow (\neg_G P \rightarrow_G (\neg_G P \rightarrow_G P)))) \rightarrow_G Q$

...

entails *Q* in $Fuzzy_G$.

25 a. Derive Q from $\neg_G P \rightarrow_G Q$ and $(\neg_G P \rightarrow_G P) \rightarrow_G Q$ in BL_{PA}.

b. Derive Q from $\neg_G P \rightarrow_G Q$ and $(\neg_G P \rightarrow_G (\neg_G P \rightarrow_G (\neg_G P \rightarrow_G P))) \rightarrow_G Q$ in BL_{PA}.

26 Prove Result 11.4, the Deduction Theorem for BL_GA. *Hint:* Review the proof in Chapter 2 of the Deduction Theorem for classical propositional logic (Result 2.5, Section 2.4).

27 Produce an example that shows that the $Fuzzy_G$ conditional is not continuous.

SECTION 13.5

28 Show that the algebraic interpretations of the axiom schemata

$\mathbf{BL_P9.}\ \neg_P\neg_P \mathbf{P} \rightarrow_P (((\mathbf{Q}\ \&_P\ \mathbf{P}) \rightarrow_P (\mathbf{R}\ \&_P\ \mathbf{P})) \rightarrow_P (\mathbf{Q} \rightarrow_P \mathbf{R}))$

and

$\mathbf{BL_P10.}\ (\mathbf{P}\ \&_P \neg_P \mathbf{P}) \rightarrow_P \mathbf{0}$

are true in every product algebra.

29 Derive the formula $\neg_P(P\ \&_P \neg_P P)$ (without any assumptions) in BL_PA.

30 Show that the set Σ^* consisting of the formulas $\neg_P\neg_P R$, $(P \vee_P R) \rightarrow_P R) \rightarrow_P (Q \vee_P R)$, and the infinitely many formulas in the series

$(((P \vee_P R) \rightarrow_P R) \rightarrow_P (P \vee_P R)) \rightarrow_P (Q \vee_P R)$

$(((P \vee_P R) \rightarrow_P R) \rightarrow_P (((P \vee_P R) \rightarrow_P R) \rightarrow_P (P \vee_P R))) \rightarrow_P (Q \vee_P R)$

$(((P \vee_P R) \rightarrow_P R) \rightarrow_P (((P \vee_P R) \rightarrow_P R) \rightarrow_P (((P \vee_P R) \rightarrow_P R) \rightarrow_P (P \vee_P R)))) \rightarrow_P (Q \vee_P R)$

$(((P \vee_P R) \rightarrow_P R) \rightarrow_P (((P \vee_P R) \rightarrow_P R) \rightarrow_P (((P \vee_P R) \rightarrow_P R) \rightarrow_P (((P \vee_P R) \rightarrow_P R) \rightarrow_P (P \vee_P R))))) \rightarrow_P (Q \vee_P R)$

...

semantically entails the formula $Q \vee_P R$ in $Fuzzy_P$ but that none of its finite subsets does.

To do this:

a. Assume that each of the formulas in Σ^* is true, and for each of the conditionals in the set explain what the value of R must be in order for the conditional's antecedent to be true.

b. Noting that the value of R cannot be 0, because of the inclusion of $\neg_P\neg_P R$ in Σ^*, explain why the antecedents of one of the conditionals in the set must be true—for it will follow that the consequent of that conditional, $Q \vee_P R$, will also be true and hence the entailment from the set Σ^* holds.

c. For each formula **S** in Σ^*, show that if **S** is excluded from a finite subset Ψ of Σ^* then Ψ does not entail $Q \vee_P R$; that is, all of the formulas in Ψ can be true while $Q \vee_P R$ is not.

31 Prove Result 13.5, the Modified Deduction Theorem for BL_PA. *Hint*: Review the proof of Result 13.3 in Section 13.1.

32 Produce an example that shows that the $Fuzzy_P$ conditional is not continuous.

SECTION 13.7

33 Produce a derivation in $F_{L}\Delta PA$ that shows that if the formula P has the value .3 in $Fuzzy_{L}$, then $\neg\Delta P \wedge \neg\Delta\neg P$ has the value 1.

34 Produce a derivation in $F_{L}\Delta PA$ that shows that if the formula P has the value .5 in $Fuzzy_{L}$, then $P \rightarrow \Delta P$ also has the value .5.

35 Produce a derivation that shows that the formula $\Delta P \vee \Delta\neg\neg P$ is a theorem in $BL_{G}\Delta A$.

36 Produce a derivation that shows that the formula $(\Delta P \& \Delta\neg P) \rightarrow \Delta\mathbf{0}$ is a theorem in $BL_{P}\Delta A$.

14 Fuzzy First-Order Logics: Semantics

14.1 Fuzzy Interpretations

Interpretations for fuzzy first-order logic use fuzzy sets to represent the meanings of vague predicates. Recall from Chapter 11 that a fuzzy set is a set to which entities belong to certain degrees, ranging from 1 (absolutely a member) to 0 (definitely not a member). An interpretation will represent each fuzzy set as a function mapping n-tuples of members of the domain to their degrees of membership in the set.

An ***interpretation I*** for fuzzy first-order logic consists of

1. A nonempty set D (the *domain*)
2. An assignment to each predicate $\mathbf{P}^n$ (of arity n) of a function mapping each n-tuple of members of D to a value in [0. .1]:

   $\mathrm{I}(\mathbf{P}^n)(<\mathrm{x}_1, \ldots, \mathrm{x}_n>) \in [0..1]$
3. An assignment of a member of D to each individual constant **a**: I(**a**) ∈ D.

Here's an example of a (partial) interpretation (where we use *T* for *tall* and *E* for *is one-eighth inch less than*):

D: set of heights between 4′ 7″ and 6′ 7″ by $^1/_8$″ increments, inclusive

$\mathrm{I(T)}(<\mathrm{x}>) = (\mathrm{x} - 4'\,7'')\ /\ 24''$

(i.e., subtract 4′ 7″ from the height x and divide the result by 24″)

$\mathrm{I(E)}(<\mathrm{x}_1,\mathrm{x}_2>) = 1$ if x_1 is $^1/_8$″ less than x_2

0 otherwise

I(a) = 6′ 7″

I(b) = 6′ 6⅞″

I(c) = 5′ 8″

I(d) = 5′ 7″

I(e) = 5′ 6⅞″

I(f) = 5′ 1″

I(g) = 4′ 7″

and here are some values that this interpretation produces for the fuzzy set of heights: $\mathrm{I}(T)(<6'\ 7''>) = {}^{24}/_{24}$ (i.e., 1); $\mathrm{I}(T)(<6'\ 6^7/_8''>) = {}^{191}/_{192}$ (approx .995);

$I(T)(<5'\,8''>) = {}^{104}/_{192}$ (approx. .54); $I(T)(<5'\,7''>) = {}^{12}/_{24}$ (.5); and $I(T)(<4'\,7''>) = {}^{0}/_{24}$ (i.e., 0). We'll refer to this interpretation as *SST* (for ***six foot seven is tall***).

The specific function assigned to *T* was chosen to produce a reasonable representation of degrees of membership in the fuzzy set of tall heights, but there certainly are other plausible ways to define this function. However, fuzzy logical systems are defined independently of specific membership functions, so we'll wait until Chapter 17 to discuss other approaches to defining membership functions. The sample function that we present for SST—which will be sufficient for investigating how fuzzy first-order logic handles vagueness—is merely meant to illustrate how we can define membership functions for fuzzy predicates like *tall*. SST also demonstrates how to interpret crisp predicates by assigning only 1 and 0 as membership values.

14.2 Łukasiewicz Fuzzy First-Order Logic

We begin our study of fuzzy first-order logics with a first order version of Fuzzy_L, which we'll call $\text{Fuzzy}_{\text{L}\forall}$.

To specify truth-conditions in $\text{Fuzzy}_{\text{L}\forall}$ we'll once again need variable assignments: a **variable assignment v** assigns a member of the domain to each individual variable **x**, and an **x-variant** of a variable assignment v is an assignment v′ such that v′(**y**) = v(**y**) for every variable **y** other than **x** (it may or may not assign the same value to **x**). As in earlier chapters we use the notation $I_v(\mathbf{P})$ to stand for the value that a formula **P** has under a variable assignment v on an interpretation I.

Truth-conditions for atomic formulas are defined in $\text{Fuzzy}_{\text{L}\forall}$ as:

1. $I_v(\mathbf{Pt}_1 \ldots \mathbf{t}_n) = I(\mathbf{P})(<I^*(\mathbf{t}_1), \ldots, I^*(\mathbf{t}_n)>)$, where $I^*(\mathbf{t}_i)$ is $I(\mathbf{t}_i)$ if $\mathbf{t}_i$ is a constant and is $v(\mathbf{t}_i)$ if $\mathbf{t}_i$ is a variable.

That is, the truth-value of an atomic formula is just the degree to which the *n*-tuple of entities denoted by the formula's terms is a member of the fuzzy set corresponding to the predicate. So, on interpretation SST (given any variable assignment) we have the following values (rounded to the nearest thousandth, a practice we'll adopt in examples to follow as well):

$I_v(Ta) = 1$
$I_v(Tb) = .995$
$I_v(Tc) = .542$
$I_v(Td) = .5$
$I_v(Te) = .495$
$I_v(Tf) = .167$
$I_v(Tg) = 0$
$I_v(Eaa) = 0$
$I_v(Eab) = 0$
$I_v(Eba) = 1$
$I_v(Ebb) = 0$

The truth-conditions for formulas governed by propositional connectives are defined as in Fuzzy$_Ł$:

2. $I_v(\neg \mathbf{P}) = 1 - I_v(\mathbf{P})$
3. $I_v(\mathbf{P} \wedge \mathbf{Q}) = \min\ (I_v(\mathbf{P}), I_v(\mathbf{Q}))$
4. $I_v(\mathbf{P} \vee \mathbf{Q}) = \max\ (I_v(\mathbf{P}), I_v(\mathbf{Q}))$
5. $I_v(\mathbf{P} \rightarrow \mathbf{Q}) = \min\ (1, 1 - I_v(\mathbf{P}) + I_v(\mathbf{Q}))$
6. $I_v(\mathbf{P} \leftrightarrow \mathbf{Q}) = \min\ (1, 1 - I_v\ (\mathbf{P}) + I_v(\mathbf{Q}), 1 - I_v(\mathbf{Q}) + I_v(\mathbf{P}))$
7. $I_v(\mathbf{P}\ \&\ \mathbf{Q}) = \max\ (0, I_v(\mathbf{P}) + I_v(\mathbf{Q}) - 1)$
8. $I_v(\mathbf{P} \nabla \mathbf{Q}) = \min\ (1, I_v(\mathbf{P}) + I_v(\mathbf{Q}))$

These clauses give the following fuzzy truth-values based on the interpretation SST:

Formula	*Value*
¬Ta	0
¬Tc	.458
¬Tg	1
Ta ∧ Tb	.995
Ta ∧ Td	.5
Ta ∧ Tg	0
Ta ∨ Tb	1
Tb ∨ Tc	.995
Tb ∨ Tf	.995
Ta → Tb	.995
Td → Te	.995
Tb → Tg	.005
Tc ↔ Tf	.625
Ta & Tb	.995
Tb & Tc	.537
Td & Te	0
Ta ∇ Tb	1
Tb ∇ Tf	1
Te ∇ Tf	.662

In Chapter 9 we used the *min* and *max* functions to define the numeric truth-conditions for quantified formulas of $Ł_3$, on the basis of the idea that universal quantification is like conjunction (the min function) and existential quantification is like disjunction (the max function). However, we can't use the min and max functions to interpret the quantifiers in Fuzzy$_{Ł\forall}$. Suppose, for example, that we interpret a fuzzy predicate *F* over the domain [0. .1] (a legitimate domain; we may wish to talk about the real numbers in the unit interval) as follows:

$$I(F)(\langle x \rangle) = \begin{cases} 1 \text{ if } x = 0 \\ {}^{x}/_{2} \text{ otherwise} \end{cases}$$

Now, what is the minimum value that *Fx* can have? There isn't one—it gets as close to 0 as you like, but never quite there. So once again we'll use the greatest lower bound (*glb*) of a set of real numbers from the unit interval; recall that a number *n* in [0. .1] is the greatest lower bound of a set R of real numbers in [0. .1] if and only if *n* is less than or equal to each member of R, and there is no *m* in [0. .1] such that *m* is also a lower bound of R but *m* is greater than *n*. For a similar reason, we'll use the dual concept of *least upper bound* (*lub*) instead of the maximum function, where a number *n* in [0. .1] is the least upper bound of a set R of real numbers in [0. .1] if and only if *n* is greater than or equal to each member of R, and there is no *m* in [0. .1] such that *m* is also an upper bound of R but *m* is less than *n*.

Here are the truth-conditions for quantified formulas in $\text{Fuzzy}_{Ł\forall}$:

9. $I_v((\forall \mathbf{x})\mathbf{P}) = \text{glb}\{I_{v'}(\mathbf{P})\text{: v}' \text{ is an } \mathbf{x}\text{-variant of v}\}$
10. $I_v((\exists \mathbf{x})\mathbf{P}) = \text{lub}\{I_{v'}(\mathbf{P})\text{: v}' \text{ is an } \mathbf{x}\text{-variant of v}\}$

The notation *glb* $\{I_{v'}(\mathbf{P})$*: v′ is an* **x***-variant of v*$\}$ stands for *the greatest lower bound of the values that the formula* **P** *can have for any* **x***-variant of v*; similarly for *lub* $\{I_{v'}(\mathbf{P})$*: v′ is an* **x***-variant of v*$\}$. Given these clauses and the interpretation

$$I(F)(<x>) = \begin{cases} 1 \text{ if } x = 0 \\ {}^{x}/_{2} \text{ otherwise} \end{cases}$$

of the predicate *F*, the formula *(∀x)Fx* has the value 0, since that is the greatest lower bound of the values that *Fx* can have, while the formula *(∃x)Fx* has the value 1.

Finally, the value assigned to a closed formula (a formula with no free variables) on an interpretation is the value that is assigned to that formula under every variable assignment on that interpretation. As in classical and three-valued first-order logic, the value assigned to a closed formula will be the same with respect to every variable assignment on a given interpretation.

We'll look at a few quantified formulas using interpretation SST. The formula *(∀x)Tx* has the value 0 on SST. Every variable assignment has an *x*-variant that assigns 4′ 7″ to *x*, and *Tx* has the value 0 on such a variant. Thus 0 is the greatest lower bound of the values that *Tx* can receive on any assignment's *x*-variants, so every variable assignment v assigns the value 0 to the quantified formula *(∀x)Tx*. On the other hand, the formula *(∃x)Tx* has the value 1 on SST. Every variable assignment has an *x*-variant that assigns 6′ 7″ to *x*, and on this variant the formula *Tx* has the value 1. So 1 is the least upper bound of the values that *Tx* can have on any *x*-variant, and this means that *(∃x)Tx* is assigned the value 1 by every variable assignment.

Now consider the weak disjunction version of the Law of Excluded Middle formula, *(∀x)(Tx ∨ ¬Tx)*. The value of *(∀x)(Tx ∨ ¬Tx)* on SST is the greatest lower bound of the values that *Tx ∨ ¬Tx* can receive on any variable assignment, which is .5. If a variable assignment v assigns 5′ 7″ to *x* then $\max(I_v(Tx), 1 - I_v(Tx)) = \max(.5, .5) = .5$, and this is the smallest value we can get for the disjunction—when we assign a taller height to *x* the value of the first disjunct increases, and when we assign a shorter height to x the value of the second disjunct increases. Thus the formula $(\forall x)(Tx \vee \neg Tx)$ has the value .5 on interpretation SST.

On the other hand, the version of the Law of Excluded Middle using bold disjunction—$(\forall x)(Tx \nabla \neg Tx)$—is a tautology in $\text{Fuzzy}_{\text{L}\forall}$; it has the value 1 on interpretation SST as well as on every other interpretation. Again, we need to ask (since the formula is universally quantified) how small the value of $Tx \nabla \neg Tx$ can get for any variable assignment. The value $I_v(Tx \nabla \neg Tx)$ is min $(1, I_v(Tx) + (1 - I_v(Tx)))$, which is min (1, 1), or 1, in every case. The universal generalization of the formula must therefore always have the value 1 as well.

The quantifiers are interdefinable in $\text{Fuzzy}_{\text{L}\forall}$ exactly as they are in the classical and three-valued first-order systems: $(\forall \mathbf{x})\mathbf{P}$ is equivalent to $\neg(\exists \mathbf{x})\neg \mathbf{P}$ for any formula $\mathbf{P}$, and $(\exists \mathbf{x})\mathbf{P}$ is equivalent to $\neg(\forall \mathbf{x})\neg \mathbf{P}$. We can show the former by verifying that the two formulas will have the same value for any variable assignment and interpretation: $I_v(\neg(\exists \mathbf{x})\neg \mathbf{P}) = 1 - I_v((\exists \mathbf{x})\neg \mathbf{P}) = 1 - \text{lub}\{I_{v'}(\neg \mathbf{P})\text{: v}' \text{ is an } \mathbf{x}\text{-variant of v}\} = 1 - \text{lub}\{1 - I_{v'}(\mathbf{P})\text{: v}' \text{ is an } \mathbf{x}\text{-variant of v}\}$. Now, the least upper bound of the values you get by subtracting a value of $\mathbf{P}$ from 1 is 1 minus the greatest lower bound of the values that $\mathbf{P}$ can have, so $1 - \text{lub}\{1 - I_{v'}(\mathbf{P})\text{: v}' \text{ is an } \mathbf{x}\text{-variant of v}\} = 1 - (1 - \text{glb}\{I_{v'}(\mathbf{P})\text{: v}' \text{ is an } \mathbf{x}\text{-variant of v}\}) = \text{glb}\{I_{v'}(\mathbf{P})\text{: v}' \text{ is an } \mathbf{x}\text{-variant of v}\} = I_v((\forall \mathbf{x})\mathbf{P})$. The equivalence of $(\exists \mathbf{x})\mathbf{P}$ and $\neg(\forall \mathbf{x})\neg \mathbf{P}$ is established similarly.

The fuzzy versions of the K^S_3 conditional and biconditional are definable in $\text{Fuzzy}_{\text{L}\forall}$ in the now familiar way:

$$I_v(\mathbf{P} \rightarrow_K \mathbf{Q}) = \max(1 - I_v(\mathbf{P}), I_v(\mathbf{Q}))$$
$$I_v(\mathbf{P} \leftrightarrow_K \mathbf{Q}) = \min(\max(1 - I_v(\mathbf{P}), I_v(\mathbf{Q})), \max(1 - I_v(\mathbf{Q}), I_v(\mathbf{P})))$$

Not surprisingly, the universally quantified formula $(\forall x)(Tx \rightarrow_K Tx)$ fails to be a tautology in $\text{Fuzzy}_{\text{L}\forall}$. In particular, it fails to be true on interpretation SST because the formula $Tx \rightarrow_K Tx$ will have the value .5 on any variable assignment that assigns 5′ 7″ to x. Indeed, because $Tx \rightarrow_K Tx$ is equivalent to $Tx \vee \neg Tx$ we know that .5 is the lowest value that any variable assignment will give to $Tx \rightarrow_K Tx$ on SST. So the formula $(\forall x)(Tx \rightarrow_K Tx)$ has the value .5 on this interpretation. On the other hand, the existentially quantified $(\exists x)(Tx \rightarrow_K Tx)$ has the value 1 on SST, since the existential quantifier looks at the least upper bound of values that the formula following the quantifier can have. That value, which occurs when either 4′ 7″ or 6′ 7″ is assigned to x, is 1.

14.3 Tautologies and Other Semantic Concepts

The semantic concepts from Chapter 11 are defined for first-order fuzzy logic on the basis of interpretations rather than truth-value assignments. A closed formula of $\text{Fuzzy}_{\text{L}\forall}$ is a **tautology** if it has the value 1 on every interpretation, and a closed formula is a **contradiction** if it has the value 0 on every interpretation. A set Γ of closed formulas **entails** a closed formula $\mathbf{P}$ in $\text{Fuzzy}_{\text{L}\forall}$ if $\mathbf{P}$ has the value 1 on every interpretation on which all of the members of Γ have the value 1, and an argument in $\mathrm{Fuzzy}_{\text{Ł}\forall}$ is **valid** if its conclusion is entailed by the set of its premises.

The major results from Chapter 11 concerning tautologies, contradictions, and entailment in $\mathrm{Fuzzy}_{\text{Ł}}$ carry over, mutatis mutandis,[1] to $\mathrm{Fuzzy}_{\text{Ł}\forall}$:

Result 14.1: Every tautology of $\mathrm{Fuzzy}_{\text{Ł}\forall}$ is a tautology in classical first-order logic, and every contradiction of $\mathrm{Fuzzy}_{\text{Ł}\forall}$ is a contradiction in classical first-order logic, but the converses do not hold.

The propositional connectives of $\mathrm{Fuzzy}_{\text{Ł}\forall}$ are normal, as are its quantifiers, and completely "crisp" fuzzy interpretations (where all predicates have crisp membership functions) are classical interpretations, so $\mathrm{Fuzzy}_{\text{Ł}\forall}$ tautologies must be classical tautologies, and similarly for contradictions. The counterexamples for the negative results in $\mathrm{Fuzzy}_{\text{Ł}}$ work here as well (using formulas of first-order rather than propositional logic). Additionally, Result 14.1 (along with14.2 and 14.3) holds if we use the bold versions of conjunction and negation in place of the weak.

Result 14.2: Every $\mathrm{Fuzzy}_{\text{Ł}\forall}$ entailment is a classical entailment, but the converse does not hold.

Result 14.3: Every tautology of $\mathrm{Fuzzy}_{\text{Ł}\forall}$ is also a tautology in $\text{Ł}_3\text{A}$, and every contradiction of $\mathrm{Fuzzy}_{\text{Ł}\forall}$ is also a contradiction in Ł_3, but the converses do not hold.

We'll put these semantic concepts to work in the next section when we evaluate $\mathrm{Fuzzy}_{\text{Ł}\forall}$'s ability to handle the problems arising from vagueness. Moving on, we define the "fuzzy" semantic concepts for $\mathrm{Fuzzy}_{\text{Ł}\forall}$ as follows: A closed formula of $\mathrm{Fuzzy}_{\text{Ł}\forall}$ is an ***n*-tautology** if *n* is the greatest lower bound of the set of truth-values that the formula can have on any interpretation. A set of closed formulas Γ **degree-entails** a closed formula **P** in $\mathrm{Fuzzy}_{\text{Ł}\forall}$ if on every fuzzy interpretation the value of **P** is greater than or equal to the greatest lower bound of the values of members of Γ on that interpretation, and an argument is **degree-valid** in $\mathrm{Fuzzy}_{\text{Ł}\forall}$ if the set of its premises degree-entail its conclusion. A closed set of formulas Γ ***n*-degree-entails** a closed formula **P** if $1 - n$ is the maximum downward distance between Γ and **P** on any interpretation, and an argument is ***n*-degree-valid** if the set of its premises *n*-degree-entails its conclusion. The relevant results carrying over from Chapter 11 are:

Result 14.4: Every degree-entailment in $\mathrm{Fuzzy}_{\text{Ł}\forall}$ is a classical entailment, but the converse does not hold.

Result 14.5: Every *n*-degree-entailment in $\mathrm{Fuzzy}_{\text{Ł}\forall}$ with $n > 0$ is also a classical entailment (whether we use the weak or the bold connectives as the

[1] That is, with the necessary changes made. The necessary changes involve replacing talk of truth-value assignments with talk of interpretations, replacing talk of maxima and minima with talk of least upper and greatest lower bounds, and so on.

counterparts to the classical ones), but some classical entailments are only n-degree-entailments for very small values of n, including 0.

We'll put these concepts to work in the next section as well.[2]

14.4 Łukasiewicz Fuzzy Logic and the Problems of Vagueness

We introduced fuzzy sets as an answer to the New Problem of the Fringe. Rather than hypothesize three sets corresponding to each vague predicate, we now talk of an infinite number of degrees of membership in one set, without any need for sharp cutoff points to distinguish (or to separate) the extension, fringe, and counterextension. Let us pause to consider a doubt that might linger. We have said that the degree of set membership 1 represents something like *clearly in the set*, 0 represents *clearly not in the set*, and the other values represent somewhere between these two extremes. Doesn't this just reintroduce the extension, counterextension, and fringe with clear cutoff points, where the extension of a predicate includes all those entities that are members of the corresponding fuzzy set to degree 1, the counterextension includes all those entities that are members of the fuzzy set to degree 0, and all the rest constitute the fringe?

Perhaps, the way we have talked so far. So let us consider what else the membership degrees might mean. A degree of membership 1 indicates that a predicate clearly applies, but we can allow that degrees *close* to 1 might also count as clear application. We're not forced to say *where* clear application begins or ends, although we have spoken as if this were clear-cut—so that we're not forced to recognize a precise extension for a fuzzy predicate interpreted as a fuzzy set. Similar comments hold for 0 and degrees of membership close to 0—the latter as well as the former might indicate the clear inapplicability of a predicate. We have a continuum of values here, and we are not forced to interpret them in any particular way, other than saying that 1 (and maybe other values) represents clear applicability and that 0 (and maybe other values) represents clear inapplicability.[3] In fact, in Chapter 16 we will explore the notion of a *fuzzy truth-value*, which formalizes the idea that degrees of truth less than 0 may still count as *true*.

So let's see how $\text{Fuzzy}_{\text{L}\forall}$ deals with vagueness. The Law of Excluded Middle using Łukasiewicz weak disjunction fails to be a tautology in $\text{Fuzzy}_{\text{L}\forall}$, as we saw in Section 14.2, although it cannot be false: it is a .5-tautology. On the other hand,

2 We can also define *fuzzy consequence* for $\text{Fuzzy}_{\text{L}\forall}$, analogous to fuzzy consequence for Fuzzy_{L}. But we don't need the concept in this chapter, so we skip the definition.

3 Dorothy Edgington (1999) has made similar suggestions: "Even fixing the context, it is unclear where clear truth leaves off and something very close to it begins: whether 1 or $1 - \varepsilon$ should be assigned" (p. 298); "There are no exactly correct numbers to assign. . . . The demand for an exact account of a vague phenomenon is unrealistic. The demand for an account which is precise enough to exhibit its important and puzzling features is not" (pp. 308–309). We note that Edgington espouses a "degree of truth" logic that differs from fuzzy logics insofar as her connectives are not generally *degree-functional* (for example, the value of a conjunction is not defined to be a function of the values of its conjuncts).

if we really wish to assert that, well, any person is either tall or not, we can use the bold disjunction version of the Law of Excluded Middle, since that is indeed a tautology in $\text{Fuzzy}_{Ł\forall}$.

Max Black's formula $(\exists x)\ (\neg Tx \wedge \neg\neg Tx)$, asserting the existence of a fringe (and thus the vagueness of a predicate), has the value .5 on interpretation SST—and this is the maximum value that this formula can have on *any* interpretation in $\text{Fuzzy}_{Ł\forall}$. The value of the formula under any variable assignment on any interpretation is min $(\text{I}_\text{V}(\neg Tx), \text{I}_\text{V}(\neg\neg Tx))$, which is min $(1 - \text{I}_\text{V}(Tx), 1 - (1 - \text{I}_\text{V}(Tx))) =$ min $(1 - \text{I}_\text{V}(Tx), \text{I}_\text{V}(Tx))$. The least upper bound of the value of the possible values here is .5, as will be the case when $\text{I}(T)(<\text{I}_\text{V}(x)>) = .5$. Because the formula has the maximum value .5, we cannot use it to assert truly that there are objects in the fringe of the vague predicate *T*.

Recall that in Chapter 7 we used Bochvar's external negation, which is definable in $Ł_3\forall$, to express the existence of borderline cases for the predicate *T* as $(\exists x)(\neg_{BE}Tx \wedge \neg_{BE}\neg Tx)$—this formula is true (and its negation is false) in $Ł_3\forall$ on any interpretation in which *T* has a nonempty fringe. But as we noted in Chapter 11 the fuzzy versions of Bochvar's external connectives are noncontinuous and therefore not definable in $\text{Fuzzy}_Ł$, and the indefinability carries over to $\text{Fuzzy}_{Ł\forall}$. So we must *add* the fuzzy external assertion operator Δ to $\text{Fuzzy}_{Ł\forall}$ in order to express the existence of borderline cases with the formula $(\exists x)(\neg\Delta Tx \wedge \neg\Delta\neg Tx)$. This formula will be true if there is at least one member of the domain whose degree of membership in the fuzzy set *T* is neither 1 nor 0. (Moreover, because the external assertion operator falls between the two negations in the right conjunct, this formula is *not* equivalent to a formula violating the Law of Noncontradiction and thus does not reintroduce the problem that Black wanted to avoid.)

But even without the external assertion operator there is another way to assert the existence of vague predicates in $\text{Fuzzy}_{Ł\forall}$. To show this, we anticipate the extension of $\text{Fuzzy}_{Ł\forall}$ with constant formulas denoting truth-values that will be used to develop a Pavelka-style axiomatic system in Chapter 15. Let us introduce the constant $^1\!/_2$, denoting .5, for our present purposes. We can then say that a formula **P** is true to degree .5 with the formula $\mathbf{P} \leftrightarrow {}^1\!/_2$. This formula has the value 1 when, and only when, **P** has the value .5. The following may then be a reasonable way to state that the predicate *T* is vague: $(\exists x)\ (Tx \leftrightarrow {}^1\!/_2)$. This formula asserts that something is T to degree .5. Or we can introduce the constants $^1\!/_4$ and $^3\!/_4$ and assert that at least one thing is T to a degree that falls in the interval between .25 and .75 inclusive thus: $(\exists x)((^1\!/_4 \rightarrow Tx) \wedge (Tx \rightarrow {}^3\!/_4))$. The formula $^1\!/_4 \rightarrow Tx$ has the value 1 only when the value of *Tx* is at least .25, and the formula $Tx \rightarrow {}^3\!/_4$ has the value 1 only when the value of *Tx* is at most .75.

This leaves the Sorites paradox. Again, here's our formulation of the argument with all premises made explicit:

Ts_1
Es_2s_1
Es_3s_2

Es_4s_3

...

$Es_{193}s_{192}$

$(\forall x)(\forall y)((Tx \wedge Eyx) \rightarrow Ty)$

Ts_{193}

The argument is valid in Fuzzy$_{L\forall}$: Assume that all of the premises have the value 1 on an interpretation I. From the truth of the first two premises, we know that $I(T)(<I(s_1)>) = 1$ and $I(E)(<I(s_2), I(s_1)>) = 1$. Given the truth of the Principle of Charity premise, the conditional formula $(Tx \wedge Eyx) \rightarrow Ty$ must have the value 1 on every variable assignment, and in particular on every assignment v such that $v(x) = I(s_1)$ and $v(y) = I(s_2)$. Since the antecedent of the conditional has the value 1 on this assignment the consequent must have the value 1 as well, and this means that $I(T)(<I(s_2)>) = 1$. Repeating this reasoning we eventually arrive at the conclusion that $I(T)(<I(s_{193})>) = 1$ and so the formula Ts_{193} has the value 1 on any interpretation on which all of the premises have the value 1.

Although the argument is valid, we can give a reasonable interpretation on which the premises are not all true and therefore on which the conclusion need not be true either. We'll add interpretations of the constants in the Sorites argument to the interpretation SST in Section 14.1, giving us the following relevant assignments:

D: set of heights between 4′ 7″ and 6′ 7″ by $^1/_8$″ increments, inclusive

$I(T)(<x>) = (x - 4'\ 7'') \;/\; 24''$

$I(E)(<x_1, x_2>) = 1$ if x_1 is $^1/_8$″ less than x_2

0 otherwise

$I(s_1) = 6'\ 7''$

$I(s_2) = 6'\ 6^7/_8''$

...

$I(s_{193}) = 4'\ 7''$

The first premise Ts_1 has the value 1 on this interpretation, since 6′ 7″ is a member to degree 1 of the fuzzy set corresponding to *T*. All premises Es_is_j have the value 1, because in each case the first specified height is $^1/_8$″ less than the second. The conclusion has the value 0 because 4′ 7″ has degree of membership 0 in the fuzzy set corresponding to *T*.

This leaves the second premise, $(\forall x)(\forall y)((Tx \wedge Eyx) \rightarrow Ty)$. The value of this premise is the minimum value that $(Tx \wedge Eyx) \rightarrow Ty$ can have for some variable assignment v. (We can speak of the minimum value rather than the greatest lower bound because the domain D is finite and so the set of different values that the formula can have—determined by the various combinations of heights that can be assigned to *x* and *y*—is also finite.) What is this minimum value? We first note that the formula will have the value 1 on any variable assignment v such that v(*y*) fails to be $^1/_8$″ less than v(*x*), because in this case *Eyx* will have the value 0 and so will $Tx \wedge Eyx$—giving the conditional the value 1. Consequently the minimum value that the formula *(Tx ∧ Eyx) → Ty* can have for some variable assignment v must be the minimum value that this formula can have on a variable assignment where v(*y*) *is* $^1/_8''$ less than v(*x*). Now, because *Eyx* has the value 1 on such assignments, the value of *(Tx ∧ Eyx) → Ty* on these assignments is therefore identical to the value of *Tx → Ty* because $I_v(Tx \wedge Eyx) = I_v(Tx)$ when $I_v(Eyx) = 1$.

So we need to find the smallest value that *Tx → Ty* can have on any variable assignment v such that v(*y*) is $^1/_8''$ less than v(*x*), or, looking at the satisfaction clause for conditionals, the smallest value that min $(1, 1 - I_v(Tx) + I_v(Ty))$ can be on such assignments. Well, when two heights differ by $^1/_8''$ their degree of membership in the fuzzy set *T* differs by $^1/_{192}$, that is, $I_v(Ty) = I_v(Tx) - {}^1/_{192}$, and so $1 - I_v(Tx) + I_v(Ty) = {}^{191}/_{192}$, or approximately .995. Thus .995 is the smallest value that *Tx → Ty*, and therefore *(Tx ∧ Eyx) → Ty*, can have on a variable assignment, and it follows that the quantified formula *(∀x)(∀y)((Tx ∧ Eyx) → Ty)* has the value .995 on the augmented interpretation SST.

This dissolves the Sorites paradox: although the argument is valid, any interpretation that captures the vagueness of the concept *tall* will assign a noncrisp interpretation to the predicate *T* in which at least two heights have different degrees of tallness and tallness never increases as height decreases; and every such interpretation of the predicate *T* (keeping the rest of SST the same) will give the Principle of Charity premise a value less than 1 (this will be established in an exercise). Thus the validity of the argument does not force us to accept its conclusion because on any reasonable interpretation the premises won't all have the value 1.

Moreover, we've addressed the first issue raised in Chapter 10: although we claim that the Principle of Charity premise is not true, we can interpret it as being very close to true—as we have just done—and so we can capture that intuition about the Principle of Charity premise. But now a new worry arises: if one premise is very close to true, and the rest of the premises are all true, shouldn't the conclusion be very close to true? Let us note that the Sorites argument is not degree-valid; the interpretation SST shows this (since the conclusion is not as true as the least true premise). Nor is it, say, .9-degree-valid, since the downward distance between the degree of truth of the least true premise—.995—and the degree of truth of the conclusion—0—is much greater than .1. But wouldn't we expect it to be *n*-degree-valid for *some* high value of *n*, given that it sure looks as if the conclusion follows from the premises?

The answer is ***no***, and for the reason we refer to Section 11.4 of Chapter 11. There we considered an argument with one simple premise *A* and then a chain of additional premises *A → B*, *B → C*, *C → D*, and so on, with the conclusion identical to the consequent of the last premise in the chain, and we explained that by increasing the number of premises in the chain we decrease the *n*-degree-validity of the argument. We call this ***decaying validity***. Decaying validity is exactly the phenomenon that we encounter with the Sorites argument, since the universally quantified conditional premise encapsulates a chain of conditional premises that lead to the conclusion. If the argument were to the effect that given the tallness of 6′ 7″

the tallness of 6′ 6″ follows, it would be plausible. If we changed the conclusion to be the claim that 6′ 5″ is tall, it would still be fairly plausible. But as we decrease the height mentioned in the conclusion the argument becomes less plausible, and it does so precisely because we have a longer chain of heights decreasing by $^1/_8$″ each. We will see this chain "in action" in Chapter 15, in the Pavelka axiomatic system for $\text{Fuzzy}_{\text{Ł}\forall}$.

Interpretation SST shows that the Sorites argument as we have symbolized it is at most $^1/_{192}$ (or approximately .005)-degree-valid in $\text{Fuzzy}_{\text{Ł}\forall}$, since the downward distance between the truth-value of the least true premise and that of the conclusion is approximately .995. This is close to the actual *n*-degree-validity of the argument, which is $^1/_{193}$. To establish this claim we'll first describe an interpretation SST* on which the downward distance between the truth-value of the least true premise and that of the conclusion in $\text{Fuzzy}_{\text{Ł}\forall}$ is $^{192}/_{193}$, showing that the argument is *at most* $^1/_{193}$-degree-valid. SST* is like SST, except that

$$\text{I (T)(<x>)} = (\text{x} - 4' \, 7'') \; / \; 24^1/_8''$$

On this modified interpretation, both Ts_1 and the Principle of Charity premise have truth-value $^{192}/_{193}$, all the other premises have the truth-value 1, and the conclusion has the truth-value 0. Thus, this interpretation shows that the argument is at most $^1/_{193}$-degree-valid.

Second, we'll show that $^1/_{193}$ is *exactly* the degree-validity of the argument as we have symbolized it in $\text{Fuzzy}_{\text{Ł}\forall}$. Consider: In order for the argument to be less than $^1/_{193}$-degree-valid, the downward distance between the truth-value of the least true premise and that of the conclusion would have to be greater than $^{192}/_{193}$, and we can show that this is impossible. Let's assume that we can increase the downward distance so that it is greater than $^{192}/_{193}$ on some interpretation I. Then each of the premises must have a truth-value that is greater than $^{192}/_{193}$, since the least value that the conclusion can have is 0. That being the case, let $\text{b}_1 \ldots \text{b}_{192}$ be the values such that

a. $\text{I}(\text{Es}_{i+1}\text{s}_i) = {}^{192}/_{193} + \text{b}_i, \; {}^1/_{193} \geq \text{b}_i > 0.$

Turning to the Principle of Charity premise, if its value is to be greater than $^{192}/_{193}$, then the unquantified conditional formula $(Tx \wedge Eyx) \rightarrow Ty$ must also have a value greater than $^{192}/_{193}$ on every variable assignment. Now, for any variable assignment v such that $\text{v}(x) = \text{I}(s_i)$ and $\text{v}(y) = \text{I}(s_{i+1})$ for some i, $1 \leq i \leq 192$, the value of the formula $(Tx \wedge Eyx) \rightarrow Ty$ will be the same as the value of the formula $(Ts_i \wedge Es_{i+1}s_i) \rightarrow Ts_{i+1}$. So for each i, $1 \leq i \leq 192$, the value of the conditional $\text{I}_\text{v}((Ts_i \wedge Es_{i+1}s_i) \rightarrow Ts_{i+1})$ must be greater than $^{192}/_{193}$. For each i let c_i be the value such that $\text{I}_\text{v}((Ts_i \wedge Es_{i+1}s_i) \rightarrow Ts_{i+1}) = {}^{192}/_{193} + \text{c}_i$. It follows from the truth-conditions for complex formulas that

b. $\min (1, 1 - \min (\text{I}_\text{v}(Ts_i), \text{I}_\text{v}(Es_{i+1}s_i)) + \text{I}_\text{v}(Ts_{i+1})) = {}^{192}/_{193} + \text{c}_i, \; {}^1/_{193} \geq \text{c}_i > 0.$

In addition, for the Sorites argument to be less than $^1/_{193}$-degree-valid we must have

c. $I_v(Ts_{193}) < b_i$ and $I_v(Ts_{193}) < c_i$, for $1 \leq i \leq 192$

—for otherwise the downward distance between the truth-value of one of the premises and that of the conclusion would fail to be greater than $^{192}/_{193}$. We will show that it follows from these inequalities that $I_v(Ts_1) < {}^{192}/_{193}$ and so the downward distance from the premises to the conclusion cannot be greater than $^{192}/_{193}$.

Plugging (a) into (b), we get

b′. $\min(1, 1 - \min(I_v(Ts_i), {}^{192}/_{193} + b_i) + I_v(Ts_{i+1})) = {}^{192}/_{193} + c_i$,
where $^{1}/_{193} \geq b_i > 0$ and $^{1}/_{193} \geq c_i > 0$.

Depending on which of the two values on the left-hand side of (b′) is the minimum, we thus have for any *i* either

d. $\min(1, 1 - \min(I_v(Ts_i), {}^{192}/_{193} + b_i) + I_v(Ts_{i+1})) = 1$, so
$1 - \min(I_v(Ts_i), {}^{192}/_{193} + b_i) + I_v(Ts_{i+1}) \geq 1$, in which case
$I_v(Ts_{i+1}) \geq \min(I_v(Ts_i), {}^{192}/_{193} + b_i)$,

or

e. $\min(1, 1 - \min(I_v(Ts_i), {}^{192}/_{193} + b_i) + I_v(Ts_{i+1})) =$
$1 - \min(I_v(Ts_i), {}^{192}/_{193} + b_i) + I_v(Ts_{i+1}) = {}^{192}/_{193} + c_i$ and so
$I_v(Ts_{i+1}) = c_i - {}^{1}/_{193} + \min(I_v(Ts_i), {}^{192}/_{193} + b_i)$ and
$I_v(Ts_{i+1}) = c_i + {}^{191}/_{193} + \min(I_v(Ts_i) - {}^{192}/_{193}, b_i)$.

Now we'll look at specific cases, beginning with the case where $i = 192$. We have either

d. $I_v(Ts_{193}) \geq \min(I_v(Ts_{192}), {}^{192}/_{193} + b_{192})$, or
e. $I_v(Ts_{193}) = c_{192} + {}^{191}/_{193} + \min(I_v(Ts_{192}) - {}^{192}/_{193}, b_{192})$.

Now, because $c_{192} \leq {}^{1}/_{193}$, $I_v(Ts_{193}) < {}^{1}/_{193}$ by (c). Since $b_{192} > 0$, it follows that $I_v(Ts_{193}) < {}^{192}/_{193} + b_{192}$. Thus, in case (d) we must have $I_v(Ts_{193}) \geq I_v(Ts_{192})$ and consequently $I_v(Ts_{192}) < {}^{1}/_{193}$. In case (e), because $I_v(Ts_{193}) < c_{192}$ we must have $^{191}/_{193} + \min(I_v(Ts_{192}) - {}^{192}/_{193}, b_{192}) < 0$. Thus $\min(I_v(Ts_{192}) - {}^{192}/_{193}, b_{192}) < -{}^{191}/_{193}$ and so, because $b_{192} > 0$, $I_v(Ts_{192}) - {}^{192}/_{193}$ must be less than $-{}^{191}/_{193}$, giving us $I_v(Ts_{192}) < {}^{1}/_{193}$. Because one of the two cases (d) and (e) must hold, we may conclude that $I_v(Ts_{192}) < {}^{1}/_{193}$.

Next, when $i = 191$, we have either

d. $I(Ts_{192}) \geq \min(I(Ts_{191}), {}^{192}/_{193} + b_{191})$, or
e. $I(Ts_{192}) = c_{191} + {}^{191}/_{193} + \min(I(Ts_{191}) - {}^{192}/_{193}, b_{191})$.

In case (d) we once again conclude that $I_v(Ts_{191}) < {}^{1}/_{193}$, because $I_v(Ts_{192}) < {}^{1}/_{193}$. In case (e) we have $c_{191} + {}^{191}/_{193} + \min(I_v(Ts_{191}) - {}^{192}/_{193}, b_{191}) < {}^{1}/_{193}$ because $I_v(Ts_{192}) < {}^{1}/_{193}$, so $c_{191} + \min(I_v(Ts_{191}) - {}^{192}/_{193}, b_{191}) < -{}^{190}/_{193}$; and because c_{191} and b_{191} are both positive, $I_v(Ts_{191}) - {}^{192}/_{193} < -{}^{190}/_{193}$, and so $I_v(Ts_{191}) < {}^{2}/_{193}$. Because either (d) or (e) holds, it follows that $I_v(Ts_{191}) < {}^{2}/_{193}$.

This reasoning can, with appropriate substitutions, be repeated to yield the conclusion that $I_v(Ts_i) < {}^{(193-i)}/_{193}$ for $191 \geq i \geq 1$. It follows that $I_v(Ts_1) < {}^{192}/_{193}$. But this contradicts the assumption that all of the premises have a value greater than ${}^{192}/_{193}$. We conclude that the Sorites argument as we have symbolized it is *exactly* ${}^{1}/_{193}$-degree-valid in $\text{Fuzzy}_{L\forall}$.

We'll now look at how the Sorites argument fares in $\text{Fuzzy}_{L\forall}$ when the Principle of Charity premise is symbolized using Łukasiewicz bold, rather than weak, conjunction:

$$(\forall x)(\forall y)((Tx \,\&\, Eyx) \rightarrow Ty))$$

First, we point out that on interpretation SST, the bold conjunction version of the Principle of Charity has the same truth-value as the weak conjunction version. This is because on that interpretation, *Eyx* has the value 1 for all of the cases that matter in determining whether the value of the Principle of Charity premise is less than 1, and as we noted in Chapter 11, all t-norms agree on the value produced when one of their arguments has the value 1. In particular, whenever *Eyx* has the value 1 on a variable assignment, the truth-value of both *Tx & Eyx* and $Tx \wedge Eyx$ will be the value of *Tx* on that assignment. Since the other premises and the conclusions are the same for both the bold and weak conjunction versions of the Sorites argument, it follows that the truth-values of the premises and the conclusion for the bold version are identical to those for the weak version on SST. They remain identical under interpretation SST*, for the same reason, so we may conclude that the bold conjunction version of the argument is at most ${}^{1}/_{193}$-degree-valid.

In fact, unlike the weak conjunction version, the bold conjunction version is strictly less than ${}^{1}/_{193}$-degree-valid; it is ${}^{1}/_{385}$-degree-valid. We show this in two steps. As the first step, the following interpretation provides a downward distance of ${}^{384}/_{385}$ between the truth-value of the least true premise and that of the conclusion in the bold version of the argument:

D: set of heights between 4′ 7″ and 6′ 7″ by ${}^{1}/_{8}$″ increments, inclusive
$I(T)(<x>) = (2 \cdot (x - 4'\,7'')) \,/\, 48^{1}/_{8}''$
$I(E)(<x_1, x_2>) = {}^{384}/_{385}$ for all x_1, x_2
$I(s_1) = 6'\,7''$
$I(s_2) = 6'\,6^{7}/_{8}''$
...
$I(s_{193}) = 4'\,7''$

On this interpretation all of the premises have the value ${}^{384}/_{385}$, while the conclusion has the value 0 (proof is left as an exercise). Thus the argument is *at most* ${}^{1}/_{385}$-degree-valid.

As the second step we will argue that ${}^{1}/_{385}$ is *exactly* the degree-validity of the Łukasiewicz bold version of the argument. Assume, contrary to what we want to prove, that we can increase the downward distance between the truth-value of the least true premise and that of the conclusion to more than $^{384}/_{385}$ on some interpretation I. Then each of the premises must have a truth-value that is greater than $^{384}/_{385}$. Let $b_1, \ldots, b_{192}$ be the values such that

a. $I_v(Es_{i+1}s_i) = {}^{384}/_{385} + b_i,\ {}^{1}/_{384} \geq b_i > 0$

In the case of the Principle of Charity premise, any variable assignment such that $v(x) = I(s_i)$ and $v(y) = I(s_{i+1})$, for some i, will give the unquantified conditional formula $(Tx \,\&\, Eyx) \rightarrow Ty$ a value greater than $^{384}/_{385}$, and $(Ts_i \wedge Es_{i+1}s_i) \rightarrow Ts_{i+1}$ will have this same value. For each i, let c_i be the value such that $I_v((Ts_i \wedge Es_{i+1}s_i) \rightarrow Ts_{i+1}) = {}^{384}/_{385} + c_i$, $c_i > 0$; it follows from the truth-conditions for complex formulas that $\min(1, 1 - \max(0, I_v(Ts_i) + I_v(Es_{i+1}s_i) - 1) + I_v(Ts_{i+1})) = {}^{384}/_{385} + c_i$. Moreover, $\min(1, 1 - \max(0, I_v(Ts_i) + I_v(Es_{i+1}s_i) - 1) + I_v(Ts_{i+1})) = \min(1, \min(1 - 0 + I_v(Ts_{i+1}), 1 - (I_v(Ts_i) + I_v(Es_{i+1}s_i) - 1) + I_v(Ts_{i+1}))) = \min(1, 1 + I_v(Ts_{i+1}), 2 - I_v(Ts_i) - I_v(Es_{i+1}s_i) + I_v(Ts_{i+1})) = \min(1, 2 - I_v(Ts_i) - I_v(Es_{i+1}s_i) + I_v(Ts_{i+1}))$ since $1 + I_v(Ts_{i+1}) \geq 0$. Thus

b. $\min(1, 2 - I_v(Ts_i) - I_v(Es_{i+1}s_i) + I_v(Ts_{i+1})) = {}^{384}/_{385} + c_i,\ {}^{1}/_{385} \geq c_i > 0$

for $1 \leq i \leq 192$. Finally, for the bold conjunction version of the Sorites argument to be less than $^{1}/_{385}$-degree-valid in $\text{Fuzzy}_{\text{L}\forall}$ we must have

c. $I_v(Ts_{193}) < b_i$ and $I_v(Ts_{193}) < c_i$, for $1 \leq i \leq 192$.

We can show that if all of these inequalities hold then $I_v(Ts_1) < {}^{384}/_{385}$ and so the argument cannot be less than $^{1}/_{385}$-degree-valid.

Again plugging (a) into (b), we get $\min(1, 2 - I_v(Ts_i) - ({}^{384}/_{385} + b_i) + I_v(Ts_{i+1})) = {}^{384}/_{385} + c_i$, or

b′. $\min(1, {}^{386}/_{385} - I_v(Ts_i) - b_i + I_v(Ts_{i+1})) = {}^{384}/_{385} + c_i$, with $^{1}/_{385} \geq b_i > 0$ and $^{1}/_{385} \geq c_i > 0$.

We thus have for any i either

d. $\min(1, {}^{386}/_{385} - I_v(Ts_i) - b_i + I_v(Ts_{i+1})) = 1$ so
$^{386}/_{385} - I_v(Ts_i) - b_i + I_v(Ts_{i+1})) \geq 1$ so
$I_v(Ts_{i+1}) \geq I_v(Ts_i) + b_i - {}^{1}/_{385}$, or

e. $\min(1, {}^{386}/_{385} - I_v(Ts_i) - b_i + I_v(Ts_{i+1})) = {}^{386}/_{385} - I_v(Ts_i) - b_i + I_v(Ts_{i+1}) = {}^{384}/_{385} + c_i$ and so
$I_v(Ts_{i+1}) = c_i + b_i + I_v(Ts_i) - {}^{2}/_{385}$.

Turning to cases, when $i = 192$ we have either

d. $I_v(Ts_{193}) \geq I_v(Ts_{192}) + b_{192} - {}^{1}/_{385}$, or
e. $I_v(Ts_{193}) = c_{192} + b_{192} + I_v(Ts_{192}) - {}^{2}/_{385}$.

$I_v(Ts_{193}) < {}^{1}/_{385}$ because $I_v(Ts_{193}) < c_{192}$ and $c_{192} \leq {}^{1}/_{385}$. In case (d) it therefore follows that $I_v(Ts_{192}) + b_{192} - {}^{1}/_{385} < {}^{1}/_{385}$ and so $I_v(Ts_{192}) + b_{192} < {}^{2}/_{385}$ and, because $b_{192} > 0$, $I_v(Ts_{192}) < {}^{2}/_{385}$. In case (e), from $I_v(Ts_{193}) < c_{192}$ it follows that $b_{192} + I_v(Ts_{192}) - {}^{2}/_{385} < 0$. But $b_{192} > 0$, so $I_v(Ts_{192}) - {}^{2}/_{385} < 0$ and $I_v(Ts_{192}) < {}^{2}/_{385}$. Either (d) or (e) holds, so we conclude that $I_v(Ts_{192}) < {}^{2}/_{385}$.

When $i = 191$, either

d. $I_v(Ts_{192}) \geq I_v(Ts_{191}) + b_{191} - {}^1/_{385}$, or
e. $I_v(Ts_{192}) = c_{191} + b_{191} + I_v(Ts_{191}) - {}^2/_{385}$.

In case (d), since $I_v(Ts_{192}) < {}^2/_{385}$ and $b_{191} > 0$ we have $I_v(Ts_{191}) < {}^3/_{385}$. In case (e) we have $c_{191} + b_{191} + I_v(Ts_{191}) - {}^2/_{385} < {}^2/_{385}$ and since $c_{191} > 0$ and $b_{191} > 0$, it follows in this case that $I_v(Ts_{191}) < {}^4/_{385}$. Either way, then, $I_v(Ts_{191}) < {}^4/_{385}$. By repeating this reasoning, we arrive at the general conclusion that $I_v(Ts_i) < {}^{(2(193-i))}/_{385}$ for $191 \geq i \geq 1$ and, in particular, that $I_v(Ts_1) < {}^{384}/_{385}$. But this contradicts our assumption that all of the premises have a value greater than ${}^{384}/_{385}$. We conclude that the Łukasiewicz bold version of the Sorites argument is exactly ${}^1/_{385}$-degree-valid.

Generally, then, both Łukasiewicz fuzzy versions of the Sorites argument agree on their analyses of the Sorites argument: the Principle of Charity premise is close to true but the argument, although valid, has a very low n-degree-validity and that is why we can go from premises that are very close to all being true to a conclusion that is false.

If we use a fuzzy Kleene conditional in the Principle of Charity premise, the Sorites argument remains valid in $Fuzzy_{LV}$. However, the truth-value of the new premise $(\forall x)(\forall y)((Tx \wedge Eyx) \rightarrow_K Ty)$ on SST is different from the version with the Łukasiewicz conditional. The value of the new premise will be the minimum value that $(Tx \wedge Eyx) \rightarrow_K Ty$ can have for different variable assignments v. What is this value? As in the case of the Łukasiewicz conditional, and for the same reasons, the minimum value that $(Tx \wedge Eyx) \rightarrow_K Ty$ can have will be the minimum value it can have on those variable assignments where v(y) is ${}^1/_8''$ less than v(x), and on such assignments $I_v((Tx \wedge Eyx) \rightarrow_K Ty) = I_v(Tx \rightarrow_K Ty)$, which is max $(1 - I_v(Tx), I_v(Ty))$. Recall that when two heights differ by ${}^1/_8''$, their membership degrees in the fuzzy set for T as defined by SST differ by ${}^1/_{192}$, so max $(1 - I_v(Tx), I_v(Ty))$ in this case is max $(1 - (I_v(Ty) + {}^1/_{192}), I_v(Ty))$, or max $({}^{191}/_{192} - I_v(Ty), I_v(Ty))$. The minimum value that $(Tx \wedge Eyx) \rightarrow_K Ty$ can have is therefore the minimum value of max $({}^{191}/_{192} - I_v(Ty), I_v(Ty))$, which is .5 (you will be asked to verify this in an exercise). It follows that the Kleenean Principle of Charity premise $(\forall x)(\forall y)((Tx \wedge Eyx) \rightarrow_K Ty)$ has the value .5 on SST. To the extent that we would like to capture the intuition that the Principle of Charity premise is *close* to true, the Kleenean conditional therefore does not compare well with the Łukasiewicz conditional (and replacing weak conjunction with bold conjunction in the premise will not change our conclusion).

With the comparison between the Łukasiewicz and Kleene conditionals at hand, we pause to address an issue that has been raised in the literature concerning an alternative formulation of Sorites paradoxes. It's been claimed (for example, by Crispin Wright [1987]) that although a fuzzy analysis of Sorites arguments can dissolve the paradoxicality by claiming that the Principle of Charity premise is *close to* true, as the Łukasiewicz analysis does, fuzzy analyses fail when the Principle of Charity is restated using conjunction and negation. Saying that *every height that's* ${}^1/_8''$ *less than a tall height is itself tall* is equivalent to saying that *there's no height that is tall while the height that is* $^1/_8''$ *less isn't.* Now if we use weak conjunction to symbolize the Principle of Charity in $\text{Fuzzy}_{\text{L}\forall}$ as $(\forall x)(\forall y)\neg((Tx \wedge Eyx) \wedge \neg Ty)$, it won't be close to true on a reasonable interpretation—because this formula is equivalent to the Kleene conditional version of the Principle of Charity premise! (The Kleene conditional formula $\mathbf{P} \rightarrow_K \mathbf{Q}$ is equivalent to $\neg(\mathbf{P} \wedge \neg\mathbf{Q})$). If we are satisfied with the Łukasiewicz conditional version of the Principle of Charity premise, and we are, the conjunctive version should use bold rather than weak conjunction when the conditional is eliminated, since $\mathbf{P} \rightarrow \mathbf{Q}$ is equivalent to $\neg(\mathbf{P}\ \&\ \neg\mathbf{Q})$ in $\text{Fuzzy}_{\text{L}\forall}$. This gives us the formula $(\forall x)(\forall y)\neg((Tx \wedge Eyx)\ \&\ \neg Ty)$, which will also be close to true on SST. So the claim that a Łukasiewiczian fuzzy analysis is inadequate for Sorites paradoxes that have a conjunctive version of the Principle of Charity premise is just plain wrong.

Returning to the Kleene conditional version of the Sorites argument, though, we should also consider the question, For what value *n* is this version of the Sorites argument *n*-degree-valid? Interpretation SST shows that it is *at most* .5-degree-valid, since the downward distance from the value of the Principle of Charity premise to the conclusion is .5. In fact, the Kleene conditional version of the Sorites paradox is *exactly* .5-degree-valid; that is, the maximum downward distance between the premises and the conclusion on any interpretation is .5.

Why? Well, assume that there's an interpretation on which this downward distance is greater than .5. Then the value of each premise must be more than .5 greater than the value of the conclusion, Ts_{193}. In particular, the value of the Principle of Charity premise must exceed the conclusion's value by more than .5. Now, the value of $(\forall x)(\forall y)((Tx \wedge Eyx) \rightarrow_K Ty)$ is the greatest lower bound of max $(1 - \min(I_v(Tx), I_v(Eyx)), I_v(Ty))$, or max $(1 - I_v(Tx), 1 - I_v(Eyx), I_v(Ty))$, for any assignment v, and so this greatest lower bound must exceed .5. Let's first consider a variable assignment v_1 such that $v_1(x) = I(s_1)$ and $v_1(y) = I(s_2)$. Because we're assuming that Ts_1 and Es_2s_1 have values greater than $I_v(Ts_{193}) + .5$, it follows that $1 - I_{v1}(Tx) < .5$ and $1 - I_{v1}(Eyx) < .5$. So if max $(1 - I_{v1}(Tx), 1 - I_{v1}(Eyx), I_{v1}(Ty)) > I_{v1}(Ts_{193}) + .5$, it must be because $I_{v1}(Ty) > I_{v1}(Ts_{193}) + .5$. Now consider a variable assignment v_2 such that $v_2(x) = I(s_2)$ and $v_2(y) = I(s_3)$. By our previous reasoning, $I_{v2}(Tx) > I_{v2}(Ts_{193}) + .5$ (because $v_2(x) = v_1(y)$), and by our assumption $I_{v2}(Es_2s_1)$ must be greater than $I_{v2}(Ts_{193}) + .5$, so for this assignment v_2 we also have both $1 - I_{v2}(Tx)$ and $1 - I_{v2}(Eyx)$ less than .5. If the value of $(Tx \wedge Eyx) \rightarrow_K Ty$ is greater than $I_{v2}(Ts_{193}) + .5$, then, it must be because $I_{v2}(Ty) > I_{v2}(Ts_{193}) + .5$.

By repeating this reasoning, we will eventually get to a variable assignment v_{192} such that $v_{192}(x) = I(s_{192})$ and $v_{192}(y) = I(s_{193})$, where $I_{v192}(Ty) > I_{v192}(Ts_{193}) + .5$. But that's impossible if $v_{192}(y) = I(s_{193})$! Thus, contrary to our assumption, the maximum downward distance from the premises to the conclusion is .5 and so the Kleenean version of the Sorites argument is exactly .5-degree-valid. Surveying this reasoning, we can also see an additional point: When we use the Kleene conditional in the Sorites argument we don't get decaying validity as we do with the Łukasiewicz conditional; in fact, the Kleenean argument remains .5-degree-valid no matter how much we increase the subscript in the conclusion.

14.5 Gödel Fuzzy First-Order Logic

Adding the same quantifier clauses as in Łukasiewicz fuzzy first-order logic, we arrive at the following truth-conditions for formulas in $\text{Fuzzy}_{G\forall}$, Gödel fuzzy first-order logic:

1. $I_v(\mathbf{Pt}_1 \ldots \mathbf{t}_n) = I(\mathbf{P})(<I^*(\mathbf{t}_1), \ldots, I^*(\mathbf{t}_n)>)$, where $I^*(\mathbf{t}_i)$ is $I(\mathbf{t}_i)$ if $\mathbf{t}_i$ is a constant and is $v(\mathbf{t}_i)$ if $\mathbf{t}_i$ is a variable.
2. $I_v(\neg_G \mathbf{P}) = 1$ if $I_v(\mathbf{P}) = 0$
   0 otherwise
3. $I_v(\mathbf{P} \,\&_G\, \mathbf{Q}) = \min(I_v(\mathbf{P}), I_v(\mathbf{Q}))$
4. $I_v(\mathbf{P} \nabla_G \mathbf{Q}) = \max(I_v(\mathbf{P}), I_v(\mathbf{Q}))$
5. $I_v(\mathbf{P} \rightarrow_G \mathbf{Q}) = 1$ if $I_v(\mathbf{P}) \leq I_v(\mathbf{Q})$
   $I_v(\mathbf{Q})$ otherwise
6. $I_v(\mathbf{P} \leftrightarrow_G \mathbf{Q}) = 1$ if $I_v(\mathbf{P}) = I_v(\mathbf{Q})$
   $\min(I_v(\mathbf{P}), I_v(\mathbf{Q}))$ otherwise
7. $I_v((\forall \mathbf{x})\mathbf{P}) = \text{glb}\{I_{v'}(\mathbf{P})\text{: v' is an } \mathbf{x}\text{-variant of v}\}$
8. $I_v((\exists \mathbf{x})\mathbf{P}) = \text{lub}\{I_{v'}(\mathbf{P})\text{: v' is an } \mathbf{x}\text{-variant of v}\}$

(Because the quantifiers are defined identically for our three fuzzy first-order systems we omit subscripts.) Recall that Gödel fuzzy logic defines its conditional as the adjunct residuum for Łukasiewicz weak conjunction, and that the negation of a formula is defined as $\neg\mathbf{P} =_{\text{def}} \mathbf{P} \rightarrow \mathbf{0}$. The general results that we examined for Gödel fuzzy propositional logic carry over to the first-order system: every formula that is a tautology in $\text{Fuzzy}_{G\forall}$ is a tautology in classical logic and every entailment in $\text{Fuzzy}_{G\forall}$ is a classical entailment, but the converses do not hold.

Black's problem—how to express the existence of borderline cases for vague predicates—can be addressed in $\text{Fuzzy}_{G\forall}$ (and in $\text{Fuzzy}_{P\forall}$, to be introduced in Section 14.6) just as it is in $\text{Fuzzy}_{L\forall}$ if we augment the language with the fuzzy external assertion operator. So let us move right along and examine the Sorites paradox with Gödel implication (which we here subscript with a *G*) and Gödel bold conjunction—recall that Gödel weak conjunction is identical to Gödel bold conjunction:

Ts_1
Es_2s_1
Es_3s_2
Es_4s_3
...
$\text{Es}_{193}\text{s}_{192}$
$\underline{(\forall x)(\forall y)((\text{Tx} \,\&_G\, \text{Eyx}) \rightarrow_G \text{Ty})}$
Ts_{193}

The argument is valid in $Fuzzy_{G\forall}$. But consider again interpretation SST:

D: set of heights between 4′ 7″ and 6′ 7″ by $^1/_8$″ increments, inclusive
I(T)(<x>) = (x – 4′ 7″) / 24″
$I(E)(<x_1,x_2>) = 1$ if x_1 is $^1/_8$″ less than x_2
0 otherwise
$I(s_1) = 6' 7''$
$I(s_2) = 6' 6^7/_8''$
...
$I(s_{193}) = 4' 7''$

All of the premises except the last have the value 1 on this interpretation. For the Principle of Charity premise we must determine the least value that $(Tx \&_G Eyx) \rightarrow_G Ty$ can have for any variable assignment. The formula has the value 1 whenever *Eyx* has the value 0, so to find the formula's minimum value we must consider variable assignments that assign the value 1 to *Eyx*—assignments such that v(*y*) is $^1/_8$″ less than v(*x*). The formula $(Tx \&_G Eyx) \rightarrow_G Ty$ has the same value as $Tx \rightarrow_G Ty$ on these variable assignments, and because $I_v(Tx) > I_v(Ty)$, $I_v(Tx \rightarrow_G Ty) = I_v(Ty)$ on these assignments. When v(*y*) = 4′ 7″, the value of $I_v(Ty)$ is 0, so that is the least value that $Tx \rightarrow_G Ty$ can have. Thus the Gödel Principle of Charity premise is false on interpretation SST; but that doesn't jibe with the intuition that the premise is close to true. In fact, the only way the principle can be close to true is to have a membership function for *tall* that assigns a high degree of membership to every height—or to have a membership function that assigns nondecreasing membership values for decreasing heights—in either case that won't be true to the facts.

It is interesting to note that the Sorites argument is also 1-degree-valid in $Fuzzy_{G\forall}$! Let's assume that it isn't; that is, we'll assume that there is at least one interpretation on which every premise has a value greater than the conclusion. Now, for the Principle of Charity premise to have a greater value than the argument's conclusion, we must have $I_v((Tx \&_G Eyx) \rightarrow_G Ty) > I_v(Ts_{193})$ for every variable assignment v. Consider first an assignment v_1 such that $v_1(x) = I(s_{192})$ and $v_1(y) = I(s_{193})$, so that $I_{v1}((Tx \&_G Eyx) \rightarrow_G Ty) = I_{v1}((Ts_{192} \&_G Es_{193}s_{192}) \rightarrow_G Ts_{193})$. By the way that Gödel implication is defined, $I_{v1}((Ts_{192} \&_G Es_{193}s_{192}) \rightarrow_G Ts_{193}) > I_{v1}(Ts_{193})$ only if $I_{v1}(Ts_{192} \&_G Es_{193}s_{192}) \leq I_{v1}(Ts_{193})$ (otherwise the value of the conditional would be $I_{v1}(Ts_{193})$). Thus either $I_{v1}(Ts_{192}) \leq I_{v1}(Ts_{193})$ or $I_{v1}(Es_{193}s_{192}) \leq I_{v1}(Ts_{193})$.) But $I_{v1}(Es_{193}s_{192}) > I_{v1}(Ts_{193})$ by our assumption, so we conclude that $I_{v1}(Ts_{192}) \leq I_{v1}(Ts_{193})$.

Next consider an assignment v_2 such that $v_2(x) = I(s_{191})$ and $v_2(y) = I(s_{192})$. In this case $I_{v2}((Tx \&_G Eyx) \rightarrow_G Ty) = I_{v2}((Ts_{191} \&_G Es_{192}s_{191}) \rightarrow_G Ts_{192})$. From the assumption that $I_{v2}((Tx \wedge_G Eyx) \rightarrow_G Ty) > I_{v2}(Ts_{193})$, then, it follows that $I_{v2}((Ts_{191} \&_G Es_{192}s_{191}) \rightarrow_G Ts_{192}) > I_{v2}(Ts_{193})$ and consequently that $I_{v2}((Ts_{191} \&_G Es_{192}s_{191})) \leq I_{v2}(Ts_{192})$ (otherwise the value of the conditional would be $I_{v2}(Ts_{192})$, which is, by the previous paragraph, less than or equal to $I_{v1}(Ts_{193})$). But again, we've assumed $I_{v2}(Es_{192}s_{191}) > I_{v2}(Ts_{193})$. So it must be that $I_{v2}(Ts_{191}) \leq I_{v2}(Ts_{193})$.

Repeating this reasoning we eventually end up concluding that even on an assignment v_{192} such that $v_{192}(x) = I(s_1)$ and $v_{192}(y) = I(s_2)$, $I_{v192}(Tx)$, which is $I_{v192}(Ts_1)$, is less than or equal to $I_{v192}(Ts_{193})$. But this contradicts the assumption that all of the premises are truer than Ts_{193}. We conclude that the Sorites argument as symbolized in $\text{Fuzzy}_{G\forall}$ is 1-degree-valid. This reasoning generalizes for longer Sorites chains as well, so the Sorites argument also fails to display decaying validity in $\text{Fuzzy}_{G\forall}$.

14.6 Product Fuzzy First-Order Logic

Again, we use the same truth-condition clauses for the quantifiers to define the system $\text{Fuzzy}_{P\forall}$:

1. $I_v(\mathbf{Pt_1 \ldots t_n}) = I(\mathbf{P})(<I^*(\mathbf{t_1}), \ldots, I^*(\mathbf{t_n})>)$, where $I^*(\mathbf{t_i})$ is $I(\mathbf{t_i})$ if $\mathbf{t_i}$ is a constant and is $v(\mathbf{t_i})$ if $\mathbf{t_i}$ is a variable
2. $I_v(\neg_P \mathbf{P}) = 1$ if $I_v(\mathbf{P}) = 0$
   0 otherwise
3. $I_v(\mathbf{P} \,\&_P\, \mathbf{Q}) = I_v(\mathbf{P}) \cdot I_v(\mathbf{Q})$
4. $I_v(\mathbf{P} \triangledown_P \mathbf{Q}) = I_v(\mathbf{P}) + I_v(\mathbf{Q}) - (I_v(\mathbf{P}) \cdot I_v(\mathbf{Q}))$
5. $I_v(\mathbf{P} \rightarrow_P \mathbf{Q}) = 1$ if $I_v(\mathbf{P}) \leq I_v(\mathbf{Q})$
   $I_v(\mathbf{Q}) \,/\, I_v(\mathbf{P})$ otherwise
6. $I_v(\mathbf{P} \leftrightarrow_P \mathbf{Q})$: left as an exercise in Chapter 11
7. $I_v((\forall \mathbf{x})\mathbf{P}) = \text{glb}\{I_{v'}(\mathbf{P})$: v′ is an **x**-variant of v}
8. $I_v((\exists \mathbf{x})\mathbf{P}) = \text{lub}\{I_{v'}(\mathbf{P})$: v′ is an **x**-variant of v}

Here bold conjunction is defined to be the algebraic product operation, bold disjunction is defined to be the algebraic sum operation, and negation turns out to be identical to Gödel negation. The main semantic results for Fuzzy_P carry over to $\text{Fuzzy}_{P\forall}$. Every formula that is a tautology in $\text{Fuzzy}_{P\forall}$ is a tautology in classical logic, but not conversely; and every entailment in $\text{Fuzzy}_{P\forall}$ is a classical entailment, but not conversely. The Sorites argument

Ts_1
Es_2s_1
Es_3s_2
Es_4s_3
. . .
$Es_{193}s_{192}$
$(\forall x)(\forall y)((Tx \wedge_P Eyx) \rightarrow_P Ty)$
Ts_{193}

is valid in $\text{Fuzzy}_{P\forall}$, as is the version using product bold conjunction in the Principle of Charity premise: *$(\forall x)(\forall y)((Tx \,\&_P\, Eyx) \rightarrow_P Ty)$*. But not all of the premises have the value 1 on interpretation SST: in fact, both versions of the Principle of Charity premise have the value 0. When $v(x) = 4'\ 7^1/_8''$ and $v(y) = 4'\ 7''$, $I_v((Tx \wedge_P Eyx) \rightarrow_P Ty) = 0$ because the consequent has the value 0 and the antecedent has the greater value $^1/_{192}$, from which it follows that the value of the conditional is the value of the consequent divided by that of the antecedent. Similarly, $I_v((Tx \&_P Eyx) \rightarrow_P Ty) = 0$ because the value of the antecedent is also $^1/_{192}$ in this case. Like the Gödel version, the product versions of the Principle of Charity compare poorly with the Łukasiewicz version on interpretation SST given the intuition that the Principle of Charity is close to true. In this connection we note that if we choose a membership function for *tall* that *doesn't* assign 0 as the degree of membership for any height, then the Sorites argument in product fuzzy logic can still have a conclusion whose value is very close to 0 while the Principle of Charity premise has a much higher value.[4]

We can approximate the n-degree-validity of the bold product conjunction version of the Sorites argument as follows (we approximate in this case because the analysis is considerably more complicated than in the other systems).[5] Let m_1, $m_2, \ldots, m_{193}$ be the values assigned to $Ts_1, Ts_2, \ldots, Ts_{193}$ on an interpretation I, and let $n_1, n_2, \ldots n_{192}$ be the values assigned to $Es_2s_1, Es_3s_2, \ldots, Es_{193}s_{192}$. We note that when $v(x) = I(s_i)$ and $v(y) = I(s_{i+1})$ for some i, $1 \leq i \leq 192$, the value p_i of $(Tx \&_P Eyx) \rightarrow_P Ty$ is 1 if $m_i \cdot n_i \leq m_{i+1}$ and it is $m_{i+1} / (m_i \cdot n_i)$ otherwise. In all other cases (i.e., all assignments v that don't assign such values x and y) we'll assume that $I_v((Tx \&_P Eyx) \rightarrow_P Ty) = 1$ – it can because $I_v(Eyx)$ can be 0 in these cases. (This assumption is insignificant given that we're trying to find a maximum downward distance.) Thus the value of the Principle of Charity premise is min $(p_1, \ldots, p_{192})$. So the maximum possible downward distance between the value of the least true premise and that of the conclusion is then the maximum value that min $(m_1, n_1, \ldots, n_{192}, p_1, \ldots, p_{192}) - m_{192}$ can be for some interpretation.

If we restrict our attention to interpretations on which each $n_1 = n_2 = \cdots = n_{192} = m_1$, then min $(m_1, n_1, \ldots, n_{192}, p_1, \ldots, p_{192}) - m_{192}$ reduces to min $(m_1, p_1, \ldots, p_{192}) - m_{192}$. (Letting the n_i values be lower than m_1 would decrease the maximum downward distance, so this restriction is also insignificant for our purposes.) If we further restrict our attention to interpretations on which $m_1 < 1$ and $m_i = {m_1}^i$ for each $i > 1$, then as i gets larger the values ${m_1}^i$ are getting smaller and smaller— and we *are* trying to make the value assigned to the conclusion as small as we can. Moreover, if $m_i = {m_1}^i$ for each $i > 1$ then every p_i is 1 because $m_i \cdot n_i = m_i \cdot m_1 = {m_1}^{i+1} \leq {m_1}^{i+1}$, and so min $(m_1, p_1, \ldots, p_{192}) - m_{192}$ now reduces to $m_1 - m_{192}$ or $m_1 - {m_1}^{193}$! (This is why we didn't want the n_i values to be greater than m_1.) So we can ask, What is the maximum value that $m_1 - {m_1}^{193}$ can be when m_1 is a value in the unit interval [0. .1]? This value is around .97, meaning the argument is at most .03-degree-valid, clearly a desirable result. The reader will be asked to do a similar

4 In fact, the examples in Goguen (1968–1969) use such a membership function—degrees of membership can get close to, but never reach 0. We'll look at this function in Chapter 17. (Goguen's membership function does not, however, increase the value of the Gödel version of the Principle of Charity.)

We hasten to make one technical point: although the choice of membership function can make a clear difference here, the *logical* results like validity and n-degree validity do not depend on any particular membership function.

5 I am grateful to my colleague Michael Albertson for pointing out that the restrictions used in this analysis yield a simple but good approximation to the n-degree-validity of the Sorites argument.

approximate analysis of the n-degree-validity of the Sorites argument when weak product conjunction is used in place of bold product conjunction, and to discuss whether either product version of the Sorites argument displays decaying validity.

14.7 The Sorites Paradox: Comparison of Fuzzy$_{Ł\forall}$, Fuzzy$_{G\forall}$, and Fuzzy$_{P\forall}$

Here is a table summarizing the different behavior of the three versions of fuzzy logic with respect to the Sorites argument as we've symbolized it:

	Principle of Charity on Interpretation SST	*Sorites argument valid?*	*N-degree-validity of Sorites argument*	*Decaying validity for Sorites argument?*
Fuzzy$_{Ł\forall}$	Close to true	Yes	Low n-degree-validity	Yes
Fuzzy$_{G\forall}$	False, and the only way the principle can be close to true is to have a membership function for *tall* that assigns a high degree of membership to every height or that assigns nondecreasing membership values for decreasing heights	Yes	1-degree-valid	No
Fuzzy$_{P\forall}$	False, but can be made close to true if every height has a degree of membership greater than 0 in the fuzzy set of tall heights	Yes	Low n-degree-validity	*Assigned as exercise*

Since Chapter 11 we've indicated our preference for Łukasewicz fuzzy logic to handle issues arising from vagueness, and these results show a good reason for this preference. But there is one more important comparison we'll need to make in Chapter 15: the extent to which the three varieties of fuzzy first-order logic are axiomatizable.

14.8 Exercises

SECTION 14.1

1 Consider the "wealth" version of the Sorites paradox:

Anyone who has a million dollars is wealthy.
Anyone who has only one dollar less than a wealthy person is also wealthy.
_____
Anyone who has ten dollars is wealthy.

a. Symbolize this argument using *Wx* to mean: *x dollars is wealth*, and letting the domain consist of the set of integers from 10 to 1,000,000.

b. Using the example for the predicate *tall* in this chapter as a model, find a formula for assigning a fuzzy value in the range [0. .1] to each member of the domain for the predicate *W*, such that

$I(W)(<10>) = 0,$

$I\,(W)(<1{,}000{,}000>) = 1,$

and all other members of the domain are assigned values between 0 and 1 as seems appropriate but subject to the stipulation that $I(W)(<x>) \leq I(W)(<y>)$ when $x < y$.

SECTION 14.2

2 Consider an interpretation I whose domain D consists of the positive integers {1, 2, 3, ...} and that makes the following assignments:

$I(M)(<x>) = {}^{1}/_{x}$

$I(N)(<x_1,x_2>) = {}^{1}/_{(x1\,+\,x2)}$

$I(a) = 1$

$I(b) = 5$

$I(c) = 20$

Using the semantic clauses for $\text{Fuzzy}_{\text{Ł}\forall}$, what is the (fuzzy) value assigned to each of the following formulas?

a. $Ma \wedge Mc$

b. $Mb \;\&\; Mc$

c. $(\exists x)Mx$

d. $(\exists x)\neg Mx$

e. $(\exists x)(Mx \;\&\; Nxx)$

f. $(\exists x)(Mx \vee Nxx)$

g. $(\exists x)(Mx \nabla Nxx)$

h. $(\forall x)(Mx \vee \neg Mx)$

i. $(\forall x)(Nxx \leftrightarrow \neg Nxx)$

j. $(\forall x)(Nxx \rightarrow Mx)$

k. $(\forall x)(\forall y)(Nxy \rightarrow Mx)$

l. $(\forall x)(\forall y)(Nxy \rightarrow (Mx \wedge My))$

m. $(\forall x)(\forall y)(Nxy \rightarrow (Mx \;\&\; My))$

3 What are the values of the formulas i–m in Exercise 2 when the Kleene conditional and biconditional are used in place of the Łukasiewicz connectives?

SECTION 14.3

4 For each of the following formulas, state the degree *n* to which it is an *n*-tautology in $\text{Fuzzy}_{\text{Ł}\forall}$, and explain your answer:

a. $(\forall x)Px \rightarrow (\exists x)Px$

b. $(\forall x)(Px \rightarrow (\exists x)Px)$

c. $(\exists x)(\forall y)Lxy \rightarrow (\forall y)(\exists x)Lxy$

d. $(\forall x)(\exists y)Lxy \rightarrow (\exists y)(\forall x)Lxy$
e. $(\forall x)Px \vee (\forall x)\neg Px$
f. $(\forall x)Px \nabla (\forall x)\neg Px$
g. $(\forall x)Px \vee (\exists x)\neg Px$
h. $(\forall x)Px \nabla (\exists x)\neg Px$

5 For each of the following arguments, state the degree *n* to which it is *n*-degree-valid in $\text{Fuzzy}_{\text{L}\forall}$, and explain your answer:

a. $\begin{array}{l}\underline{(\forall x)Px}\\ (\exists x)Px\end{array}$

b. $\begin{array}{l}\underline{(\exists x)Px}\\ (\forall x)Px\end{array}$

c. $\begin{array}{l}(\forall x)Px\\ \underline{(\forall x)Rx}\\ (\exists x)(Px \wedge Rx)\end{array}$

d. $\begin{array}{l}(\forall x)Px\\ \underline{(\forall x)Rx}\\ (\exists x)(Px \,\&\, Rx)\end{array}$

e. $\begin{array}{l}\underline{(\forall x)Px}\\ Qa \rightarrow Pa\end{array}$

SECTION 14.4

6 Show that the formulas $(\exists x)(\neg\Delta Tx \wedge \neg\Delta\neg Tx)$ and $(\exists x)(\neg\Delta Tx \wedge \Delta Tx)$ are not equivalent in $\text{Fuzzy}_{\text{L}\forall}$ by producing an interpretation on which the formulas have different values.

7 We claimed that any variation of SST that assigns to the predicate *T* in our Sorites argument a noncrisp interpretation in which at least two heights have different degrees of tallness and tallness never increases as height decreases (but that keeps the rest of SST the same) will give the Principle of Charity premise a value less than 1 in $\text{Fuzzy}_{\text{L}\forall}$. Prove this claim.

8 Explain why both Ts_1 and the weak conjunction version of the Principle of Charity premise, $(\forall x)(\forall y)((Tx \wedge Eyx) \rightarrow Ty)$, have the truth-value $^{192}/_{193}$ on interpretation SST*.

9 Prove that on the interpretation

D: set of heights between 4′ 7″ and 6′ 7″ by $^{1}/_{8}$″ increments, inclusive
$I(T)(<x>) = (2 \cdot (x - 4'\,7'')) \,/\, 48^{1}/_{8}''$
$I(E)(<x_1,x_2>) = {}^{384}/_{385}$ for all x_1, x_2
$I(s_1) = 6'\,7''$
$I(s_2) = 6'\,6^{7}/_{8}''$
. . .
$I(s_{193}) = 4'\,7''$

$I(Ts_1) = I(Es_2s_1) = I(Es_3s_2) = \cdots = I(Es_{193}s_{192}) = I((\forall x)\,(\forall y)\,((Tx \,\&\, Eyx) \rightarrow Ty)) = {}^{384}/_{385}$ in $\text{Fuzzy}_{\text{L}\forall}$, while $I(Ts_{193}) = 0$.

10 Assume that v is a variable assignment (for an interpretation) such that the values of *Tx* and *Eyx* are identical. What is the relation between the values of $(Tx \wedge Eyx) \rightarrow Ty$ and $(Tx \,\&\, Eyx) \rightarrow Ty$ on v? Be as specific as you can.

11 Prove that .5 is the minimum value that max $({}^{191}/_{192} - \mathrm{I_v}(Ty), \mathrm{I_v}(Ty))$ can have for any variable assignment v on interpretation SST.

12 a. We claimed that when we use the Kleene conditional in a Sorites argument we don't get decaying validity as we increase the subscript in the conclusion, but that rather the argument remains .5-degree-valid. Prove this claim. (When we increase the subscript we will automatically add the necessary additional $Es_{i+1}Es_i$ premises.)

b. A Kleenean Sorites argument is also .5-degree valid if we decrease the subscript in the conclusion—to a point. What is the smallest subscript for which the Kleenean argument is .5-degree-valid? Prove that you are right.

SECTION 14.5

13 What are the values of the formulas in Exercise 2 when the Łukasiewicz connectives are replaced with Gödel connectives?

14 What is the value of the Gödel conjunctive version of the Principle of Charity, $(\forall x)(\forall y)\neg_G((Tx \,\&_G\, Eyx) \,\&_G\, \neg_G Ty)$, on interpretation SST?

15 Show that the Gödel conjunctive version of the Sorites argument (using the Principle of Charity $(\forall x)(\forall y)\neg_G((Tx \,\&_G\, Eyx) \,\&_G\, \neg_G Ty)$) is not valid in $\mathrm{Fuzzy}_{G\forall}$.

16 Show that if we modify the conclusion of the Gödel conjunctive version of the Sorites argument to be $\neg_G\neg_G Ts_{193}$, the resulting argument *is* valid in $\mathrm{Fuzzy}_{G\forall}$.

17 Analyze Max Black's fringe formula $(\exists x)(\neg_G Tx \wedge_G \neg_G\neg_G Tx)$ in $\mathrm{Fuzzy}_{G\forall}$: what is the least value that this formula can have? What is the greatest value?

SECTION 14.6

18 What are the values of the formulas in Exercise 2 when the Łukasiewicz connectives are replaced with product connectives?

19 We claimed that if we choose a membership function for *tall* that doesn't assign 0 as the degree of membership for any height, then a Sorites argument in $\mathrm{Fuzzy}_{P\forall}$ can still have a conclusion whose value is very close to 0 while the Principle of Charity premise has a much higher value. Show that this is so by modifying SST's definition for the fuzzy set denoted for *T* to be:

$$\mathrm{I(T)(<x>)} = \begin{cases} (x - 4'\,7'')\,/\,24'' & \text{if } x > 4'\,7'' \\ {}^{1}/_{192} & \text{if } x = 4'\,7'' \end{cases}$$

and then determining the value of the Principle of Charity premise (with either bold or weak conjunction in the antecedent).

20 What is the value of the bold product conjunctive version of the Principle of Charity, $(\forall x)(\forall y)\neg_P((Tx \,\&_P\, Eyx) \,\&_P\, \neg_P Ty)$, on interpretation SST?

21 Show that the bold product conjunctive version of the Sorites argument (using the Principle of Charity $(\forall x)(\forall y)\neg_P((Tx \,\&_P\, Eyx) \,\&_P\, \neg_P Ty)$) is not valid in $\mathrm{Fuzzy}_{P\forall}$.

22 Show that if we modify the conclusion of the bold product conjunctive version of the Sorites argument to be $\neg_P\neg_P Ts_{193}$, the resulting argument *is* valid in $\text{Fuzzy}_{P\forall}$.

23 Analyze the n-degree-validity of the (conditional version of the) Sorites argument in $\text{Fuzzy}_{P\forall}$ when weak product conjunction is used instead of bold product conjunction in the Principle of Charity premise. To simplify, you may restrict your attention to interpretations on which the $Es_{i+1}s_i$ premises all have the value 1—but be sure to explain why this restriction won't produce a wildly inaccurate anaysis. Try to think of other simplifications that may help in the analysis.

24 Does either product version of the Sorites argument display decaying validity?

25 Analyze Max Black's fringe formula $(\exists x)\ (\neg_P Tx \wedge_P \neg_P\neg_P Tx)$ in $\text{Fuzzy}_{P\forall}$: what is the least value that this formula can have? What is the greatest value?

15 Derivation Systems for Fuzzy First-Order Logic

15.1 Axiomatic Systems for Fuzzy First-Order Logic: Overview

Of the three varieties of fuzzy first-order logic that we presented in Chapter 14, only $Fuzzy_{G\forall}$ can be adequately axiomatized for tautologies and entailment *simpliciter*—specifically, only $Fuzzy_{\forall G}$ is *recursively axiomatizable* for tautologies and entailment. A system is ***recursively axiomatizable*** if we can produce a sound and complete set of axioms such that it can be mechanically determined whether a formula of the language is (or is not) an axiom. All of our axiomatizations thus far have been recursive axiomatizations. In all of the systems other than $F_ŁPA$ we've listed all of the axiom schemata and so checking whether a formula is an axiom amounts to checking whether its form instantiates one of the finitely many axiom schemata, a mechanical procedure for sure. And in the Pavelka-style system $F_ŁPA$ for $Fuzzy_{Ł\forall}$, it can in addition be mechanically determined whether or not formulas meet the specifications of axioms $F_ŁP5.1$–$F_ŁP6.2$ involving truth-value constants and their values, specifications like:

$$p = min\,(1, 1 - m + n).$$

$Fuzzy_{Ł\forall}$ and $Fuzzy_{\forall P}$ are not recursively axiomatizable for tautologies and entailment. This is a significant limitation, one that for example negatively affects our ability to define tautology and entailment conditions computationally for these systems.[1] On the other hand, $Fuzzy_{Ł\forall}$ (but not the other two systems) can be axiomatized in a Pavelka-style system that captures *fuzzy* consequence.[2] $Fuzzy_{G\forall}$ and $Fuzzy_{P\forall}$ are not axiomatizable in Pavelka-style systems for the same reason that the

[1] The negative result for $Fuzzy_{Ł\forall}$ was established in Scarpellini (1962). Gottwald (2001) has an excellent discussion of this negative result and presents weaker positive results; for example, if one allows for inference rules that work on *infinite* numbers of formulas, one can get a weakly complete and sound system for $Fuzzy_{Ł\forall}$. We'll see another weaker positive result in Section 15.2.

[2] Fuzzy consequence is now defined in terms of interpretations rather than truth-values: each fuzzy interpretation that makes every formula in a *fuzzy* set Φ of closed formulas of $Fuzzy_{Ł\forall}$ at least as true as its degree of membership in Φ is a ***consonant interpretation*** for the set Φ, and the **fuzzy consequence** of a fuzzy set of formulas Φ is defined to be another fuzzy set, the fuzzy set in which the degree of membership for any closed formula **P** is the greatest lower bound of the truth-values that **P** can have on any consonant interpretation for Φ.

propositional Gödel and product logics aren't: they have noncontinuous propositional operations. Here is a table summarizing the situation:

	Weakly complete axiomatization	*Strongly complete axiomatization*	*Fuzzily complete axiomatization*
Fuzzy$_{L\forall}$	No	No	Yes
Fuzzy$_{G\forall}$	Yes	Yes	No
Fuzzy$_{P\forall}$	No	No	No

15.2 A Pavelka-Style Derivation System for Fuzzy$_{Ł\forall}$

In this section we present a Pavelka-style system for Fuzzy$_{Ł\forall}$, which we call $F_Ł\forall PA$[3]. We must add atomic formulas denoting the rational values in the unit interval to the language Fuzzy$_{Ł\forall}$—we call the expanded system RFuzzy$_{Ł\forall}$. The axioms of $F_Ł\forall$PA include the axiom schemata for the propositional system $F_Ł$PA:

$F_Ł\forall$P1. [**P** → (**Q** → **P**), 1]
$F_Ł\forall$P2. [(**P** → **Q**) → ((**Q** → **R**) → (**P** → **R**)), 1]
$F_Ł\forall$P3. [(¬**P** → ¬**Q**) → (**Q** → **P**), 1]
$F_Ł\forall$P4. [((**P** → **Q**) → **Q**) → ((**Q** → **P**) → **P**), 1]
$F_Ł\forall$P5.1. *Includes every graded formula* [(**m** → **n**) → **p**, 1] *where **m**, **n**, and **p** are atomic formulas denoting rational truth-values m, n, and p in the unit interval such that* $p = min\ (1, 1 - m + n)$
$F_Ł\forall$P5.2. *Includes every graded formula* [**p** → (**m** → **n**), 1] *where **m**, **n**, and **p** are as in $F_Ł\forall$P5.1*
$F_Ł\forall$P6.1. *Includes every graded formula* [¬**m** → **p**, 1] *where **m** and **p** are atomic formulas denoting rational truth-values m and p such that* $p = 1 - m$
$F_Ł\forall$P6.2. *Includes every graded formula* [**p** → ¬**m**, 1] *where **m** and **p** are as in $F_Ł\forall$P6.1*
$F_Ł\forall$P7. *Includes* [**m**, *m*] *for any rational value m in the unit interval, where **m** is the atomic formula that denotes the value m*

$F_Ł\forall$PA also has two axiom schemata for quantifiers:

$F_Ł\forall$P8. [(∀**x**)(**P** → **Q**) → (**P** → (∀**x**)**Q**), 1]
where **P** is a formula in which **x** does not occur free
$F_Ł\forall$P9. [(∀**x**)**P** → **P(a/x)**, 1]
where **a** is any individual constant and the expression **P(a/x)** means: *the result of substituting the constant **a** for the variable **x** wherever **x** occurs free in* **P**.

3 Vilém Novák (1990) generalized Pavelka-syle systems for Łukasiewicz fuzzy propositional logic to *first-order* systems. $F_Ł\forall$PA is from Novák, Perfilieva, and Močkoř (1999), which owes partly to Hájek (1997).

The rules of $\text{F}_\text{L}\forall$PA are

> **MP**. From $[\mathbf{P}, m]$ and $[\mathbf{P} \to \mathbf{Q}, n]$, infer $[\mathbf{Q}, p]$, where $p = \max(0, m + n - 1)$
>
> **TCI.** From $[\mathbf{P}, m]$ infer $[\mathbf{m} \to \mathbf{P}, 1]$,
> where **m** is the atomic formula that denotes the value m
>
> **UG.** From $[P(a/x), m]$ **infer** $[(\forall x)P, m]$
> where **x** is any individual variable, provided that no assumption contains the constant **a** and that **P** does not contain the constant **a**

The rule UG tells us that if an arbitrary instance of a universal quantification has at least the value m, we may infer that the quantified formula also has at least the value m. This is acceptable because what is true of an arbitrary instance must be true of all instances, so every instance of the quantified formula must have at least the value that an arbitrary instance has.

Using the definition of existential quantification in $\text{Fuzzy}_{\text{L}\forall}$:

$$(\exists \mathbf{x})\mathbf{P} =_{\text{def}} \neg(\forall \mathbf{x})\neg\mathbf{P}$$

we have the following derived axiom schemata (we start numbering at 20 because we have all of the derived axioms—as well as the derived rules—from the axiomatic system F_LPA):

> **$\text{F}_\text{L}\forall$PD20.** $[(\forall \mathbf{x})(\mathbf{P} \to \mathbf{Q}) \to ((\exists \mathbf{x})\mathbf{P} \to \mathbf{Q}), 1]$
> where **Q** is a formula in which **x** does not occur free

Justification. We derive the formula $[(\forall \mathbf{x})(\mathbf{P} \to \mathbf{Q}) \to (\neg(\forall \mathbf{x})\neg\mathbf{P} \to \mathbf{Q}), 1]$, where **x** does not occur free in **Q** and **a** is a constant that does not occur in **P** or **Q**:

1	$[(\forall x)(P \to Q) \to (P(a/x) \to Q), 1]$	$\text{F}_\text{L}\forall$P9, with $(\forall x)(P \to Q)$ / $(\forall x)P$, a/a
	Note that $(P \to Q)(a/x)$ is just $P(a/x) \to Q$, because x does not occur free in Q	
2	$[(\forall x)(P \to Q) \to (\neg Q \to \neg P(a/x)), 1]$	1, GCON
3	$[(\forall x)((\forall x)(P \to Q) \to (\neg Q \to \neg P)), 1]$	2, UG
4	$[(\forall x)((\forall x)(P \to Q) \to (\neg Q \to \neg P)) \to$ $((\forall x)(P \to Q) \to (\forall x)(\neg Q \to \neg P)), 1]$	$\text{F}_\text{L}\forall$P8, with $(\forall x)((\forall x)(P \to Q) \to$ $(\neg Q \to \neg P))$ / $(\forall x)(P \to Q)$
5	$[(\forall x)(P \to Q) \to (\forall x)(\neg Q \to \neg P), 1]$	3,4 MP
6	$[(\forall x)(\neg Q \to \neg P) \to (\neg Q \to (\forall x)\neg P), 1]$	$\text{F}_\text{L}\forall$P8, with $(\forall x)(\neg Q \to \neg P)$ / $(\forall x)(P \to Q)$
7	$[(\forall x)(P \to Q) \to (\neg Q \to (\forall x)\neg P), 1]$	5,6 HS
8	$[(\forall x)(P \to Q) \to (\neg(\forall x)\neg P \to \neg\neg Q), 1]$	7, GCON
9	$[\neg\neg Q \to Q, 1]$	$\text{F}_\text{L}\forall$PD8, with Q / P
10	$[(\forall x)(P \to Q) \to (\neg(\forall x)\neg P \to Q), 1]$	8,9 GHS

> **$\text{F}_\text{L}\forall$PD21.** $[\mathbf{P}(\mathbf{a}/\mathbf{x}) \to (\exists \mathbf{x})\mathbf{P}, 1]$

Justification: Justification is left as an exercise.

We mention these derived axioms because they will appear as explicit axioms in the system for $\text{Fuzzy}_{\text{G}\forall}$ in the next section.

The following derivation produces a graded formula asserting that the universally quantified Law of Excluded Middle using bold disjunction has at least the value 1 in $\text{Fuzzy}_{\text{Ł}\forall}$. Again we express $Ta \,\underline{\vee}\, \neg Ta$ as $\neg Ta \rightarrow \neg Ta$:

1	$[\neg Ta \rightarrow \neg Ta, 1]$	$F_{\text{Ł}}\forall$PD7, with $\neg$Ta / **P**
2	$[(\forall x)(\neg Tx \rightarrow \neg Tx), 1]$	1, UG

On the other hand, the Law of Excluded Middle using weak disjunction can only be proved to have at least the value .5. Here is a derivation analogous to that in Chapter 8, where we rewrite disjunction in terms of negation and the conditional so that *$(\forall x)(Tx \vee \neg Tx)$* becomes *$(\forall x)((Tx \rightarrow \neg Tx) \rightarrow \neg Tx)$*:

1	$[\neg Ta \rightarrow ((Ta \rightarrow \neg Ta) \rightarrow \neg Ta), 1]$	$F_{\text{Ł}}\forall$P1, with $\neg$Ta / **P**, Ta $\rightarrow \neg$Ta / **Q**
2	$[({}^{1}/_{2} \rightarrow \neg Ta) \rightarrow ((\neg Ta \rightarrow ((Ta \rightarrow \neg Ta) \rightarrow \neg Ta)) \rightarrow$ $({}^{1}/_{2} \rightarrow ((Ta \rightarrow \neg Ta) \rightarrow \neg Ta))), 1]$	$F_{\text{Ł}}\forall$P2, with ${}^{1}/_{2}$ / **P**, $\neg$Ta / **Q**, (Ta $\rightarrow \neg$Ta) $\rightarrow \neg$Ta / **R**
3	$[({}^{1}/_{2} \rightarrow \neg Ta) \rightarrow ({}^{1}/_{2} \rightarrow ((Ta \rightarrow \neg Ta) \rightarrow \neg Ta))), 1]$	1,2 GMP
4	$[Ta \rightarrow ((Ta \rightarrow \neg Ta) \rightarrow \neg Ta), 1]$	$F_{\text{Ł}}\forall$PD3, with Ta / **P**, $\neg$Ta / **Q**
5	$[({}^{1}/_{2} \rightarrow Ta) \rightarrow ((Ta \rightarrow ((Ta \rightarrow \neg Ta) \rightarrow \neg Ta)) \rightarrow$ $({}^{1}/_{2} \rightarrow ((Ta \rightarrow \neg Ta) \rightarrow \neg Ta))), 1]$	$F_{\text{Ł}}\forall$P2, with ${}^{1}/_{2}$ / **P**, Ta / **Q**, (Ta $\rightarrow \neg$Ta) $\rightarrow \neg$Ta / **R**
6	$[({}^{1}/_{2} \rightarrow Ta) \rightarrow ({}^{1}/_{2} \rightarrow ((Ta \rightarrow \neg Ta) \rightarrow \neg Ta)),]$	4,5 GMP
7	$[({}^{1}/_{2} \rightarrow Ta) \vee (Ta \rightarrow {}^{1}/_{2}), 1]$	$F_{\text{Ł}}\forall$PD13, with ${}^{1}/_{2}$ / **P**, Ta / **Q**
8	$[({}^{1}/_{2} \rightarrow Ta) \vee (\neg{}^{1}/_{2} \rightarrow \neg Ta), 1]$	7, GCON
9	$[\neg{}^{1}/_{2} \rightarrow {}^{1}/_{2}, 1]$	$F_{\text{Ł}}\forall$P6.1
10	$[{}^{1}/_{2} \rightarrow \neg{}^{1}/_{2}, 1]$	$F_{\text{Ł}}\forall$P6.2
11	$[({}^{1}/_{2} \rightarrow Ta) \vee ({}^{1}/_{2} \rightarrow \neg Ta), 1]$	8,9,10 SUB
12	$[{}^{1}/_{2} \rightarrow ((Ta \rightarrow \neg Ta) \rightarrow \neg Ta), 1]$	3,6,11 DS
13	$[{}^{1}/_{2}, .5]$	$F_{\text{Ł}}\forall$P7
14	$[(Ta \rightarrow \neg Ta) \rightarrow \neg Ta, .5]$	12,13 MP
15	$[(\forall x)((Tx \rightarrow \neg Tx) \rightarrow \neg Tx), .5]$	14, UG

Next we'll construct a derivation for the weak conjunction version of the Sorites argument that concludes that if the Principle of Charity premise has at least the value ${}^{191}/_{192}$ and the other premises have the value 1, then the conclusion has at least the value 0. The derivation here generalizes a derivation from Section 13.3 of Chapter 13. (As with that derivation, this derivation is longer than it need be; we can derive $[Ts_{193}, 0]$ as a theorem using $F_{\text{Ł}}\forall$PD19, without any of the Sorites premises. But we include the longer derivation because it illustrates decaying validity.) For the derivation we'll use the rule

WCI (Weak Conjunction Introduction). From [**P**, *m*] and [**Q**, *n*] infer [**P** $\wedge$ **Q**, *p*] where $p = \min(m, n)$

which is derivable in $F_{\text{Ł}}$PA and therefore also in $F_{\text{Ł}}\forall$PA.

1	$[Ts_1, 1]$	Assumption
2	$[Es_2s_1, 1]$	Assumption
3	$[Es_3s_2, 1]$	Assumption
...	...	
193	$[Es_{193}s_{192}, 1]$	Assumption
194	$[(\forall x)(\forall y)((Tx \wedge Eyx) \rightarrow Ty), {}^{191}/_{192}]$	Assumption
195	$[(\forall x)(\forall y)((Tx \wedge Eyx) \rightarrow Ty) \rightarrow$	$F_Ł\forall P9$, with $(\forall x)(\forall y)((Tx \wedge Eyx) \rightarrow Ty)$ / $(\forall \mathbf{x})\mathbf{P}$,
	$(\forall y)((Ts_1 \wedge Eys_1) \rightarrow Ty), 1]$	s_1 / **a**
196	$[(\forall y)((Ts_1 \wedge Eys_1) \rightarrow Ty), {}^{191}/_{192}]$	194,195 MP
197	$[(\forall y)((Ts_1 \wedge Eys_1) \rightarrow Ty) \rightarrow$	$F_Ł\forall P9$, with $(\forall y)((Ts_1 \wedge Eys_1) \rightarrow Ty)$ / $(\forall \mathbf{x})\mathbf{P}$,
	$((Ts_1 \wedge Es_2s_1) \rightarrow Ts_2), 1]$	s_2 / **a**
198	$[(Ts_1 \wedge Es_2s_1) \rightarrow Ts_2, {}^{191}/_{192}]$	196,197 MP
199	$[Ts_1 \wedge Es_2s_1, 1]$	1,2 WCI
200	$[Ts_2, {}^{191}/_{192}]$	198,199 MP
201	$[(\forall x)(\forall y)((Tx \wedge Eyx) \rightarrow Ty) \rightarrow$	$F_Ł\forall P9$, with $(\forall x)(\forall y)((Tx \wedge Eyx) \rightarrow Ty)$ / $(\forall \mathbf{x})\mathbf{P}$,
	$(\forall y)((Ts_2 \wedge Eys_2) \rightarrow Ty), 1]$	s_2 / **a**
202	$[(\forall y)((Ts_2 \wedge Eys_2) \rightarrow Ty), {}^{191}/_{192}]$	194,201 MP
203	$[(\forall y)((Ts_2 \wedge Eys_2) \rightarrow Ty) \rightarrow$	$F_Ł\forall P9$, with $(\forall y)((Ts_2 \wedge Eys_2) \rightarrow Ty)$ / $(\forall \mathbf{x})\mathbf{P}$,
	$((Ts_2 \wedge Es_3s_2) \rightarrow Ts_3), 1]$	s_3 / **a**
204	$[(Ts_2 \wedge Es_3s_2) \rightarrow Ts_3, {}^{191}/_{192}]$	202,203 MP
205	$[Ts_2 \wedge Es_3s_2, {}^{191}/_{192}]$	3,200 WCI
206	$[Ts_3, {}^{190}/_{192}]$	204,205 MP
...	... *{repeating 201–206 with appropriate substitutions we arrive at}*	
1346	$[Ts_{193}, 0]$	1344,1345 MP

With obvious substitutions we can also produce a derivation for the *bold* conjunction version of the Sorites argument, using the rule

BCI (Bold Conjunction Introduction). From $[\mathbf{P}, m]$ and $[\mathbf{Q}, n]$ infer $[\mathbf{P} \,\&\, \mathbf{Q}, k]$ where $k = \max(0, m + n - 1)$

which was also derived in $F_ŁPA$.

It is left as an exercise to produce derivations that show that if the Principle of Charity premise in either version of the Sorites argument has at *most* the value ${}^{191}/_{192}$ and the other premises have at most the value 1, then the conclusion has at *most* the value 0. Together these derivations show that if the Principle of Charity premise in the Sorites argument (either version) has *exactly* the value ${}^{191}/_{192}$ and all of the other premises are true, the conclusion is false.

Recall that our Kleenean version of the Sorites argument is .5-valid. It should therefore be possible to produce derivations in which the premises are all graded with values greater than .5 and the grade of the conclusion is .5 less than the least of these. In fact, in some cases we can infer a higher grade for the conclusion (this doesn't contradict .5-validity, which gives an *upper bound* of .5 on the distance between the values of the premises and that of the conclusion). For example, if the premises Ts_1, Es_2s_1, $Es_3s_2, \ldots,$ $Es_{193}s_{192}$, and $(\forall x)(\forall y)((Tx \wedge_K Eyx) \rightarrow_K Ty)$ all have the value .6, then so must the conclusion. The Kleenean Principle of Charity is equivalent to $(\forall x)(\forall y)(\neg(Tx \wedge Eyx) \vee Ty)$ using only Łukasiewicz's connectives, and if both Ts_1 and Es_2s_1 have the value .6 then $\neg(Tx \wedge Eyx)$ has the value .4 and so Ts_2 must have the value .6 to give the disjunction the value .6. This reasoning may be repeated over and over again, finally showing that Ts_{193} must also have the value .6 as well. Here is a derivation showing that if all the premises have at least the value .6, so must the conclusion (we switch freely between the equivalent forms $\mathbf{P} \vee \mathbf{Q}$ and $(\mathbf{P} \rightarrow \mathbf{Q}) \rightarrow \mathbf{Q}$, depending on whether we are using the formula as a disjunction or as a conditional). We begin by deriving the formula $(Ts_2 \rightarrow {}^4/_{10}) \rightarrow {}^8/_{10}$, which contains only Ts_2 and truth-value constants:

1	$[Ts_1, .6]$	Assumption
2	$[Es_2s_1, .6]$	Assumption
3	$[Es_3s_2, .6]$	Assumption
…	…	
193	$[Es_{193}s_{192}, .6]$	Assumption
194	$[(\forall x)(\forall y)(\neg(Tx \wedge Eyx) \vee Ty), .6]$	Assumption
195	$[(\forall x)(\forall y)(\neg(Tx \wedge Eyx) \vee Ty) \rightarrow (\forall y)(\neg(Ts_1 \wedge Eys_1) \vee Ty), 1]$	$F_L\forall$P9, with $(\forall x)(\forall y)(\neg(Tx \wedge Eyx) \vee Ty)$ / $(\forall \mathbf{x})\mathbf{P}$, s_1 / $\mathbf{a}$
196	$[(\forall y)(\neg(Ts_1 \wedge Eys_1) \vee Ty), .6]$	194,195 MP
197	$[(\forall y)(\neg(Ts_1 \wedge Eys_1) \vee Ty) \rightarrow (\neg(Ts_1 \wedge Es_2s_1) \vee Ts_2), 1]$	$F_L\forall$P9, with $(\forall y)(\neg(Ts_1 \wedge Eys_1) \vee Ty)$ / $(\forall \mathbf{x})\mathbf{P}$, s_2 / $\mathbf{a}$
198	$[\neg(Ts_1 \wedge Es_2s_1) \vee Ts_2, .6]$	196,197 MP
199	$[{}^6/_{10} \rightarrow (\neg(Ts_1 \wedge Es_2s_1) \vee Ts_2), 1]$	198, TCI
200	$[Ts_1 \wedge Es_2s_1, .6]$	1,2 WCI
201	$[{}^6/_{10} \rightarrow (Ts_1 \wedge Es_2s_1), 1]$	200, TCI
202	$[\neg(Ts_1 \wedge Es_2s_1) \rightarrow \neg{}^6/_{10}, 1]$	201, GCON
203	$[\neg{}^6/_{10} \rightarrow {}^4/_{10}, 1]$	$F_L\forall$P6.1
204	$[\neg(Ts_1 \wedge Es_2s_1) \rightarrow {}^4/_{10}, 1]$	202,203 HS
205	$[{}^4/_{10} \rightarrow (Ts_2 \vee {}^4/_{10}), 1]$	$F_L\forall$P1, with ${}^4/_{10}$ / $\mathbf{P}$, $Ts_2 \rightarrow {}^4/_{10}$ / $\mathbf{Q}$
	{Note: $Ts_2 \vee {}^4/_{10}$ is defined to be $(Ts_2 \rightarrow {}^4/_{10}) \rightarrow {}^4/_{10}$}	
206	$[\neg(Ts_1 \wedge Es_2s_1) \rightarrow (Ts_2 \vee {}^4/_{10}), 1]$	204,205 HS
207	$[Ts_2 \rightarrow (Ts_2 \vee {}^4/_{10}), 1]$	$F_L\forall$PD3, with Ts_2 / $\mathbf{P}$, ${}^4/_{10}$ / $\mathbf{Q}$
208	$[(\neg(Ts_1 \wedge Es_2s_1) \vee Ts_2) \rightarrow (Ts_2 \vee {}^4/_{10}), 1]$	206,207 DC
209	$[{}^6/_{10} \rightarrow ((Ts_2 \rightarrow {}^4/_{10}) \rightarrow {}^4/_{10}), 1]$	199,208 HS
210	$[(Ts_2 \rightarrow {}^4/_{10}) \rightarrow ({}^6/_{10} \rightarrow {}^4/_{10}), 1]$	209, TRANS
211	$[({}^6/_{10} \rightarrow {}^4/_{10}) \rightarrow {}^8/_{10}, 1]$	$F_L\forall$P5.1
212	$[(Ts_2 \rightarrow {}^4/_{10}) \rightarrow {}^8/_{10}, 1]$	210,211 HS

At this point we know that Ts_2 must have at least the value .6, for if it didn't, the antecedent $Ts_2 \rightarrow {}^4/_{10}$ of the formula on line 212 would have a value greater than .8 and the conditional's value would be less than 1, not 1 as it is graded. Our task now is to derive the formula ${}^6/_{10} \rightarrow Ts_2$ that *says* that Ts_2 has at least the value .6. We will do this using a strategy that we saw in the derivation of rule CV in Chapter 13, first showing that Ts_2 has at least the value .2, then at least .4, then at least .6:

213	$[\mathbf{1}, 1]$	$\mathrm{F_L}\forall$P7
214	$[\mathbf{1} \rightarrow (\mathbf{0} \rightarrow {}^{4}/_{10}), 1]$	$\mathrm{F_L}\forall$P5.2
215	$[\mathbf{0} \rightarrow {}^{4}/_{10}, 1]$	213,214 MP
216	$[(\mathrm{Ts}_2 \rightarrow \mathbf{0}) \rightarrow ((\mathbf{0} \rightarrow {}^{4}/_{10}) \rightarrow (\mathrm{Ts}_2 \rightarrow {}^{4}/_{10})), 1]$	$\mathrm{F_L}\forall$P2, with Ts_2 / **P**, **0** / **Q**, ${}^{4}/_{10}$ / **R**
217	$[(\mathrm{Ts}_2 \rightarrow \mathbf{0}) \rightarrow (\mathrm{Ts}_2 \rightarrow {}^{4}/_{10}), 1]$	215,216 GMP
218	$[(\mathrm{Ts}_2 \rightarrow \mathbf{0}) \rightarrow {}^{8}/_{10}, 1]$	212,217 HS
219	$[{}^{8}/_{10} \rightarrow (({}^{8}/_{10} \rightarrow \mathbf{0}) \rightarrow \mathbf{0}), 1]$	$\mathrm{F_L}\forall$PD3, with ${}^{8}/_{10}$ / **P**, **0** / **Q**
220	$[(\mathrm{Ts}_2 \rightarrow \mathbf{0}) \rightarrow (({}^{8}/_{10} \rightarrow \mathbf{0}) \rightarrow \mathbf{0}), 1]$	218,219 HS
221	$[({}^{8}/_{10} \rightarrow \mathbf{0}) \rightarrow ((\mathrm{Ts}_2 \rightarrow \mathbf{0}) \rightarrow \mathbf{0}), 1]$	220, TRANS
222	$[((\mathrm{Ts}_2 \rightarrow \mathbf{0}) \rightarrow \mathbf{0}) \rightarrow ((\mathbf{0} \rightarrow \mathrm{Ts}_2) \rightarrow \mathrm{Ts}_2), 1]$	$\mathrm{F_L}\forall$P4, with Ts_2 / **P**, **0** / **Q**
223	$[({}^{8}/_{10} \rightarrow \mathbf{0}) \rightarrow ((\mathbf{0} \rightarrow \mathrm{Ts}_2) \rightarrow \mathrm{Ts}_2), 1]$	221,222 HS
224	$[\mathrm{Ts}_2, \mathbf{0}]$	$\mathrm{F_L}\forall$PD19, with Ts_2 / **P**
225	$[\mathbf{0} \rightarrow \mathrm{Ts}_2, 1]$	224, TCI
226	$[({}^{8}/_{10} \rightarrow \mathbf{0}) \rightarrow \mathrm{Ts}_2, 1]$	223,225 GMP
227	$[{}^{2}/_{10} \rightarrow ({}^{8}/_{10} \rightarrow \mathbf{0}), 1]$	$\mathrm{F_L}\forall$P5.2
228	$[{}^{2}/_{10} \rightarrow \mathrm{Ts}_2, 1]$	226,227 HS
229	$[\mathbf{1} \rightarrow ({}^{2}/_{10} \rightarrow {}^{4}/_{10}), 1]$	$\mathrm{F_L}\forall$P5.2
230	$[{}^{2}/_{10} \rightarrow {}^{4}/_{10}, 1]$	213, 229 MP
231	$[(\mathrm{Ts}_2 \rightarrow {}^{2}/_{10}) \rightarrow (({}^{2}/_{10} \rightarrow {}^{4}/_{10}) \rightarrow (\mathrm{Ts}_2 \rightarrow {}^{4}/_{10})), 1]$	$\mathrm{F_L}\forall$P2, with Ts_2 / **P**, ${}^{2}/_{10}$ / **Q**, ${}^{4}/_{10}$ / **R**
232	$[(\mathrm{Ts}_2 \rightarrow {}^{2}/_{10}) \rightarrow (\mathrm{Ts}_2 \rightarrow {}^{4}/_{10}), 1]$	230,231 GMP
233	$[(\mathrm{Ts}_2 \rightarrow {}^{2}/_{10}) \rightarrow {}^{8}/_{10}, 1]$	212,232 HS
234	$[{}^{8}/_{10} \rightarrow (({}^{8}/_{10} \rightarrow {}^{2}/_{10}) \rightarrow {}^{2}/_{10}), 1]$	$\mathrm{F_L}\forall$PD3, with ${}^{8}/_{10}$ / **P**, ${}^{2}/_{10}$ / **Q**
235	$[(\mathrm{Ts}_2 \rightarrow {}^{2}/_{10}) \rightarrow (({}^{8}/_{10} \rightarrow {}^{2}/_{10}) \rightarrow {}^{2}/_{10}), 1]$	233,234 HS
236	$[({}^{8}/_{10} \rightarrow {}^{2}/_{10}) \rightarrow ((\mathrm{Ts}_2 \rightarrow {}^{2}/_{10}) \rightarrow {}^{2}/_{10}), 1]$	235, TRANS
237	$[((\mathrm{Ts}_2 \rightarrow {}^{2}/_{10}) \rightarrow {}^{2}/_{10}) \rightarrow (({}^{2}/_{10} \rightarrow \mathrm{Ts}_2) \rightarrow \mathrm{Ts}_2), 1]$	$\mathrm{F_L}\forall$P4, with Ts_2 / **P**, ${}^{2}/_{10}$ / **Q**
238	$[({}^{8}/_{10} \rightarrow {}^{2}/_{10}) \rightarrow (({}^{2}/_{10} \rightarrow \mathrm{Ts}_2) \rightarrow \mathrm{Ts}_2), 1]$	236,237 HS
239	$[({}^{8}/_{10} \rightarrow {}^{2}/_{10}) \rightarrow \mathrm{Ts}_2, 1]$	228,238 GMP
240	$[{}^{4}/_{10} \rightarrow ({}^{8}/_{10} \rightarrow {}^{2}/_{10}), 1]$	$\mathrm{F_L}\forall$P5.2
241	$[{}^{4}/_{10} \rightarrow \mathrm{Ts}_2, 1]$	239,240 HS
242	$[{}^{8}/_{10} \rightarrow (({}^{8}/_{10} \rightarrow {}^{4}/_{10}) \rightarrow {}^{4}/_{10}), 1]$	$\mathrm{F_L}\forall$PD3, with ${}^{8}/_{10}$ / **P**, ${}^{4}/_{10}$ / **Q**
243	$[(\mathrm{Ts}_2 \rightarrow {}^{4}/_{10}) \rightarrow (({}^{8}/_{10} \rightarrow {}^{4}/_{10}) \rightarrow {}^{4}/_{10}), 1]$	212,242 HS
244	$[({}^{8}/_{10} \rightarrow {}^{4}/_{10}) \rightarrow ((\mathrm{Ts}_2 \rightarrow {}^{4}/_{10}) \rightarrow {}^{4}/_{10}), 1]$	243, TRANS
245	$[((\mathrm{Ts}_2 \rightarrow {}^{4}/_{10}) \rightarrow {}^{4}/_{10}) \rightarrow (({}^{4}/_{10} \rightarrow \mathrm{Ts}_2) \rightarrow \mathrm{Ts}_2), 1]$	$\mathrm{F_L}\forall$P4, with Ts_2 / **P**, ${}^{4}/_{10}$ / **Q**
246	$[({}^{8}/_{10} \rightarrow {}^{4}/_{10}) \rightarrow (({}^{4}/_{10} \rightarrow \mathrm{Ts}_2) \rightarrow \mathrm{Ts}_2), 1]$	244,245 HS
247	$[({}^{8}/_{10} \rightarrow {}^{4}/_{10}) \rightarrow \mathrm{Ts}_2, 1]$	241,246 GMP
248	$[{}^{6}/_{10} \rightarrow ({}^{8}/_{10} \rightarrow {}^{4}/_{10}), 1]$	$\mathrm{F_L}\forall$P5.2
249	$[{}^{6}/_{10} \rightarrow \mathrm{Ts}_2, 1]$	247,248 HS
...	... *{lines 195–202, 204–210, 212, 216–218, 220–226, 228, 231–233, 235–239, 241, 243–247, and 249 repeat 191 times (we don't need to repeat the formulas that contain only truth-value constants), with 1 added to the subscripts each time, ending with}*	
8271	$[{}^{6}/_{10} \rightarrow \mathrm{Ts}_{193}, 1]$	8269,8270 HS
8272	$[{}^{6}/_{10}, .6]$	$\mathrm{F_L}\forall$P7
8273	$[\mathrm{Ts}_{193}, .6]$	8271,8272 MP

As we did for F_LPA, we will say that a formula **P** is a ***theorem to degree n*** in $F_L\forall PA$ if *n* is the least upper bound of the values *m* such that there is a derivation of the graded formula [**P**, *m*], and that a formula **P** is **derivable to degree *n*** from a set Γ of graded formulas if *n* is the least upper bound of the degrees *m* such that [**P**, *m*] is derivable from the graded formulas in Γ.

The system $F_L\forall PA$ is **fuzzy sound**: every formula that is a theorem to degree *n* in $F_L\forall PA$ is an *n*-tautology of $RFuzzy_{L\forall}$, and if a formula **P** is derivable to degree *n* from a graded set of formulas then *n* is the greatest lower bound of the values that **P** can have in $RFuzzy_{L\forall}$ given the graded values of the set of formulas. (As earlier, interpretations for $RFuzzy_{L\forall}$ are like those for $Fuzzy_{L\forall}$ except that, in addition, each of the special formulas added to denote rational truth-values is assigned the truth-value that it denotes.) $F_L\forall PA$ is also **fuzzy complete**: every formula that is an *n*-tautology in $RFuzzy_{L\forall}$ is a theorem to degree *n* in $F_L\forall PA$, and a formula **P** is derivable to degree *n* from a graded set of formulas if *n* is the least upper bound of the values that **P** can have in $RFuzzy_{L\forall}$ given the graded values of the set of formulas (see Novák et al. 1999, pp. 147–150).

As a consequence of the non-recursive axiomatizability of $Fuzzy_{L\forall}$ (and $RFuzzy_{L\forall}$), soundness and (weak) completeness in the traditional sense must fail for $F_L\forall PA$ (and for any other axiomatic system for $Fuzzy_{L\forall}$). Specifically, it is traditional completeness, not soundness, that is problematic: not every formula of $RFuzzy_{L\forall}$ that has the value 1 on every interpretation has a derivation with graded value 1. But Hájek (1998b) has proved that a weaker result does hold: that a formula of $RFuzzy_{L\forall}$ will have a derivation with graded value 1 in $F_L\forall PA$ if that formula has the value *unit* on every linear MV-algebraic interpretation in which the algebra includes the rationals in the unit interval along with every glb and lub that is required by the truth-conditions for quantified formulas. (An algebra is linearly ordered if for any two elements x and y in its domain either $x \leq y$ or $y \leq x$.) Hájek also proved that a related weaker result holds for a non-Pavelka axiomatization of $Fuzzy_{L\forall}$ based on F_LA with additional axioms and rules for the quantifiers. The reason that these are weaker results is that there are tautologies over interpretations based on the unit interval [0. .1] that do not evaluate to *unit* on every algebraic interpretation in the broader classes, and so these formulas are not guaranteed by the weaker results to have derivations with graded value 1.

15.3 An Axiomatic Derivation System for $Fuzzy_{G\forall}$

Here is an axiomatic system $BL_G\forall A$ from Hájek (1997) that is both sound and complete (in the traditional sense) for $Fuzzy_{G\forall}$, the only one of our three fuzzy first-order systems for which this is possible. $BL_G\forall A$ includes the axiom schemata for BL_GA from Chapter 13, which we here list with the additional prefix ∀ since we are now working with a first-order system:

$BL_G\forall 1$. $(\mathbf{P} \rightarrow_G \mathbf{Q}) \rightarrow_G ((\mathbf{Q} \rightarrow_G \mathbf{R}) \rightarrow_G (\mathbf{P} \rightarrow_G \mathbf{R}))$
$BL_G\forall 2$. $(\mathbf{P} \,\&_G\, \mathbf{Q}) \rightarrow_G \mathbf{P}$

$BL_G\forall 3$. $(P \mathbin{\&_G} Q) \rightarrow_G (Q \mathbin{\&_G} P)$
$BL_G\forall 4$. $(P \mathbin{\&_G} (P \rightarrow_G Q)) \rightarrow_G (Q \mathbin{\&_G} (Q \rightarrow_G P))$
$BL_G\forall 5$. $(P \rightarrow_G (Q \rightarrow_G R)) \rightarrow_G ((P \mathbin{\&_G} Q) \rightarrow_G R)$
$BL_G\forall 6$. $((P \mathbin{\&_G} Q) \rightarrow_G R) \rightarrow_G (P \rightarrow_G (Q \rightarrow_G R))$
$BL_G\forall 7$. $((P \rightarrow_G Q) \rightarrow_G R) \rightarrow_G (((Q \rightarrow_G P) \rightarrow_G R) \rightarrow_G R)$
$BL_G\forall 8$. $0 \rightarrow_G P$
$BL_G\forall 9$. $P \rightarrow_G (P \mathbin{\&_G} P)$

along with the following axiom schemata for quantifiers:

$BL_G\forall 9$. $(\forall x)(P \rightarrow_G Q) \rightarrow_G (P \rightarrow_G (\forall x)Q)$
where **P** is a formula in which **x** does not occur free

$BL_G\forall 10$. $(\forall x)P \rightarrow_G P(a/x)$
where **a** is any individual constant

$BL_G\forall 11$. $(\forall x)(P \rightarrow_G Q) \rightarrow ((\exists x)P \rightarrow_G Q)$
where **Q** is a formula in which **x** does not occur free

$BL_G\forall 12$. $P(a/x) \rightarrow_G (\exists x)P$
where **a** is any individual constant

$BL_G\forall 13$. $(\forall x)(P \vee_G Q) \rightarrow_G ((\forall x)P \vee_G Q)$
where **Q** is a formula in which **x** does not occur free

(For the last axiom, recall that weak disjunction is identical to strong disjunction in Fuzzy_G and hence in $\text{Fuzzy}_{G\forall}$.) The rules are

MP. From **P** and $P \rightarrow_G Q$, infer **Q**

and

UG. From **P(a/x)**, infer **(∀x)P**
where **x** is any individual variable, provided no assumption contains the constant **a** and that **P** does not contain the constant **a**.

Note that $BL_G\forall 9$ and $BL_G\forall 10$ are (ungraded versions of) $F_Ł\forall PA$'s axiom schemata $F_Ł\forall G8$ and $F_Ł\forall G9$ in Section 15.2. We showed there that the next two axioms, $BL_G\forall 11$ and $BL_G\forall 12$, are derivable in $F_Ł\forall GA$. But here we need explicitly to include the remaining axioms to capture quantified claims because the operations are Gödel operations. In Łukasiewicz fuzzy logic we can define the existential quantifier in terms of the universal quantifier and negation—that means we don't need special axioms for the existential quantifier in addition to those for the universal quantifier. But we cannot similarly define the existential quantifier in Gödel fuzzy logic because Gödel negation behaves differently from Łukasiewicz negation (this will be further explored in an exercise).

We shall show that the conclusion of the Sorites argument using Gödel bold conjunction and the Gödel conditional:

Ts_1
Es_2s_1
Es_3s_2
Es_4s_3
...
$Es_{193}s_{192}$
$(\forall x)(\forall y)((Tx \&_G Eyx) \rightarrow_G Ty)$
———
Ts_{193}

is derivable from the premises in $B_LG\forall A$. (Recall that weak and bold conjunction are identical in $Fuzzy_{G\forall}$, so the derivation also establishes derivability when the Principle of Charity uses weak rather than bold conjunction.) First, we'll derive the rule:

BCI (Bold Conjunction Introduction). From **P** and **Q** infer **P** $\&_G$ **Q**

The rule is derived as follows (recall that HS is derivable rule in all of the BL-axiomatic systems):

1	**P**	Assumption
2	**Q**	Assumption
3	$(\mathbf{P} \&_G \mathbf{Q}) \rightarrow ((\mathbf{P} \&_G \mathbf{Q}) \&_G (\mathbf{P} \&_G \mathbf{Q}))$	$BL_G\forall 9$, with $\mathbf{P} \&_G \mathbf{Q}$ / **P**
4	$((\mathbf{P} \&_G \mathbf{Q}) \&_G (\mathbf{P} \&_G \mathbf{Q})) \rightarrow (\mathbf{P} \&_G \mathbf{Q})$	$BL_G\forall 2$, with $(\mathbf{P} \&_G \mathbf{Q})$ / **P**, $(\mathbf{P} \&_G \mathbf{Q})$ / **Q**
5	$(\mathbf{P} \&_G \mathbf{Q}) \rightarrow_G (\mathbf{P} \&_G \mathbf{Q})$	3,4 HS
6	$((\mathbf{P} \&_G \mathbf{Q}) \rightarrow_G (\mathbf{P} \&_G \mathbf{Q})) \rightarrow_G (\mathbf{P} \rightarrow_G (\mathbf{Q} \rightarrow_G (\mathbf{P} \&_G \mathbf{Q})))$	$BL_G\forall 6$, with **P** / **P**, **Q** / **Q**, $\mathbf{P} \&_G \mathbf{Q}$ / **R**
7	$\mathbf{P} \rightarrow_G (\mathbf{Q} \rightarrow_G (\mathbf{P} \&_G \mathbf{Q}))$	5,6 MP
8	$\mathbf{Q} \rightarrow_G (\mathbf{P} \&_G \mathbf{Q})$	1,7 MP
9	$\mathbf{P} \&_G \mathbf{Q}$	2,8 MP

And here's the Sorites derivation:

1	Ts_1	Assumption
2	Es_2s_1	Assumption
...	...	
193	$Es_{193}s_{192}$	Assumption
194	$(\forall x)(\forall y)((Tx \&_G Eyx) \rightarrow_G Ty)$	Assumption
195	$(\forall x)(\forall y)((Tx \&_G Eyx) \rightarrow_G Ty) \rightarrow_G (\forall y)((Ts_1 \&_G Eys_1) \rightarrow_G Ty)$	$BL_G\forall 10$, with $(\forall x)(\forall y)((Tx \&_G Eyx) \rightarrow_G Ty)$ / $(\forall \mathbf{x})\mathbf{P}$, s_1 / **a**
196	$(\forall y)((Ts_1 \&_G Eys_1) \rightarrow_G Ty)$	194,195 MP
197	$(\forall y)((Ts_1 \&_G Eys_1) \rightarrow_G Ty) \rightarrow_G ((Ts_1 \&_G Es_2s_1) \rightarrow_G Ts_2)$	$BL_G\forall 10$, with $(\forall y)((Ts_1 \&_G Eys_1) \rightarrow_G Ty)$ / $(\forall \mathbf{x})\mathbf{P}$, s_2 / **a**
198	$(Ts_1 \&_G Es_2s_1) \rightarrow_G Ts_2$	196,197 MP
199	$Ts_1 \&_G Es_2s_1$	1,2 BCI
200	Ts_2	198,199 MP
...	... *{repeating 195–200 with appropriate substitutions we arrive at}*	
1346	Ts_{193}	1344,1345 MP

which should look familiar—the reasoning is exactly the same as the reasoning displayed in the Sorites derivations in Section 15.2!

Finally, we note that the set of theorems of $B_LG\forall A$ is not decidable, not surprising given the undecidability of classical and three-valued first-order systems.

15.4 Combining Fuzzy First-Order Logical Systems; External Assertion

Our bias in favor of Łukasiewicz fuzzy logic has been clear: it (augmented with an external assertion operation) deals quite well with the issues of vagueness that have concerned us, and we also have a Pavelka-style axiomatic system to examine n-degree-validity, decaying validity, and so forth syntactically. On the other hand, Gödel negation, which is not definable in $Fuzzy_{L\forall}$, has some interest, and so do the t-norm of product logic and its residuum. We may decide, in the end, that we would also like to have these additional operations available—just as we decided that external assertion was an important operator to have. Mindful of this, and particularly keeping in mind that $Fuzzy_{L\forall}$ isn't recursively axiomatizable, researchers have produced axiomatic systems that combine these three basic fuzzy systems. For example, Esteva, Godo, and Montagna (2001) present complete axiomatizations for a fuzzy propositional system that includes both Łukasiewicz and product connectives (recall here that external assertion is definable as long as we have Łukasiewicz negation and product/Gödel negation), and Hájek (1998b) has developed an axiomatic system for fuzzy first-order logic that includes all of the Łukasiewicz, Gödel, and product connectives, based on work in Takeuti and Titani (1984).

We can also augment either of the axiomatic systems in this chapter with Matthias Baaz's external assertion axiom schemata, which we repeat here:

Δ1. $\mathbf{\Delta P \vee \neg\Delta P}$
Δ2. $\mathbf{\Delta(P \vee Q) \rightarrow (\Delta P \vee \Delta Q)}$
Δ3. $\mathbf{\Delta P \rightarrow P}$
Δ4. $\mathbf{\Delta P \rightarrow \Delta\Delta P}$
Δ5. $\mathbf{\Delta(P \rightarrow Q) \rightarrow (\Delta P \rightarrow \Delta Q)}$

and the rule

EA (External Assertion). From **P** infer **ΔP**.

In Chapter 13 we illustrated using these axioms in the Pavelka system $F_L\Delta PA$. We did note that the resulting system was not fuzzy complete for $Fuzzy_L$ augmented with external assertion, and similarly here. However, the system is fuzzy sound, so derivations that we can produce will not lead us astray. Here we will give an example using the graded system $F_L\forall\Delta PA$. The Δ axiom schemata will all be graded with the value 1, of course. And here is a graded version of EA:

EA (External Assertion). From [**P**, 1] infer [**ΔP**, 1].

Note that if a formula **P** has any value less than 1, **ΔP** will have the value 0. But we don't bother to include this as part of the rule EA because we can derive *any* formula with graded value 0.

We'll show that if *Ta* has at least the value .5, in Fuzzy_L augmented with the external assertion operator then *(∀x)¬Tx* is clearly not clearly true (recall the reading of the external assertion operator as *clearly*):

1	[Ta, .5]	Assumption
2	$[{}^1/_2 \rightarrow \text{Ta}, 1]$	1, TCI
3	$[\neg\text{Ta} \rightarrow \neg{}^1/_2, 1]$	2, GCON
4	$[\neg{}^1/_2 \rightarrow {}^1/_2, 1]$	$F_L\forall\Delta P6.1$
5	$[\neg\text{Ta} \rightarrow {}^1/_2, 1]$	3,4 HS
6	$[(\forall x)\neg\text{Tx} \rightarrow \neg\text{Ta}, 1]$	$F_L\forall\Delta P9$, with $(\forall x)\neg Tx$ / $(\forall \mathbf{x})\mathbf{Px}$, a / **a**
7	$[(\forall x)\neg\text{Tx} \rightarrow {}^1/_2, 1]$	5,6 HS
8	$[\Delta(\forall x)\neg\text{Tx} \rightarrow (\forall x)\neg\text{Tx}, 1]$	Δ3, with $\Delta(\forall x)\neg Tx$ / **P**
9	$[\Delta(\forall x)\neg\text{Tx} \rightarrow {}^1/_2, 1]$	7,8 HS
10	$[\neg{}^1/_2 \rightarrow \neg\Delta(\forall x)\neg\text{Tx}, 1]$	9, GCON
11	$[{}^1/_2 \rightarrow \neg{}^1/_2, 1]$	$F_L\forall\Delta P6.2$
12	$[{}^1/_2 \rightarrow \neg\Delta(\forall x)\neg\text{Tx}, 1]$	10,11 HS
13	$[\Delta(\forall x)\neg\text{Tx} \rightarrow \neg\Delta(\forall x)\neg\text{Tx}, 1]$	9,12 HS
14	$[(\Delta(\forall x)\neg\text{Tx} \rightarrow \neg\Delta(\forall x)\neg\text{Tx}) \rightarrow \neg\Delta(\forall x)\neg\text{Tx}, 1]$	Δ1, with $(\forall x)\neg Tx$ / **P**
15	$[\neg\Delta(\forall x)\neg\text{Tx}, 1]$	13,14 MP
16	$[\Delta\neg\Delta(\forall x)\neg\text{Tx}, 1]$	15, EA

15.5 Exercises

SECTION 15.2

1 Construct a derivation that justifies the derived axiom schema

$F_L\forall$PD21. [P(a/x) → (∃x)P, 1]

in $F_L\forall$PA.

2 Present a derivation in $F_L\forall$PA based on the method of reasoning in the corresponding "chain" derivation in Section 13.3 of Chapter 13, that shows that if the Principle of Charity premise in the weak conjunction version of the Sorites argument has at *most* the value $^{191}/_{192}$ and the other premises have at most the value 1, then the conclusion has at most the value 0.

3 Present a derivation in $F_L\forall$PA that shows that if the Principle of Charity premise in the bold conjunction version of the Sorites argument has at *most* the value $^{191}/_{192}$ and the other premises have at most the value 1, then the conclusion has at most the value 0.

4 Present a derivation that shows that if the premises of the Kleenean version of the Sorites argument all have at most the value .6, then so does the conclusion.

5 Consider the non-Pavelka axiomatic system $\mathrm{F}_{\text{Ł}}\forall\mathrm{A}$ that consists of axiom schemata $\mathrm{F}_{\text{Ł}}\forall\mathrm{P1}$–$\mathrm{F}_{\text{Ł}}\forall\mathrm{P4}$, $\mathrm{F}_{\text{Ł}}\forall\mathrm{P8}$ and $\mathrm{F}_{\text{Ł}}\forall\mathrm{P9}$, and the rules MP and UG, all with grades removed from the formulas. We know that this system must be incomplete, as explained in Section 15.1. Nevertheless, the Łukasiewicz weak conjunction Sorites argument, the Łukasiewicz strong conjunction Sorites argument, and the Kleene conditional Sorites arguments are all valid in this system. Show this.

SECTION 15.3

6 Explain why we cannot define the existential quantifier in Gödel fuzzy first-order logic as we did in classical logic; that is, explain why $(\exists \mathbf{x})\mathbf{P}$ and $\neg\,(\forall \mathbf{x})\neg_G\mathbf{P}$ are not in general equivalent in $\mathrm{Fuzzy}_{\mathrm{G}\forall}$. You may do this by giving an instance of these formulas and an interpretation on which these instances have different truth-values.

7 Construct a derivation of the modified conclusion of the Gödel conjunctive version of the Sorites paradox from its premises in $\mathrm{BL}_{\mathrm{G}}\forall\mathrm{A}$ (remember that $\neg\mathbf{P}$ is defined as $(\mathbf{P} \rightarrow \mathbf{0}) \rightarrow \mathbf{0}$):

$$\begin{array}{l} \mathrm{Ts1} \\ \mathrm{Es2s1} \\ \mathrm{Es3s2} \\ \ldots \\ \mathrm{Es193s192} \\ (\forall x)(\forall y)\neg_G((\mathrm{Tx}\ \&_G\ \mathrm{Eyx})\ \&_G\ \neg_G\mathrm{Ty}) \\ \hline \neg_G\neg_G\mathrm{Ts}_{193} \end{array}$$

SECTION 15.4

8 Construct a derivation of the graded formula $[\Delta(\forall x)\mathrm{Tx} \rightarrow \Delta(\exists x)\mathrm{Tx}, 1]$ in $\mathrm{F}_{\text{Ł}}\forall\Delta\mathrm{PA}$.

9 Construct a derivation of the graded formula $[\Delta(\forall x)\mathrm{Tx} \rightarrow (\exists x)\Delta\mathrm{Tx}, 1]$ in $\mathrm{F}_{\text{Ł}}\forall\Delta\mathrm{PA}$.

10 Construct a derivation that shows that $[\neg\Delta\Delta(\forall x)\mathrm{Tx}, 1]$ follows from $[\Delta\neg\Delta(\exists x)\mathrm{Tx}, 1]$ in $\mathrm{F}_{\text{Ł}}\forall\Delta\mathrm{PA}$.

11 Construct a derivation that shows that $[\Delta\mathrm{Ra}, 1]$ follows from $[\neg\mathrm{Pa}, .5]$ and $[(\forall x)\Delta(\mathrm{Px} \vee \mathrm{Rx}), .5]$ in $\mathrm{F}_{\text{Ł}}\forall\Delta\mathrm{PA}$.

16 Extensions of Fuzziness

16.1 Fuzzy Qualifiers: Hedges

So far we have studied fuzziness in connection with vague predicates, with our main concern the Sorites paradoxes and other logical puzzles. In this chapter we present two extensions of fuzziness. The formula apparatus that we present can be used to augment any of the fuzzy systems we've studied: Łukasiewicz, Gödel, or product.

We'll begin with fuzzy qualifiers, known as *hedges*.[1] Consider the adverb *very*. This adverb combines with vague predicates to form new vague predicates. So, for example, *very tall* is a vague predicate, as are *very bald* and *very big*. The fact that *very* combines with a given predicate implies that there are degrees of membership in the predicate's extension: we do not, for example, talk of natural numbers that are *very* prime; a number either is prime or is not.

Not only does *very* require that the predicates it qualifies be vague—it systematically produces new vague concepts by *raising the threshold* for membership in fuzzy sets. Recall the fuzzy membership function for *tall* in interpretation SST in Chapter 15, where the domain consists of heights between 4′ 7″ and 6′ 7″:

$$I(T)(<x>) = (x - 4'\,7'') \,/\, 24''$$

So I(*T*)(<6′ 7″>) = 1; I(*T*)(<6′ 5″>) = (roughly) .92; and I(*T*)(<5′ 8″>) = .54. When we say that *very* raises the threshold for membership in this fuzzy set we mean that in general the degree of membership of a height in the fuzzy set *very tall* will be less than that height's degree of membership in the fuzzy set *tall*; that is, it's harder to have a high degree of membership in the fuzzy set *very tall*. So, for example, if 5′ 8″ is tall to degree .54, then it is *very* tall to a lesser degree, perhaps to degree .25. Now, the threshold need not be lowered in every case. For example, 6′ 7″ may be not only tall to degree 1 but also very tall to degree 1. Moreover, the lowering need not be a *linear* function in those cases where it does occur. In our example, 5′ 8″ is a member of the fuzzy set *very tall* to a degree that is .29 less than its degree of membership in *tall*, but the degrees to which 6′ 5″ is a member of the fuzzy sets *tall* and *very tall* need not differ by the same amount – 6′ 5″ may be very tall to a higher degree than .63 (= .92 – .29), perhaps to degree .8 or .9.

[1] The term was coined by the linguist George Lakoff (1973).

To capture this, we'll interpret *very* (along with other hedges) as a function mapping membership degrees (values in the unit interval [0. .1]) to membership degrees (in [0. .1]), because when *very* attaches to a predicate it takes an objects's degree of membership in the predicate's fuzzy set and produces a new degree. Put differently, the function maps one fuzzy set into another—it maps the fuzzy set *tall* into the fuzzy set *very tall* by showing how the membership degrees change.

The particular function that's usually used for *very* is the square function.[2] It preserves 1 as the degree to which 6′ 7″ is very tall, and it produces .85 and .29 as the respective degrees to which 6′ 5″ and 5′ 8″ are very tall. Note that interpreting *very* as a function from membership degrees to membership degrees supports iterated application of the function. *Very* can intensify *very tall*, so that a height's degree of membership in the fuzzy set *very very tall* is its degree of membership in *tall* raised to the fourth power—a much higher threshold. Building on our preceding examples, 6′ 7″ is very very tall to degree 1; 6′ 5″ is very very tall to degree .72; and 5′ 8″ is very very tall to degree .08.

Unlike *very*, the qualifier *close to* generally serves to lower membership thresholds. While 5′ 8″ may only be tall to degree .54, it is *close to* tall to a higher degree, say, .75. (One exception is that we might not want to say that a height that is tall to degree 1 is *close to tall* because the latter carries the suggestion that the height is not exactly tall; the reader will be asked to explore this possibility in the exercises.) *Close to* also seems more like a linear qualifier. So perhaps the corresponding function maps a degree of membership to a degree that is .1 higher—with the obvious special case that we can't map to a degree higher than 1, so every degree of membership between .9 and 1 inclusive is mapped to 1.

We need to add a supply of qualifiers to the language of first-order logic to symbolize hedges—we'll use lowercase Greek letters (to which we may add subscripts to guarantee an infinite supply of qualifiers). We now define *predicates* to include single uppercase roman letters and all expressions formed by placing one or more occurrences of qualifiers in front of a predicate, for example, αT, $\alpha\alpha T$, $\beta\alpha\delta T$. (In English we are not allowed to mix qualifiers quite so freely; for example, while a height may be *very* tall, *very very* tall, *very very very* tall, *close to* tall, *close to close to* tall, *very close to* tall, *close to very* tall, or *somewhat* tall, it may not be *somewhat somewhat* tall (well, maybe not?), *very somewhat* tall, or *somewhat very* tall.)

Interpretations must now assign functions as the values of qualifiers, so the definition of interpretations is modified to include

An assignment of a function I(α) to each qualifier α mapping [0. .1] to [0. .1]:

$$\mathrm{I}(\alpha) \in [0..1]^{[0..1]}$$ [3]

Given the preceding analysis, I(*very*) is the function that maps each member n of [0. .1] to n^2, that is, I(*very*)(n) = n^2; and I(*close-to*)(n) = min (1, n + .1). (Here we

[2] This was first suggested by L. A. Zadeh. See, for example, Zadeh (1975).
[3] This is standard notation for the set of functions mapping the unit interval to the unit interval.

have taken the obvious liberty of using the English expressions rather than Greek letters.)

We must also add a semantic basis clause defining the fuzzy sets corresponding to predicates formed with qualifiers:

$$\mathrm{I}(\alpha\mathbf{P}^n)(<\mathrm{x}_1,\ldots,\mathrm{x}_n>) = \mathrm{I}(\alpha)(\mathrm{I}(\mathbf{P}^n)(<\mathrm{x}_1,\ \ldots,\mathrm{x}_n>)).$$

That is, the degree to which $<\mathrm{x}_1,\ \ldots,\mathrm{x}_n>$ is $\alpha\mathbf{P}^n$ is the result of applying the function α to $<\mathrm{x}_1,\ldots,\ \mathrm{x}_n>$'s degree of membership in the fuzzy set $\mathbf{P}^n$. Thus, where T is interpreted as in SST, we have

$\mathrm{I}(\mathit{very}\ T)(<\mathrm{x}>) = \mathrm{I}(\mathit{very})(\mathrm{I}(T)(<\mathrm{x}>)) = \mathrm{I}(\mathit{very})((\mathrm{x}-4'\,7'')\ /\ 24'') = ((\mathrm{x}-4'\,7'')/24')^2$

$\mathrm{I}(\mathit{very\ very}\ T)(<\mathrm{x}>) = \mathrm{I}(\mathit{very})(\mathrm{I}\ (\mathit{very}\ T)(<\mathrm{x}>)) = \mathrm{I}(\mathit{very})(((\mathrm{x}-4'\,7'')\ /\ 24'')^2) = (((\mathrm{x}-4'\,7'')\ /\ 24'')^2)^2$, or $((\mathrm{x}-4'\,7'')/24'')^4$

$\mathrm{I}(\mathit{close\text{-}to}\ T)(<\mathrm{x}>) = \mathrm{I}(\mathit{close\text{-}to})(\mathrm{I}(T)(<\mathrm{x}>)) = \mathrm{I}(\mathit{close\text{-}to})((\mathrm{x}-4'\,7'')/24'') = \min(1,\ ((\mathrm{x}-4'\,7'')/24'') + .1)$

$\mathrm{I}(\mathit{close\text{-}to\ close\text{-}to}\ T)(<\mathrm{x}>) = \mathrm{I}(\mathit{close\text{-}to})(\mathrm{I}(\mathit{close\text{-}to}\ T)(<\mathrm{x}>)) = \mathrm{I}(\mathit{close\text{-}to})(\min(1,\ ((\mathrm{x}-4'\,7'')/24'') +.1)) = \min(1,\ \min(1,\ (\mathrm{x}-4'\,7'')/24'' + .1) + .1)$, which is $\min(1,\ (\mathrm{x}-4'\,7'')/24'' + .2)$

$\mathrm{I}(\mathit{close\text{-}to\ very}\ T)(<\mathrm{x}>) = \mathrm{I}(\mathit{close\text{-}to})(\mathrm{I}(\mathit{very}\ T)(<\mathrm{x}>)) = \mathrm{I}(\mathit{close\text{-}to})(((\mathrm{x}-4'\,7'')\ /\ 24'')^2) = \min(1,\ ((\mathrm{x}-4'\,7'')\ /\ 24'')^2 + .1)$

$\mathrm{I}(\mathit{very\ close\text{-}to}\ T)(<\mathrm{x}> = \mathrm{I}(\mathit{very})(\mathrm{I}(\mathit{close\text{-}to}\ T)(<\mathrm{x}>)) = \mathrm{I}(\mathit{very})(\min(1,\ (\mathrm{x}-4'\,7'')\ /\ 24'' + .1)) = (\min(1,\ (\mathrm{x}-4'\,7'')/24'') + .1)^2$

(For perspicuity in our illustration we've used English in place of the Greek letters.) The rest of the semantic clauses remain the same, since qualified predicates work the same way as simple predicates when embedded in formulas.

Note that we can also treat *not* as a predicate qualifier to form expressions like *not tall* and *not very tall.* A reasonable interpretation for *not* in this context is

$$\mathrm{I}(\mathit{not})(n) = 1 - n$$

which produces

$$\mathrm{I}(\mathit{not}\ T)(<\mathrm{x}>) = \mathrm{I}(\mathit{not})(\mathrm{I}(\mathrm{T})(<\mathrm{x}>)) = \mathrm{I}(\mathit{not})((\mathrm{x}-4'\,7'')/24'') = 1 - ((\mathrm{x}-4'\,7'')/24'')$$

and

$$\mathrm{I}(\mathit{not\ very}\ T)(<\mathrm{x}>) = \mathrm{I}(\mathit{not})(\mathrm{I}(\mathit{very}\ T)(<\mathrm{x}>)) = \mathrm{I}(\mathrm{not})((\mathrm{x}-4'\,7'')/24'')^2 = (1 - ((\mathrm{x}-4'\,7'')/24'').^2$$

Note that *not* is modifying *very T* rather than *very*; that is, *not very T* is *not (very T)*, not *(not very) T*—the latter would require hedges that modify hedges. Obviously, with this interpretation of *not*, *not Tx* will be equivalent to $\neg Tx$ in $\text{Fuzzy}_{\text{Ł}\forall}$, but not in $\text{Fuzzy}_{\text{G}\forall}$ or $\text{Fuzzy}_{\text{P}\forall}$.[4]

[4] For further reading on hedges in fuzzy logic see Lakoff (1973), Zádeh (1975), and Novák (2001).

16.2 Fuzzy "Linguistic" Truth-Values

In addition to combining with ordinary predicates in English, hedges can modify truth-value attributions. For example, we've said often that the Principle of Charity premise in the Sorites paradox is *close to true*, and we may also say that a particular statement is *very true, very close to true, not very true, somewhat true*, and so on, or that it is *very false, close to false*, and so forth. In this book we have also talked about statements that are *clearly true, clearly not true*, and *clearly false* rather than just *true* or *false*. Lotfi Zadeh (1975) first explored these "linguistic truth-values" (natural language truth-value attributions) in 1975,[5] somewhat informally; more recently theoreticians have begun to incorporate linguistic truth-values into formal axiomatic systems (e.g., Hájek [2001].

Following Zadeh, we will interpret linguistic truth-values as fuzzy sets of truth-values. For example, the interpretation of *true* might be the fuzzy set (over the unit interval [0. .1]) defined by the function

$$\begin{aligned}&\mathrm{I(true)}(n) = 0 \text{ if } n \leq .8\\&2((n - .8)(.2)^2 \text{ if } .8 < n \leq .9\\&1 - 2((n - 1)/.2)^2 \text{ if } n > .9\end{aligned}$$

(this example is Zadeh's). Note that this function makes 1 true to degree 1 and 0 true to degree 0, a minimal requirement we might impose on such a function. The value .81 is true to degree .05, .9 is true to degree .5, .91 is true to degree .595, and .99 is true to degree .995. The linguistic truth-value *true* is a propositional operator that combines with formulas just as negation does:

If **Q** is a formula, so is *true* **Q**

and the semantic clause

$$\mathrm{I}(\mathit{true}\ \mathbf{Q}) = \mathrm{I}(\mathit{true})(\mathrm{I}(\mathbf{Q}))$$

gives the truth-conditions for formulas formed with the operator *true*. If *Tj* symbolizes *John is tall*, then *true Tj* symbolizes *It is true that John is tall*. If *Tj* has the value .91, then by the function we have assigned to *true*, *true Tj* has the value .595. More generally, interpretations will now include

An assignment to each primitive truth-value **t** of a function I(**t**) mapping [0. .1] to [0. .1]:

$$\mathrm{I}(\mathbf{t}) \in [0..1]^{[0..1]}$$

to assign fuzzy sets to all primitive truth-values, and the semantic truth-condition clauses will include

$$\mathrm{I}(\mathbf{tQ}) = \mathrm{I}(\mathbf{t})(\mathrm{I}(\mathbf{Q}))$$

to determine the values of formulas formed with linguistic truth-values.

[5] Zadeh (1975).

Because *true* is interpreted as a fuzzy set we can combine hedges with *true* (or other linguistic truth-values) to form further, complex linguistic truth-values. Formally we need to specify that

If α is a qualifier and **t** is a linguistic truth-value, then α**t** is a linguistic truth-value,

and a semantic clause to assign functions to complex linguistic truth values:

$$I(\alpha\mathbf{t})(n) = I(\alpha)(I(\mathbf{t})(n))$$

For example, the value of *very true* for any truth-value is the square of the *true* function applied to that degree. In particular, given our sample interpretations, we have

$$\begin{aligned} I(\textit{very true})(n) &= I(\textit{very})(I(\textit{true})(n)) \\ &= 0 \text{ if } n \le .8 \\ &\quad 4((n - .8)/.2)^4 \text{ if } .8 < n \le .9 \\ &\quad 1 - 4((n-1) / .2)^2 + 4((n-1) / .2)^4 \text{ if } n > .9 \\ I(\textit{not very true})(n) &= I(\textit{not})(I(\textit{very true})(n)) \\ &= 1 \text{ if n} \le .8 \\ &\quad 1 - 4((n - .8)/.2)^4 \text{ if } .8 < n \le .9 \\ &\quad 4((n-1)/.2)^2 - 4((n-1)/.2)^4 \text{ if } n > .9 \\ I(\textit{close-to true})(n) &= I(\textit{close-to})(I(\textit{true})(n)) \\ &= .1 \text{ if } n \le .8 \\ &\quad 2((n - .8)/.2)^2 + .1 \text{ if } .8 < n \le .9 \\ &\quad \min(1, 1.1 - 2((n-1)/.2)^2) \text{ if } n > .9 \end{aligned}$$

So if $I(Tj) = .91$, then

$I(\textit{very true } Tj) = I(\textit{very true})(I(Tj)) = $ (approximately) .354
$I(\textit{not very true } Tj) = I(\textit{not very true})(I(Tj)) = .646$
$I(\textit{close-to true } Tj) = I(\textit{close-to true})(I(Tj)) = .695.$

Zadeh proposed defining *false* as

$$I(\textit{false})(n) = I(\textit{true})(1 - n)$$

rather than as *not true*. This allows us to say, for example, *not false and not true* without danger that the expression will reduce to *not not true and not true*, which is equivalent to *true and not true*. Note that the expression *not false and not true* treats *and* like our other qualifiers. For this we might define

$$I(\textit{and})(m, n) = \min(m, n).$$

Thus, *not false and not true Tk* will have the value 1 when *Tk* has the value .5—just as we would hope.

As we suggested in Section 14.4 of Chapter 14, we might use linguistic truth-values to address a concern that's been raised about fuzzy logic: that fuzzy solutions to the Sorites paradox, whatever their detail, seem to assume that there is a clear cutoff for *true* insofar as the value 1 counts as true and all values less than 1 do not count as true, and similarly for *false*. The introduction of fuzzy linguistic truth-values allows that truth may be a matter of degree rather than black and white. The value 1 may be the only one that counts as *true* to degree 1, but other values can count as *true* to high degrees as well. Or perhaps values less than, but close to, 1 may even count as true to degree 1. There's a lot of flexibility here.

16.3 Other Fuzzy Extensions of Fuzzy Logic

There are other fuzzy ways that fuzzy logic has been extended. Although they are beyond the scope of this text, we'll mention three examples as further avenues of study for the interested reader.

Quantifiers can be fuzzy as well as crisp. The universal and existential quantifiers are paradigmatic crisp quantifiers, as are quantifiers representing specific cardinalities: *two people, fifty-eight heartbeats,...* They are crisp in the sense that they tell us exactly how many things we are talking about. (This may seem counterintuitive in the case of the existential quantifier, which we read as: *at least one*. But *at least one* requires exactly a positive integer.) By contrast, *few* and *many* are fuzzy quantifiers: How many people, for example, are *many people*? In a group of 100 people there are surely many people, but are there many people in a group of 20? Of 10? There's no specific cutoff point between *many* and *not many*, just as there is no specific cutoff point between *tall* and *not tall*.

Modalities, studied in modal logic, can also be fuzzy. The standard crisp modalities are *necessary* and *possible* (although it can certainly be argued that there are degrees of necessity and possibility). *Probable* is a fuzzy modality. To be sure, we can give specific probabilities for many things, and there are very precise logics of probability. But *probable* is a fuzzy modality because there is no specific probability or range of probabilities that counts as being *probable*.[6]

Finally, there is yet another issue of vagueness that we haven't explored in this text, which we'll now describe. We've noted that if we add the fuzzy external assertion operator Δ to Łukasiewicz, Gödel, or product fuzzy logic, we can use the formula $(\exists x)(\neg\Delta Tx \wedge \neg\Delta\neg Tx)$ to express the existence of borderline cases for the predicate *tall*. But now recall that Bertrand Russell posited "higher-order" fringes so that we are not forced to recognize a precise cutoff point between a vague predicate's extension and its fringe. Objects in the first higher-order fringes for a predicate, for example, are not in the predicate's extension (or counterextension), nor are they in the fringe. The issue, first raised by Crispin Wright (1987), is finding a way to express the existence of higher-order fringes.

Wright proposed expressing the existence of first-order fringes thus: *there is no pair of heights that differ by* $^1\!/_8''$ *such that one is definitely tall and the other is*

6 The reader interested in exploring these two extensions—fuzzy quantifiers and fuzzy modalities—would do well to start with Hájek (1998b).

definitely not tall. Using **D** for Wright's *definitely* operator, we can symbolize this as $\neg(\exists x)(\exists y)((DTx \wedge Eyx) \wedge D\neg Ty)$. The existence of a second-order fringe between the predicate's extension and first-order fringe would be expressed as $\neg(\exists x)(\exists y)((DDTx \wedge Eyx) \wedge D\neg DTy)$: *there is no pair of heights that differ by* 1/8″ *such that one is definitely definitely tall and the other is definitely not definitely tall*, and the existence of a third-order fringe between the predicate's extension and second-order fringe would be expressed as $\neg(\exists x)(\exists y)((DDDTx \wedge Eyx) \wedge D\neg DDTy)\ldots$[7] Because these formulas are supposed to express the existence of distinct fringes, the formulas *T, DT, DDT, DDDT, DDDDT,...*, must all be nonequivalent to one another. And this means that the external assertion operator cannot do the work of the *definitely* operator, because any formula that prefixes one or more external assertion operators to the formula **ΔP** is equivalent to **ΔP**.

Wright didn't propose a semantics for the *definitely* operator, but Richard Heck (1993) satisfactorily analyzed it outside fuzzy logic as a *modal* operator. In response to a concern that fuzzy logic cannot adequately represent higher-order vagueness, Libor Behounek ("A Model of Higher-Order Vagueness") has begun formally to explore higher-order vagueness in fuzzy logic using such a *definitely* operator. Behounek doesn't analyze the operator as a modality but rather uses fuzzy higher-order (higher than first-order) logic in which we can quantify over, and say things about, fuzzy sets.

16.4 Exercises

SECTION 16.1

1 We noted one possible exception to our function for the hedge *close to*: we may want to say that a height that is tall to degree 1 is not *close to tall* at all. Do you agree with this intuition? If so, define a function for *close to* that captures the intuition. If not, explain why you believe the intuition is incorrect.

2 We claimed that a height may be *very* tall, *very very* tall, *very very very* tall, *close to* tall, *close to close to* tall, *very close to* tall, *close to very* tall, or *somewhat* tall, but that it may not be *somewhat somewhat* tall, *very somewhat* tall, or *somewhat very* tall. Can you provide rules governing which combinations of the three hedges *very, close to,* and *somewhat* are permissible in English?

3 Given the following interpretation:

D: set of heights between 5′ and 6′ 2″ by 1/8″ increments, inclusive

$I(T)(\langle x\rangle) = (x - 5')/14''$

$I(very)(n) = n^2$

$I(close\text{-}to)(n) = \min(1, n + .1)$

$I(not)(n) = 1 - n$

[7] Wright and subsequent analysts have had a lot more to say about this operator than this brief exposition suggests, but that literature would take us too far afield. For an overview see Keefe and Smith (1997).

$I(a) = 6' 7''$
$I(b) = 6' 1''$
$I(c) = 5' 5''$
$I(d) = 5''$
$I(e) = 4' 7''$

determine the truth-value of each of the following formulas:

a. *very Ta*
b. *very Tb*
c. *very Te*
d. *close-to Ta*
e. *close-to Tb*
f. *close-to Tc*
g. *close-to Td*
h. *very very Ta*
i. *close-to very Ta*
j. *very very Td*
k. *very close-to Tb*
l. *very close-to Tc*
m. *very very close-to Tb*
n. *very very close-to Tc*
o. *close-to very very Tb*
p. *close-to very very Tc*
q. *not close-to Te*
r. *not close-to very Td*

4 Provide reasonable interpretations for the following hedges:

a. *somewhat*
b. *extremely*
c. *slightly*
d. *infinitesimally*
e. *hardly*
f. *clearly*

SECTION 16.2

5 Assuming that *Tj* has the value .9 and *Tk* has the value .2, and given the sample definitions in this chapter, determine the truth-value of each of the following formulas:

a. *true Tj*
b. *not true Tj*
c. *false Tj*
d. *not false Tj*
e. *true Tk*
f. *not true Tk*
g. *false Tk*

h. *not false Tk*
i. *very true Tj*
j. *very false Tk*
k. *not very very false Tk*
l. *not very true and not very false Tk*
m. *close-to true Tj*
n. *close-to false Tk*
o. *not very close-to true Tj*

6 Using your definition of *clearly* in Exercise 4, what degrees of truth will be *clearly true*? Does this seem correct?

17 Fuzzy Membership Functions

17.1 Defining Membership Functions

There are various shapes that membership functions for vague predicates might have. The definitions of the fuzzy set *tall* in our sample interpretations—SST and the like—have assumed that the membership function is **linear**; that is, it defines a straight line:

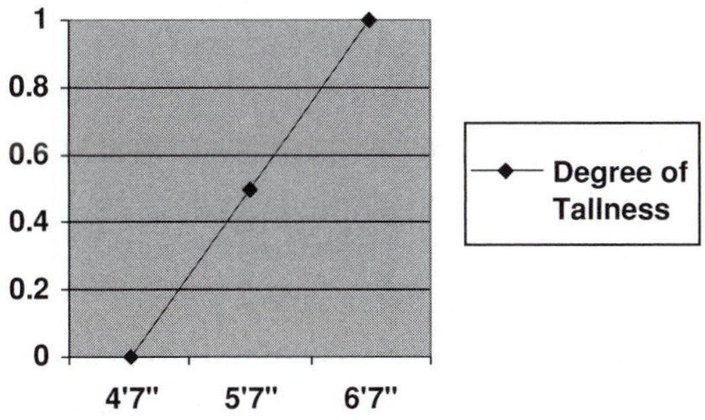

Indeed, Max Black (1937) conjectured that very vague concepts would exhibit such curves, in contrast to nearly crisp, or precise, concepts, which would have long flat portions representing degrees of membership 1 and 0, with a nearly vertical rise (or drop) connecting the two:

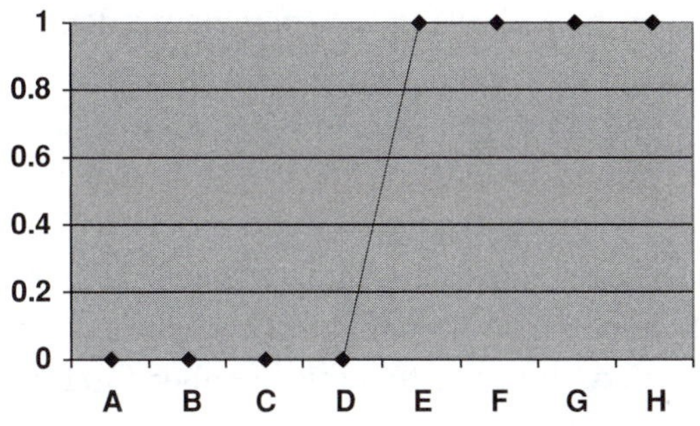

In contrast, Joseph Goguen (1968–1969) suggested that the membership curve for *short*—and by implication, the membership curve for *tall*—would be **continuous** (no big jumps), **decreasing** (the degree of shortness lessens as heights increase), and

asymptotic to 0, but not necessarily linear. As an example he offered the function I(*short*)(<*x*>) $= {}^1/_{(1+x)}$ where *x* is some quantitative measurement of height:

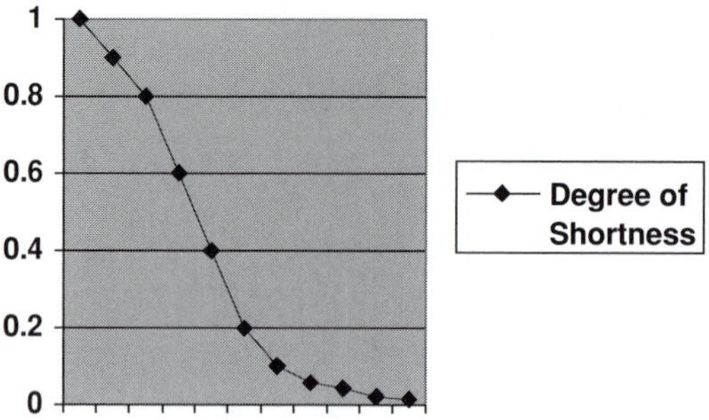

In Section 14.6 of Chapter 14 we noted that the product Principle of Charity premise for tall heights is false on interpretation SST, and the reason that it is false is that one height in the domain (but not all) is a member of the fuzzy set *tall* to degree 0. Goguen's sample function for *short* does not have a similar problem because ${}^1/_{(1+x)}$ never reaches 0. So the product Principle of Charity for short heights—$(\forall x)(\forall y)((Sx \mathbin{\&_P} Gyx) \rightarrow_P Sy)$, symbolizing *A height that is* ${}^1/_8''$ *greater than a short height is also short*—would have a value strictly greater than 0.

But we have to work to find a specific measure *x* of heights that gives plausible degrees of membership in *short*. If we set *x* to be a height's excess over 4′ 7″ in 1″ units, we have $x = 0$ for 4′ 7″, so that 4′ 7″ is short to degree 1 by Goguen's function—which looks good. But the measure $x = 1$ for 4′ 8″, which makes 4′ 8″ short to degree .5, doesn't look so good. And if we measure a height's excess over 4′ 7″ in ${}^1/_8''$ units, the situation gets even worse: $x = 8$ for 4′ 8″, which makes that height short to degree ${}^1/_9$. Moreover, the product Principle of Charity $(\forall x)(\forall y)((Sx \mathbin{\&_P} Gyx) \rightarrow_P Sy)$ has the value ${}^1/_2$ here (letting *Gyx* have the value 1 when y is ${}^1/_8''$ greater than x and the value 0 otherwise)—when v(*x*) = 4′ 7″ and v(*y*) = 4′ $7^1/_8''$, the value of $(Sx \mathbin{\&_P} Gyx) \rightarrow_P Sy$ is the ratio ${}^{1/2}/_1$ (and that's as small as it can get). But ${}^1/_2$ isn't "close to true."

To see that Goguen's function isn't that bad, here's an example of a better way to flesh out the numbers for the *short* Sorites (and others are certainly possible). Assuming a range of heights from 3′ 4″ to 8′ 4″, using ${}^1/_8''$ increments, define *x* for any height *h* to be $h - 3'$, expressed as half-foot units. Using Goguen's function, then, 3′ 4″ is short to degree 1; 3′ $4^1/_8''$ is short to degree ${}^{48}/_{49}$; 3′ $4^2/_8''$ is short to degree ${}^{24}/_{25}$; 4′ 4″ is short to degree ${}^1/_3$; 5′ 4″ is short to degree ${}^1/_5$; 6′ 4″ is short to degree ${}^1/_7$; 7′ 4″ is short to degree ${}^1/_9$; and 8′ 4″ is short to degree ${}^1/_{11}$. Moreover, the smallest value of the ratio between two successive heights is the ratio between the shortness of 3′ $4^1/_8''$ and the shortness of 3′ 4″, which is ${}^{48}/_{49}$, so that is the smallest value that the conditional can have and is therefore the value that this membership function gives the product Principle of Charity for the predicate *short*.

Not all vague predicates are best represented with membership functions that are strictly increasing or strictly decreasing. For example, some vague predicates, such as the predicate *medium heat* as used in the context of cooking (e.g., *cook over medium heat for two minutes, stirring constantly*), suggest **trapezoidal** functions (having trapezoidal shapes when charted). A membership function for *medium heat,* where the temperature x is measured in degrees Fahrenheit, might be defined as

$$
\mathrm{I}(\textit{medium heat})(<\mathrm{x}>) = \begin{cases} 0 \text{ if } x < 200, \\ {}^{(x-200)}/_{125} \text{ if } 200 \leq x \leq 325, \\ 1 \text{ if } 325 < x < 375, \\ {}^{(500-x)}/_{125} \text{ if } 375 \leq x \leq 500, \\ 0 \text{ if } x > 500 \end{cases}
$$

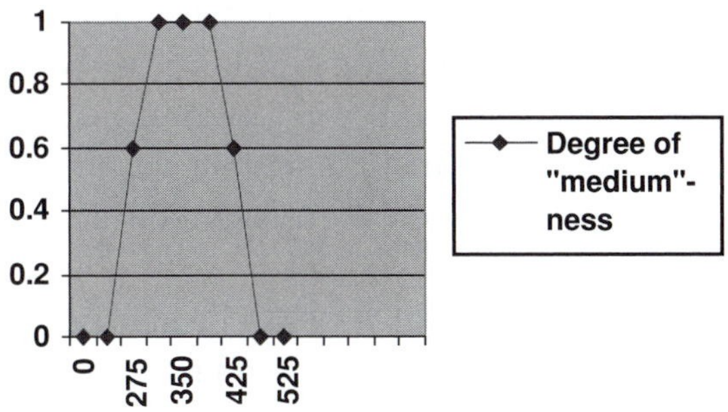

And there are more possibilities. See Pedrycz and Gomide (1998) for a good discussion of the variety of shapes that membership functions can take.

There is another important issue to be addressed in designing membership functions: sometimes degrees of membership are a function of several measurements rather than just one. We've considered tallness and shortness to be functions of height and medium-heatedness to be a function of temperature. But consider, for example, the claim that the air on a midsummer day is *comfortable.* Comfort here is (minimally) a function of temperature and humidity, so in this case fuzzy set membership should be defined as a function of pairs of values consisting of a temperature and a measure of humidity. Another more complicated example requiring several measurements is from Goguen (1998–1999): "A 'good' computer should be (at least): cheap; small; fast; reliable; easy to repair; of large storage capacity; inexpensive to run; equipped with good input and output; and very flexible" (p. 352). It is interesting that in this case each of the criteria is itself vague—for instance, what is *cheap* for a computer? Goguen suggests that we can define a fuzzy membership function over this collection of vague criteria by assigning a weight to each criterion (reflecting that criterion's relative importance in the concept of a good computer), such that the weights add up to 1, and then adding the weighted membership values under each of the criteria. As an example, given the fuzzy degrees of membership and weights in the chart

	Cheap	Small	Fast	Easy	Large	Inexpensive	Equipped	Flexible
Degree	.8	.9	.5	.3	.8	1	1	.9
Weight	.05	.1	.2	.05	.15	.1	.15	.2

the final weighted value for this computer would be $(.05 \cdot .8) + (.1 \cdot .9) + (.2 \cdot .5) + (.05 \cdot .3) + (.15 \cdot .8) + (.1 \cdot 1) + (.15 \cdot 1) + (.2 \cdot .9)$, which is .795—a fairly *good* computer according to this example.

17.2 Empirical Construction of Membership Functions

Our examples of membership functions characterize 6′ 7″ as a definitely tall height, 310 degrees as close to medium heat, and so on. Where does this information come from? In this text I've used my own intuitions to construct examples. But there are more objective ways to construct membership functions. In each case empirical data are collected and then translated into a membership function, perhaps coercing the data to get some specific type of membership curve.

As an example, Max Black proposed a method of establishing what he called a *consistency profile* for a vague predicate "based on the assumptions that while the vagueness of a word involves variations in its application by the users of the language in which it occurs, such variations must themselves be systematic and obey statistical laws if one symbol is to be distinguished from another" (1937, p. 442). We'll use the example concept of *tallness* to explain what a consistency profile is. For each height in our study, we ask each member of some group G of users of the language whether the height should be classified as *tall* or as *not tall* (they don't get a third choice). For each height h, let $m(h)$ be the number of people in G who say that h should be classified as *tall* and $n(h)$ the number of people in G who say that h should be classified as *not tall*. Then the consistency profile is the function C(*tall*) defined as $C(tall)(<h>) = m(h) / n(h)$.[1] Now, this profile will give values outside the range [0. .1] and it can also have undefined values (when $n(h) = 0$), so we need to adjust the profile to produce defined values within the unit interval. Here is one way:

$$I(tall)(<h>) = m(h) / (m(h) + n(h)),$$

That is, the percentage of people in G who say that h should be classified as *tall*. This use of consistency profiles to define fuzzy membership is an example of the **horizontal** or **polling method** of determining membership functions. Alternatively, membership values can be determined by **direct rating**: rather than ask members of G whether a given height h should be classified as *tall* or as not *tall*, we ask members of G to tell us the *degree* to which h is tall. We would then construct a membership function based on the direct ratings by averaging the degrees that members of G gave for each height or by some other method of combining the rankings for each height into a single membership degree.

[1] Actually, Black went a step further than we have and suggested using the limit that the ratio $m(h)/n(h)$ approaches as the number of heights and the number of members of the ranking group G increases.

Yet another method for determining a membership function, the **vertical** or **reverse rating method**, asks users of the language to identify, for some selected values v_i in [0. .1], the range of heights that are tall to degree v_i. The membership function is then constructed from these responses. Pedrycz and Gomide (1998) and Turksen (1991) contain excellent discussions of these and other methods of gathering data from which membership functions can be constructed.

17.3 Logical Relevance?

Does it matter, logically speaking, what shape our membership functions have, or what data we use to construct them? Not really. Logic is concerned with concepts like *tautology* and *entailment*, and fuzzy logic is additionally concerned with concepts like *n-tautology* and *n-degree entailment*. These logical concepts, fuzzy or not, are defined with respect to *all possible* interpretations, not just the specific ones we have chosen. From a logical point of view it doesn't matter whether a specific membership function arbitrarily assigns degrees of membership, just as in classical logic it doesn't matter how a specific interpretation distributes truth-values. This is our final response to the concern that fuzzy logic relies on assuming that, for any vague predicate, there is a clear cutoff point between cases to which the predicate clearly applies (to degree 1) and the borderline cases (degrees strictly between 1 and 0). *N*-tautology and *n*-degree entailment give us measures of tautologousness and entailment that don't require setting any particular clear cutoff point. So, for example, the results in the table in Section 14.7 of Chapter 14 characterizing the validity, *n*-degree-validity, and decaying validity of our Sorites argument in the three fuzzy systems are independent of the way the membership function for *tall* is defined. And that's the way it should be in logic.

17.4 Exercises

SECTION 17.1

1 Using Goguen's membership function for *short*, a domain consisting of heights between 3′ 4″ and 8′ 4″ inclusive, by increments of $^1/_8$″, x recording a height in half-foot units, and the obvious interpretation of the predicate G,

   a. what is the value of the Principle of Charity for shortness when the $\text{Fuzzy}_{\text{L}\forall}$ conditional and bold conjunction are substituted for the product conditional and bold conjunction?

   b. what is the value of the Principle of Charity for shortness when the $\text{Fuzzy}_{\text{G}\forall}$ conditional and bold conjunction are substituted for the product conditional and bold conjunction?

2 Can you find a plausible nonlinear membership function for *short* that will give the Principle of Charity for shortness a high degree of truth in $\text{Fuzzy}_{\text{G}\forall}$?

3 Suggest other plausible membership functions for *short*.
4 Find a vague *noun* whose applicability depends on several criteria. What are these criteria, and how would you weight each criterion when defining a membership function for the fuzzy set corresponding to the noun? Defend your answer.

SECTION 17.2

5 Of the methods for empirical construction of membership functions presented in Section 17.2, which do you prefer and why?
6 Propose another plausible method for empirical consruction of membership functions, and explain why it is plausible.

APPENDIX

Basics of Countability and Uncountability

How many natural numbers (0, 1, 2, 3, 4, . . .) are there? Infinitely many, of course. There are also infinitely many real numbers. But there is an important difference between these infinite sizes, which we will now describe.

There are *countably* many natural numbers. We say that a set, or collection, has ***countably*** many members (is ***countable***) either if it has finitely many members, or there is a 1-1 correspondence between that set and the set of the positive integers. (An infinite countable set is also set to be ***denumerable***.) A 1-1 correspondence between two sets is a pairing of the members of the two sets such that each member of one of the sets is paired off with exactly one of the members of the other set, and vice versa. By definition, then, there are countably many positive integers—because the identity pairing is a 1-1 correspondence between any set and itself. Here is a 1-1 correspondence that pairs each positive integer with the natural number that is 1 less (and hence each natural number with the positive integer that is greater by 1):

1	0
2	1
3	2
4	3
5	4
.	.
.	.
.	.

Note that there may be other 1-1 correspondences between two sets—the important point in establishing countability is to show that there is at least one such pairing. There are countably many even positive integers: we can correlate each positive integer n with the even positive integer $2n$ (conversely, each even positive integer $2n$ is paired with the positive integer n):

1	2
2	4
3	6
4	8

5	10
6	12
.	.
.	.
.	.

The totality of integers, positive, negative, and 0, is also countable. We will list the integers "emanating" from 0, alternating the positive ones with the negative ones:

1	0
2	1
3	−1
4	2
5	−2
6	3
7	−3
.	.
.	.
.	.

Here each positive integer n is paired with $n/2$ if it is even, $(1 - n)/2$ if it is odd. Conversely, each integer m gets paired with $2m$ if m is positive, and $2(1 - m) - 1$ if m is 0 or negative.

Although we have produced formulas showing how the numbers are paired in each of our examples, it's sufficient in establishing countability to describe how to generate a sequence of the members of the set (our right-hand columns) such that each member of the set must occur exactly once in the sequence. The position of a member in the sequence then gives us the positive integer with which it is paired.

A set is **uncountable** (has **uncountably** many members) if it is not countable. Because all finite sets are countable, an uncountable set must at a minimum be infinite. The set of real numbers, as well as the subset of real numbers in the unit interval, are both uncountable. We will prove the latter using an ingenious type of argument developed by the German mathematician Georg Cantor and known as ***Cantor's diagonal argument***. More specifically, we'll focus on the real numbers that lie strictly between 0 and 1 (we call this set the *open unit interval*), and at the end will introduce 0 and 1 into the picture. Every real number in the open unit interval can be written as an infinitely long decimal expression $0.\mathrm{d}_1\mathrm{d}_2\mathrm{d}_3\mathrm{d}_4\mathrm{d}_5\ldots$, where each d_i is a single decimal digit. (Note that at some point the trailing digits may all be 0, as in .23000000 . . . – the 0's allow an elegant presentation of the proof.) Now assume, contrary to what we want to show, that there is a 1-1 correspondence between the positive integers and the open unit interval, and that a sequence of decimal expressions displaying the correspondence begins as follows:

$$0.d_{1,1}d_{1,2}d_{1,3}d_{1,4}d_{1,5}d_{1,6}d_{1,7}d_{1,8}d_{1,9}\ldots$$
$$0.d_{2,1}d_{2,2}d_{2,3}d_{2,4}d_{2,5}d_{2,6}d_{2,7}d_{2,8}d_{2,9}\ldots$$
$$0.d_{3,1}d_{3,2}d_{3,3}d_{3,4}d_{3,5}d_{3,6}d_{3,7}d_{3,8}d_{3,9}\ldots$$
$$0.d_{4,1}d_{4,2}d_{4,3}d_{4,4}d_{4,5}d_{4,6}d_{4,7}d_{4,8}d_{4,9}\ldots$$
$$0.d_{5,1}d_{5,2}d_{5,3}d_{5,4}d_{5,5}d_{5,6}d_{5,7}d_{5,8}d_{5,9}\ldots$$
$$0.d_{6,1}d_{6,2}d_{6,3}d_{6,4}d_{6,5}d_{6,6}d_{6,7}d_{6,8}d_{6,9}\ldots$$
$$0.d_{7,1}d_{7,2}d_{7,3}d_{7,4}d_{7,5}d_{7,6}d_{7,7}d_{7,8}d_{7,9}\ldots$$
$$0.d_{8,1}d_{8,2}d_{8,3}d_{8,4}d_{8,5}d_{8,6}d_{8,7}d_{8,8}d_{8,9}\ldots$$
$$0.d_{9,1}d_{9,2}d_{9,3}d_{9,4}d_{9,5}d_{9,6}d_{9,7}d_{9,8}d_{9,9}\ldots$$

We will now show how to find a number $0.e_1e_2e_3e_4e_5\ldots$ in the open unit interval that appears nowhere in this sequence. We are going to do this by looking at the digits on the diagonal

$$0.\boldsymbol{d_{1,1}}d_{1,2}d_{1,3}d_{1,4}d_{1,5}d_{1,6}d_{1,7}d_{1,8}d_{1,9}\ldots$$
$$0.d_{2,1}\boldsymbol{d_{2,2}}d_{2,3}d_{2,4}d_{2,5}d_{2,6}d_{2,7}d_{2,8}d_{2,9}\ldots$$
$$0.d_{3,1}d_{3,2}\boldsymbol{d_{3,3}}d_{3,4}d_{3,5}d_{3,6}d_{3,7}d_{3,8}d_{3,9}\ldots$$
$$0.d_{4,1}d_{4,2}d_{4,3}\boldsymbol{d_{4,4}}d_{4,5}d_{4,6}d_{4,7}d_{4,8}d_{4,9}\ldots$$
$$0.d_{5,1}d_{5,2}d_{5,3}d_{5,4}\boldsymbol{d_{5,5}}d_{5,6}d_{5,7}d_{5,8}d_{5,9}\ldots$$
$$0.d_{6,1}d_{6,2}d_{6,3}d_{6,4}d_{6,5}\boldsymbol{d_{6,6}}d_{6,7}d_{6,8}d_{6,9}\ldots$$
$$0.d_{7,1}d_{7,2}d_{7,3}d_{7,4}d_{7,5}d_{7,6}\boldsymbol{d_{7,7}}d_{7,8}d_{7,9}\ldots$$
$$0.d_{8,1}d_{8,2}d_{8,3}d_{8,4}d_{8,5}d_{8,6}d_{8,7}\boldsymbol{d_{8,8}}d_{8,9}\ldots$$
$$0.d_{9,1}d_{9,2}d_{9,3}d_{9,4}d_{9,5}d_{9,6}d_{9,7}d_{9,8}\boldsymbol{d_{9,9}}\ldots$$

If $d_{i,i}$ is 0 we define e_i to be 1; otherwise we define e_i to be 0. The number $0.e_1e_2e_3e_4e_5\ldots$ so defined is different from the first number $0.d_{1,1}d_{1,2}d_{1,3}d_{1,4}d_{1,5}d_{1,6}$ $d_{1,7}d_{1,8}d_{1,9}\ldots$ in the list, because we have defined the first decimal digit e_1 to be different from $d_{1,1}$. Similar reasoning shows that $0.e_1e_2e_3e_4e_5\ldots$ must be different from every other number in the sequence. So on the basis of the assumption that there is a way to sequence members of the open unit interval we can define a member of the open unit interval that *doesn't* appear in the sequence. This is sufficient to show that there is no way to construct a sequence that includes every real number in the open unit interval. No matter how we try to order them in a sequence, we can always define a real number that's different from every number in the sequence. We conclude that there are uncountably many members of the open unit interval.

It is a simple matter to show that the closed unit interval *including* 0 and 1 is also uncountable. If there *were* a 1-1 correspondence between the positive integers and the members of the closed unit interval, then there would be a way to sequence members of the closed interval such that every member occurs exactly once, and removing 0 and 1 from this sequence would leave a sequence of the members of the open unit interval. But we have just shown that there is no such sequence, so we conclude that the closed unit interval is also uncountable. Generalizing this argument, we may conclude that *any* set that includes the open (or closed) unit interval must be uncountable—so the set of all real numbers is also uncountable.

The set of *rational* real numbers in the unit interval is, however, countable. Here's the beginning of a 1-1 correspondence (that we will describe later) between the positive integers and those rational numbers, expressed as fractions:

1	$^{0}/_{1}$
2	$^{1}/_{1}$
3	$^{1}/_{2}$
4	$^{1}/_{3}$
5	$^{2}/_{3}$
6	$^{2}/_{4}$
7	$^{3}/_{4}$
8	$^{3}/_{5}$
9	$^{2}/_{5}$
10	$^{3}/_{5}$
11	$^{4}/_{5}$
12	$^{1}/_{6}$
13	$^{5}/_{6}$
14	$^{1}/_{7}$
.	.
.	.
.	.

In the right-hand column we list the beginning of a sequence of the rational numbers in the unit interval, generated as follows: We begin with the denominator 1 and list all fractions representing values in the unit interval that have this denominator—said fractions being ordered by increasing value of the numerator: $^{0}/_{1}$, $^{1}/_{1}$. Then we do the same for the numerator 2, except that we skip those fractions that equal values already listed—so we skip $^{0}/_{2}$ and $^{2}/_{2}$. Then we do the same for the numerator 3, and so on. Clearly every rational number in the unit interval will be represented by some fraction in this sequence, and skipping the indicated fraction ensures that no rational number is represented more than once in the sequence.

Finally, we can show that a language contains a countable number of formulas by explaining how to generate a sequence in which each of the formulas of the language occurs exactly once. We'll first do this for the language Fuzzy_{L}, which is defined as:

1. Every uppercase roman letter, with or without an integer subscript, is a formula.
2. If **P** is a formula, so is ¬**P**.
3. If **P** and **Q** are formulas, so are **(P ∧ Q)**, **(P ∨ Q)**, **(P → Q)**, **(P ↔ Q)**, **(P & Q)**, and **(P ∇ Q)**.

We (arbitrarily) impose the following alphabetical order on all of the symbols used in formulas of Fuzzy_L:

A
B
C
.
.
.
Z
0
1
2
.
.
.
9
(
)
$\neg$
$\wedge$
$\vee$
$\rightarrow$
$\leftrightarrow$
&
∇

The formulas may now be ordered in a sequence that is ordered overall by their length, the shortest formulas first, and within a single length by their alphabetical ordering. Thus the sequence begins with the twenty-six (nonsubscripted) uppercase roman letters in alphabetical order, these being the shortest formulas, followed by the following sequence:

A_1
A_2
...
A_9
B_1
B_2
...
B_9
C_1
...
...

Z_9
$\neg A$
$\neg B$
...
$\neg Z$

—these being all of the formulas that contain exactly two symbols, and so on. Clearly each formula of $Fuzzy_L$ will appear in this sequence, so the language contains a countable number of formulas.[1]

The language $RFuzzy_L$ is also countable—this can be shown quite simply by adding the symbol / to our alphabetical list of symbols, then defining the sequence as we did in the case of $Fuzzy_L$. However, we *cannot* do the same for $RealFuzzy_L$—precisely because the totality of real numbers in the unit interval is uncountable. If we could produce such a sequence for $RealFuzzy_L$, containing the names of all values in the unit interval, then we could also produce such a sequence consisting of just those real values, by removing all the formulas except the constant atomic formulas that denote real values.

1 We need to note an important point here. We have to be careful about how we specify the sequence. We know that every formula will occur in the sequence we've described because every formula has a finite length, and we have a finite alphabet, so only finitely many formulas can occur before a given formula **P**. If we had instead tried to generate a sequence that begins with all formulas—of any length—that begin with the letter *A*; then all formulas that begin with the letter *B*, . . . ; then all formulas that begin with the negation operator; then all formulas that begin with a left parenthesis; . . . ; we'd be in trouble. There are *infinitely* many formulas that begin with the negation operator, and so the sequence would go on forever with these formulas and would never get to the formulas that begin with a left parenthesis!

Bibliography

Ackermann, Robert. 1967. *Introduction to Many-Valued Logics*. London: Routledge & Kegan Paul.

Aguzzoli, S., and A. Ciabattoni. 2000. "Finiteness in Infinite-Valued Łukasiewicz Logic." *Journal of Logic, Language, and Information* 9, pp. 5–29.

Baaz, Matthias. 1996. "Infinite-Valued Gödel Logics with 0–1-Projections and Relativizations." In ed. Petr Hájek, *Gödel '96: Logical Foundations of Mathematics, Computer Science and Physics—Kurt Gödel's Legacy*. New York: Springer, pp. 23–33.

Baaz, Matthias, Christian G. Fermüller, and Richard Zach. 1993. "Systematic Construction of Natural Deduction Systems for Many-Valued Logics." *Proceedings of the 23rd International Symposium on Multiple Valued Logic*. Los Alamitos, CA: IEEE Computer Society Press, pp. 208–213.

Baaz, Mathias, Petr Hájek, Jan Kraníček, and David Švejda. 1998. "Embedding Logics into Product Logic." *Studia Logica* 61, pp. 35–47.

Baaz, M., and R. Zach. 1998. "Compact Propositional Gödel Logics." *Proceedings of the 28th International Symposium on Multiple-Valued Logic*. Los Alamitos, CA: IEEE Computer Society Press, pp. 108–113.

Balbes, Raymond, and Philip Dwinger. 1974. *Distributive Lattices*. Columbia: University of Missouri Press.

Beall, J. C., and Mark Colyvan. 2001. "Heaps of Gluts and Hyde-ing the Sorities." *Mind* 110, pp. 401–408.

Beall, J. C., and Bas C. van Fraassen. 2003. *Possibilities and Paradox: An Introduction to Modal and Many-Valued Logic*. New York: Oxford University Press.

Behounek, Libor. "A Model of Higher-Order Vagueness in Higher-Order Fuzzy Logic." http://atlas-conferences.com/cgi-bin/abstract/casu-34.

Bergmann, Merrie, James H. Moor, and Jack Nelson. 2004. *The Logic Book*, 4th ed. New York: McGraw-Hill.

Black, Max. 1937. "Vagueness: An Exercise in Logical Analysis." *Philosophy of Science* 4, pp. 427–455.

Bochvar, D. A. 1937. "Ob odnom Tréhznačnom Isčislénii i égo Priménénii k Analizu Paradoksov Klassičéskogo Rasširénnogo Funkcional'nogo Isčislénià." *Matématčéskij Sbornik* 4 (46), pp. 287–308. (English translation by Merrie Bergmann, "On a Three-Valued Calculus and Its Application to the Analysis of the Paradoxes of the Classical Extended Functional Calculus." *History and Philosophy of Logic* 2, 1981, pp. 87–112.)

Bolc, Leonard, and Piotr Borowik. 1992. *Many-Valued Logics. I: Theoretical Foundations*. New York: Springer-Verlag.

Chang, C. C. 1958a. "Proof of an Axiom of Łukasiewicz." *Transactions of the American Mathematical Society* 87, 55–56.

Chang, C. C. 1958b. "Algebraic Analysis of Many Valued Logics." *Transactions of the American Mathematical Society* 88, pp. 476–490.

Chang, C. C. 1959. "A New Proof of the Completeness of the Łukasiewicz Axioms." *Transactions of the American Mathematical Society* 93, pp. 74–80.

Church, Alonzo. 1936. "A Note on the Entscheidungsproblem." *Journal of Symbolic Logic* 1, pp. 40–41.

Cignoli, Roberto L. O., Itala M. L. D'Ottaviano, and Daniele Mundici. 2000. *Algebraic Foundations of Many-Valued Reasoning*. Boston: Kluwer.

Delong, Howard. 1970. *A Profile of Mathematical Logic*. Reading, MA: Addison-Wesley.

Dilworth, R. P., and M. Ward. 1939. "Residuated Lattices." *Transactions of the American Mathematical Society* 45, pp. 335–354.

Dunn, J. Michael, and Gary M. Hardegree. 2001. *Algebraic Methods in Philosophical Logic*. New York: Oxford University Press.

Edgington, Dorothy. 1999. "Vagueness by Degrees." In eds. Rosanna Keefe and Peter Smith, *Vagueness: A Reader*. Cambridge, MA: MIT Press, pp. 294–316.

Esteva, Francesc, Lluis Godo, Petr Hájek, and Mirko Navara. 2000. "Residuated Fuzzy Logics with an Involutive Negation." *Archive for Mathematical Logic* 39, pp. 103–124.

Esteva, Francesc, Lluis Godo, and Franco Montagna. 2001. "The ŁΠ and ŁΠ$^1/_2$ Logics: Two Complete Fuzzy Systems Joining Lukasiewicz and Product Logics." *Archive for Mathematical Logic* 40, pp. 39–67.

Fine, Kit. 1975. "Vagueness, Truth and Logic." *Synthese* 30, pp. 265–300.

Frege, Gottlob. 1879. *Begriffsschrift, eine der arithmetischen nachgebildete Formelsprache des reinen Denkens*. Halle: Verlag von Louis Nebert.

Gödel, Kurt. 1932. "Zum Intuitionistischen Aussagenkalkül." *Anzeiger der Akademie der Wissenschaften Wien, mathematisch, naturwissenschaftliche Klasse* 69, pp. 65–66.

Goguen, Joseph A. 1967. "L-Fuzzy Sets." *Journal of Mathematical Analysis and Applications* 18, pp. 145–174.

Goguen, Joseph A. 1968–1969. "The Logic of Inexact Concepts." *Synthese* 19, pp. 325–373.

Goldberg, H., H. LeBlanc, and G. Weaver. 1974. "A Strong Completeness Theorem." *Notre Dame Journal of Formal Logic* 15 (2), pp. 325–331.

Gottwald, Siegfried. 2001. *A Treatise on Many-Valued Logics*. Philadelphia: Research Studies Press.

Gottwald, Siegfried, and Petr Hájek. 2005. "Triangular Norm-based Mathematical Fuzzy Logics." In eds. E. P. Klement and R. Mesiar, *Logical, Algebraic, Analytic, and Probabilistic Aspects of Triangular Numbers*. New York: Elsevier, pp. 275–299.

Haack, Susan. 1979. "Do We Need Fuzzy Logic?" *International Journal of Man-Machine Studies* 11, pp. 437–445.

Hájek, Petr. 1995a. "Fuzzy Logic and Arithmetical Hierarchy." *Fuzzy Sets and Systems* 73 (3), pp. 359–363.

Hájek, Petr. 1995b. "Fuzzy Logic from the Logical Point of View." In eds. M. Bartošek, J. Staudek, and J. Wiedermann, *SOFSEM '95: Theory and Practice of Informatics; Lecture Notes in Computer Science 1012*. New York: Springer-Verlag, pp. 31–49.

Hájek, Petr. 1997. "Fuzzy Logic and Arithmetical Hierarchy II." *Studia Logica* 58, pp. 129–141.

Hájek, Petr. 1998a. "Basic Fuzzy Logic and BL-algebras." *Soft Computing* 2, pp. 124–128.

Hájek, Petr. 1998b. *Metamathematics of Fuzzy Logic*. Boston: Kluwer.

Hájek, Petr. 2001. "On Very True." *Fuzzy Sets and Systems* 124, pp. 329–333.

Hájek, Petr, Jeff Paris, and John Shepherdson. 2000. "Rational Pavelka Predicate Logic Is a Conservative Extension of Łukasiewicz Predicate Logic." *The Journal of Symbolic Logic* 65 (2), pp. 669–682.

Heck, Richard G. 1993. "A Note on the Logic of (Higher-Order) Vagueness." *Analysis* 53, pp. 201–208.

Hirota, K., ed. 1993. *Industrial Applications of Fuzzy Technology* (translated by H. Solomon). New York: Springer-Verlag.

Hunter, Geoffrey. 1971. *Metalogic: An Introduction to the Metatheory of Standard First-Order Logic.* Los Angeles: University of California Press.

Hyde, Dominic. 1997. "From Heaps and Gaps to Heaps of Gluts." *Mind* N. S. 106, pp. 641–660.

Kearns, John T. 1979. "The Strong Completeness of a System for Kleene's Three-Valued Logic." *Zeitschrift für mathematische Logik und Grundlagen der Mathematik* 25, pp. 61–68.

Keefe, Rosanna, and Peter Smith, eds. 1997. *Vagueness: A Reader.* Cambridge, MA: MIT Press.

Kleene, Stephen C. 1938. "On a Notation for Ordinal Numbers." *The Journal of Symbolic Logic* 3, pp. 150–155.

Klir, George J., and Bo Yuan. 1995. *Fuzzy Sets and Fuzzy Logic: Theory and Applications.* Saddle River, NJ: Prentice Hall.

Lakoff, George. 1973. "Hedges: A Study in Meaning Criteria," *Journal of Philosophical Logic* 2, pp. 459–508.

LeBlanc, Hugues. 1977. "A Strong Completeness Theorem for 3-Valued Logic: Part II." *Notre Dame Journal of Formal Logic* 18 (1), pp. 107–116.

Lee, R. C.T., and C.-L. Chang. 1971. "Some Properties of Fuzzy Logic." *Information and Control* 19, pp. 417–431.

Łukasiewicz, Jan. 1930. "Philosophische Bemerkungen zu mehrwertigen Systemen des Aussagenkalküls." *Comptes rendus des séances de la Société des Sciences et des Lettres de Varsovie* 23, cl. iii, pp. 51–77. (English translation by H. Weber, "Philosophical Remarks on Many-Valued Systems of Propositional Logic." In ed. Storrs McCall, *Polish Logic: 1920–1939,* New York: Oxford University Press, 1967, pp. 40–65.)

Łukasiewicz, Jan. 1934. "Z historii logiki zdań." *Przeglad Filozoficzny* 37, pp. 417–437. (English translation by Storrs McCall, "On the History of the Logic of Propositions." In ed. Storrs McCall, *Polish Logic 1920–1939,* New York: Oxford University Press, 1967, pp. 66–87.)

Łukasiewicz, J., and A. Tarski. 1930. "Untersuchungen über den Aussagenkalkül." *Comptes rendus des séances de la Société des Sciences et des Lettres de Varsovie* 23, cl. iii, pp. 39–50. (English translation by J. H. Woodger, "Investigations into the Sentential Calculus." In Alfred Tarski, *Logic, Semantics, Metamathematics: Papers from 1923 to 1938,* 2nd ed., Indianapolis: Hackett Publishing Co., 1983, pp. 38–59.

Machina, Kenton F. 1976. "Truth, Belief, and Vagueness." *Journal of Philosophical Logic* 5, pp. 47–78.

MacLane, Saunders, and Garrett Birkhoff. 1999. *Algebra,* 3rd ed. Providence, RI: American Mathematical Society.

Mangani, P. 1973. "Su Certe Algebre Connesse con Logiche a Piú Valori (On Certain Algebras Related to Many-Valued Logics)." *Bollettino dell'Unione Matematica Italiana (Series 4)* 8, pp. 68–78.

Martin, Robert L., ed. 1970. *Paradox of the Liar.* New Haven, CT: Yale University Press.

Martin, Robert L., ed. 1984. *Recent Essays on Truth and the Liar Paradox.* New York: Oxford University Press.

McNaughton, Robert. 1951. "A Theorem about Infinite-Valued Sentential Logic." *Journal of Symbolic Logic* 16, pp. 1–13.

Menger, K. 1942. "Statistical Metrics." *Proceedings of the National Academy of Sciences in the USA* 8, pp. 535–537.

Meredith, C. A. 1928. "The Dependence of an Axiom of Łukasiewicz." *Transactions of the American Mathematical Society* 87, p. 54.

Minari, Pierluigi. 2003. "A Note on Łukasiewicz's Three-Valued Logic." http://eprints.unifi.it/archive/00001106/02/08_Minari.pdf.

Morgan, Charles G., and Francis J. Pelletier. 1977. "Some Notes Concerning Fuzzy Logics." *Linguistics and Philosophy* 1, pp. 79–97.

Novák, Vilém. 1990. "On the Syntactico-Semantical Completeness of First-Order Fuzzy Logic" (Parts I and II). *Kybernetika* 26, pp. 47–66, 134–154.

Novák, Vilém. 2001. "Antonyms and Linguistic Quantifiers in Fuzzy Logic." *Fuzzy Sets and Systems* 124, pp. 335–351.

Novák, Vilém, Irina Perfilieva, and Jiří Močkoř. 1999. *Mathematical Principles of Fuzzy Logic.* Boston: Kluwer.

Pavelka, J. M. 1979. "On Fuzzy Logic" (Parts I, II, and III). *Zeitschrift für mathematische Logik und Grundlagen der Mathematik* 25, pp. 45–52, 119–134, 447–464.

Pedrycz, Witold, and Fernando Gomide. 1998. *An Introduction to Fuzzy Sets: Analysis and Design.* Cambridge, MA: MIT Press.

Pogorzelski, W. A. 1964. "The Deduction Theorem for Łukasiewicz Many-Valued Propositional Calculi." *Studia Logica* 15, pp. 7–23.

Priest, Graham. 2001. *An Introduction to Non-Classical Logic.* Cambridge: Cambridge University Press, p. 215.

Quine, W. V. O. 1960. *Word and Object.* Cambridge, MA: MIT Press.

Rescher, Nicholas. 1969. *Many-Valued Logic.* New York: McGraw-Hill.

Robinson, J. A. 1965. "A Machine-Oriented Logic Based on the Resolution Principle." *Journal of the Association for Computational Machinery* 12, pp. 23–41.

Rose, Alan, and J. Barkley Rosser. 1958. "Fragments of Many-Valued Statement Calculi." *Transactions of the American Mathematical Society* 87 (1), pp. 1–53.

Ruspini, Enrique H., Piero P. Bonissone, and Witold Predrycz, eds. 1998. *Handbook of Fuzzy Computation.* Philadelphia: Institute of Physics Publishing.

Russell, Bertrand. 1923. "Vagueness." *Australasian Journal of Philosophy* 1, pp. 84–92.

Scarpellini, B. 1962. "Die Nichtaxiomatisierbarkeit des undenlichwertigen Prädikatenkalküls von Łukasiewicz." *Journal of Symbolic Logic* 27, pp. 159–170.

Słupecki, Jerzy. 1936. "Der volle dreivertige Aussagenkalkül." *Comptes rendus des séances de la Société des sciences et des lettres de Varsovie* 3 (29), pp. 9–11. (English translation by Storrs McCall, "The Full Three-Valued Propositional Calculus." In ed. Storrs McCall, *Polish Logic: 1920–1939.* New York: Oxford University Press, 1967, pp. 335–337.)

Smullyan, Raymond M. 1968. *First-Order Logic.* New York: Springer-Verlag.

Stoll, Robert R. 1961. *Set Theory and Logic.* San Francisco: W. H. Freeman.

Takeuti, Gaisi, and Satoko Titani. 1984. "Intuitionistic Fuzzy Logic and Intuitionistic Fuzzy Set Theory." *The Journal of Symbolic Logic* 49, pp. 851–866.

Tarski, Alfred. 1936. "Der Wahrheitsbegriff in den formalisierten Sprachen." *Studia Philosophica* 1, pp. 261–405.

Turksen, I. B. 1991. "Measurement of Membership Functions and Their Acquisition." *Fuzzy Sets and Systems* 40, pp. 5–38.

van Fraassen, Bas C. 1966. "Singular Terms, Truth-Value Gaps, and Free Logic." *Journal of Philosophy* 63, pp. 481–495.

Wajsberg, Mordchaj. 1931. "Aksjomatyzacja trójwartościowego rachunku zdań." *Comptes rendus des séances de la Société des Sciences et des Lettres de Varsovie* 24, cl. iii, pp. 126–145. (English translation by B. Gruchman and S.McCall, "Axiomatization of the Three-Valued Propositional Calculus." In ed. Storrs McCall, *Polish Logic: 1920–1939.* New York: Oxford University Press, 1967, pp. 264–284.)

Wright, Crispin. 1987. "Further Reflections on the Sorites Paradox." *Philosophical Topics* 15, pp. 227–290.
Wu, Olivia. 2003. "Rice Goes Digital Cooked the Fuzzy Logic Way: Side-by-Side Tests Show Appliance Makes a Difference." *San Francisco Chronicle*, December 10, p. E1.
Zadeh, Lotfi A. 1965. "Fuzzy Sets." *Information and Control* 8, pp. 338–353.
Zadeh, Lotfi A. 1975. "Fuzzy Logic and Approximate Reasoning." *Synthese* 30, pp. 407–428.

Index